IET TELECOMMUNICATIONS SERIES 83

Microstrip and Printed Antenna Design

Other volumes in this series:

Volume 9 **Phase Noise in Signal Sources** W.P. Robins
Volume 12 **Spread Spectrum in Communications** R. Skaug and J.F. Hjelmstad
Volume 13 **Advanced Signal Processing** D.J. Creasey (Editor)
Volume 19 **Telecommunications Traffic, Tariffs and Costs** R.E. Farr
Volume 20 **An introduction to Satellite Communications** D.I. Dalgleish
Volume 26 **Common-channel Signalling** R.J. Manterfield
Volume 28 **Very Small Aperture Terminals (VSATs)** J.L. Everett (Editor)
Volume 29 **ATM: The broadband telecommunications solution** L.G. Cuthbert and J.C. Sapanel
Volume 31 **Data Communications and Networks, 3rd Edition** R.L. Brewster (Editor)
Volume 32 **Analogue Optical Fibre Communications** B. Wilson, Z. Ghassemlooy and I.Z. Darwazeh (Editors)
Volume 33 **Modern Personal Radio Systems** R.C.V. Macario (Editor)
Volume 34 **Digital Broadcasting** P. Dambacher
Volume 35 **Principles of Performance Engineering for Telecommunication and Information Systems** M. Ghanbari, C.J. Hughes, M.C. Sinclair and J.P. Eade
Volume 36 **Telecommunication Networks, 2nd Edition** J.E. Flood (Editor)
Volume 37 **Optical Communication Receiver Design** S.B. Alexander
Volume 38 **Satellite Communication Systems, 3rd Edition** B.G. Evans (Editor)
Volume 40 **Spread Spectrum in Mobile Communication** O. Berg, T. Berg, J.F. Hjelmstad, S. Haavik and R. Skaug
Volume 41 **World Telecommunications Economics** J.J. Wheatley
Volume 43 **Telecommunications Signalling** R.J. Manterfield
Volume 44 **Digital Signal Filtering, Analysis and Restoration** J. Jan
Volume 45 **Radio Spectrum Management, 2nd Edition** D.J. Withers
Volume 46 **Intelligent Networks: Principles and applications** J.R. Anderson
Volume 47 **Local Access Network Technologies** P. France
Volume 48 **Telecommunications Quality of Service Management** A.P. Oodan (Editor)
Volume 49 **Standard Codecs: Image compression to advanced video coding** M. Ghanbari
Volume 50 **Telecommunications Regulation** J. Buckley
Volume 51 **Security for Mobility** C. Mitchell (Editor)
Volume 52 **Understanding Telecommunications Networks** A. Valdar
Volume 53 **Video Compression Systems: From first principles to concatenated codecs** A. Bock
Volume 54 **Standard Codecs: Image compression to advanced video coding, 3rd Edition** M. Ghanbari
Volume 59 **Dynamic Ad Hoc Networks** H. Rashvand and H. Chao (Editors)
Volume 60 **Understanding Telecommunications Business** A. Valdar and I. Morfett
Volume 65 **Advances in Body-Centric Wireless Communication: Applications and state-of-the-art** Q.H. Abbasi, M.U. Rehman, K. Qaraqe and A. Alomainy (Editors)
Volume 67 **Managing the Internet of Things: Architectures, theories and applications** J. Huang and K. Hua (Editors)
Volume 68 **Advanced Relay Technologies in Next Generation Wireless Communications** I. Krikidis and G. Zheng
Volume 69 **5G Wireless Technologies** A. Alexiou (Editor)
Volume 70 **Cloud and Fog Computing in 5G Mobile Networks** E. Markakis, G. Mastorakis, C.X. Mavromoustakis and E. Pallis (Editors)
Volume 71 **Understanding Telecommunications Networks, 2nd Edition** A. Valdar
Volume 72 **Introduction to Digital Wireless Communications** Hong-Chuan Yang
Volume 73 **Network as a Service for Next Generation Internet** Q. Duan and S. Wang (Editors)

Volume 74 **Access, Fronthaul and Backhaul Networks for 5G &** Beyond M.A. Imran, S.A.R. Zaidi and M.Z. Shakir (Editors)

Volume 76 **Trusted Communications with Physical Layer Security for 5G and Beyond** T.Q. Duong, X. Zhou and H.V. Poor (Editors)

Volume 77 **Network Design, Modelling and Performance Evaluation** Q. Vien

Volume 78 **Principles and Applications of Free Space Optical Communications** A.K. Majumdar, Z. Ghassemlooy, A.A.B. Raj (Editors)

Volume 79 **Satellite Communications in the 5G Era** S.K. Sharma, S. Chatzinotas and D. Arapoglou

Volume 80 **Transceiver and System Design for Digital Communications, 5th Edition** Scott R. Bullock

Volume 81 **Applications of Machine Learning in Wireless Communications** R. He and Z. Ding (Editors)

Volume 84 **Low Electromagnetic Emission Wireless Network Technologies: 5G and beyond** M.A. Imran, F. Héliot and Y.A. Sambo (Editors)

Volume 86 **Advances in Communications Satellite Systems Proceedings of the 36th International Communications Satellite Systems Conference (ICSSC-2018)** I. Otung, T. Butash and P. Garland (Editors)

Volume 905 **ISDN Applications in Education and Training** R. Mason and P.D. Bacsich

Microstrip and Printed Antenna Design

3rd Edition

Randy Bancroft

The Institution of Engineering and Technology

Published by The Institution of Engineering and Technology, London, United Kingdom

The Institution of Engineering and Technology is registered as a Charity in England & Wales (no. 211014) and Scotland (no. SC038698).

First published 2004
Second edition 2009
Third edition 2019

The Institution of Engineering and Technology
Michael Faraday House
Six Hills Way, Stevenage
Herts, SG1 2AY, United Kingdom

www.theiet.org

British Library Cataloguing in Publication Data
A catalogue record for this product is available from the British Library

ISBN 978-1-78561-854-3 (hardback)
ISBN 978-1-78561-855-0 (PDF)

Typeset in India by MPS Limited

Contents

Preface

The first and second edition of this book were written for designers of planar microstrip antennas; the third edition expands on this. It includes details and subjects that are useful to the physical realization of microstrip and printed antennas that are generally not found in more academic works. An appendix on balun design and analysis has been added, as a number of printed antennas require a balanced feed and impedance matching. This book is written with commercial applications in mind, but is also of use for aerospace and other fields. The designs selected have been useful to the author over his career, along with ones that are clearly of utility.

There are many very complex microstrip antenna designs in the literature, but this volume emphasizes simplicity, and accessible mathematical designs, rather than offering a geometry, and rationalizing the resonances and properties in an ad hoc manner. This text includes the most recent work available from researchers in the field of microstrip and printed antennas. This book is intended to be a succinct, accessible, handbook, which provides useful, practical, simple, and manufacturable antenna designs. It also includes references which allow the reader to investigate more complex designs.

The third edition has a considerable number of additions to the material in earlier editions, which I hope will make concepts presented clearer. The efficiency analysis of rectangular and circular microstrip antennas is expanded upon. Full-wave analysis software (HFSS) made it possible to separate out radiative, conductor, and dielectric losses, and an extra loss that is thought to be some manner of surface wave loss. The fields inside of copper conductors were computed to produce as accurate modeling as possible. The impedance bandwidth of quarter-wave microstrip antennas versus relative permittivity is more thoroughly examined. Series/parallel feed structures in microstrip arrays have been added to this edition. The Vivaldi antenna section has been rewritten, demonstrating that common textbook design geometry limits the antenna's intrinsic impedance bandwidth. A number of useful designs have been added, and the appendices expanded.

This edition of the text uses updated nomenclature. First, the term dielectric constant has been eliminated. The author has measured the relative permittivity of numerous materials in transmission lines (waveguide and coax), cavities, and with microstrip methods. Dielectric permittivity generally changes with frequency and is *not constant*. The *IEEE Standard Definitions of Terms for Radio Wave Propagation* (IEEE Std 211-1997) no longer accepts the term dielectric constant.* It allows the terms permittivity,

*Thiel, D.V., "Using the Right Terms," *IEEE Antennas and Propagation Magazine*, October 2010, Vol. 52, No. 5, p. 192.

electric permittivity and dielectric permittivity. I will also use relative permittivity for permittivity normalized to free space.

Another nomenclature change is increased, and more effective use of metric quantities. Like the use of dielectric constant, centimeters are eschewed from this text. The use of metric prefixes separated by 1000 (metric triads) is implemented. The use of commas as triad separators has been eliminated, and thinspaces replace them instead. In the case of a four digit number, either a space or no space is implemented depending on context. If a table contains values up to four digits, the space is generally suppressed. If the table has entries with five digits or more, then usually the space is inserted into the four digit values. The prefix cluster around unity: centi, deci, deca, and hecto are atavistic prefixes and should not be used in a modern context.[†] It was realized that in metric building construction, the use of millimeters as the everyday small metric unit essentially eliminated the use of a decimal point on drawings, decreasing errors. The side of a house can be 23 000 mm and also immediately understood as 23 m. The use of centimeters is *forbidden* in metric construction standards. It was noted by metrication expert Pat Naughtin, that when comparing numbers in a table, choosing a metric prefix that allows integer representation produces easy cognitive comparison of numbers in tables. Microwave engineers have done this when they use 2450 MHz rather than switching to 2.450 GHz. If a frequency sweep is from say 850 MHz to 2450 MHz, it makes no sense to introduce a cognitive discontinuity by switching to GHz at 1000 MHz. There is no reason to immediately change metric prefixes at 1000. In everyday work, millimeters, grams, and milliliters allow one to use integers to express length, mass, and volume. When possible, the thickness of dielectric substrates will be given in micrometers (μm). This allows substrates to be described with integers over the microwave and millimeter wave regions of the electromagnetic spectrum.

The anachronistic term mil, is not a formally accepted unit of length, and will not be used. The appropriate metric unit: micrometers or nanometers will be employed. When an antenna design is articulated in millimeters, the dimensions will all be in millimeters. The thickness of microwave substrates are best designated in micrometers μm as it produces integer values (similar to the pre-metric integer values 10 mil, 20 mil, etc.). I have done this with comparison tables in this book, but when substrate thicknesses are part of a design in millimeters, they are presented as millimeters. For high-frequency work, micrometers can be easier to work with than millimeters and should be used appropriately. The term micron is a slang term for micrometer (μm) and should not be used in technical papers. The term micron hides the dimensions under discussion with an argot. Micron is sometimes thought to be an inch-based pre-metric quantity, introducing a source of potential error.

When describing insertion loss, the use of the pre-metric dB/inch will not be used. It is far preferable to use dB/100 mm. This produces larger numerical values than dB/inch, which often have more utility. An insertion loss of 4 dB/100 mm is effortlessly

[†]Bancroft, R., *The Dimensions of the Cosmos*, Outskirts Press, 2016.

converted to 40 dB/1 000 mm or 40 dB/m. Converting 4 dB/in to dB/foot, or dB/yard, requires more cognitive effort.

In the second edition of this book, in Appendix A, Table A.1, on p. 237, the thickness of copper foil in terms of its weight per square foot is given. Typical textbooks express 0.5 ounce copper as possessing a thickness of 0.0007 inches, or 0.01778 mm. Generally, the smallest unit of pre-metric measure in the US is the inch,[‡] which is then illegitimately assigned a metric prefix to produce a similar number of decimal places, rather than thinking in, and using, the metric system. In this edition, Table A.1 on page 291, is a metric-only table that relates grams per square meter to foil thickness in micrometers. For instance 150 grams per square meter is equivalent to a foil thickness of 16.87 μm (0.000 664 inches). The mixing of feet and inches used in the US designation is avoided, and the potential confusion it creates. Manufacturers weigh the amount of copper foil they have deposited, rather than measuring its thickness. The US designations are presented in Table A.2.

When I was an undergraduate, I saw a directional coupler with a calibration graph on it in KiloMegaCycles. We have made some progress, but there is much, much room for metric improvement.

[‡]The smallest is the barleycorn which is 1/3 inch and used to define shoe sizes.

Chapter 1
Microstrip antennas

1.1 The origin of microstrip radiators

The use of coaxial cable and parallel two wire (or "twin lead") as a transmission line can be traced to at least the 19th century. The realization of radio frequency (RF) and microwave components using these transmission lines required considerable mechanical effort in their construction. Paul Eiser envisioned using copper foil to interconnect electronic components in the 1930s. The printed circuit board (PCB) concept remained dormant until engineers required electronic circuits which would withstand a shock of around 20 000 Gs in 1941. The application of interest was the proximity fuse. The proximity fuse uses an antenna at the top of a shell to detect its proximity with respect to a target. The internal electronics of the proximity fuse utilized the first PCB [1].

The advent of commercially viable PCB techniques in the mid-twentieth century led to the realization that printed circuit versions of RF transmission lines could be developed. This would allow for a much simpler mass production of microwave components. The printed circuit analog of a coaxial cable became known as stripline. With a groundplane image providing a virtual second conductor, the printed circuit analog of two wire ("parallel plate") transmission line became known as microstrip. For those not familiar with the details of this transmission line, they can be found in Appendix C at the end of this book.

Microstrip geometries which radiate electromagnetic waves were originally contemplated in the 1950s. The realization of radiators that are compatible with microstrip transmission line is nearly contemporary with its introduction by Grieg and Englemann in 1952 [2]. The earliest known realization of a microstrip-like antenna, integrated with a microstrip transmission line was developed in 1953 by Deshamps [3,4] (Figure 1.1). By 1955, Gutton and Baissinot patented a microstrip antenna design [5].

Early microstrip lines and radiators were specialized devices developed in laboratories. No commercially available PCBs with controlled dielectric permittivity were developed during this period. The investigation of microstrip resonators that were also efficient radiators languished. The theoretical basis of microstrip transmission lines continued to be the object of academic inquiry [6].

Stripline received more interest as a planar transmission line at the time because it supports a transverse electromagnetic wave and allowed for easier analysis, design,

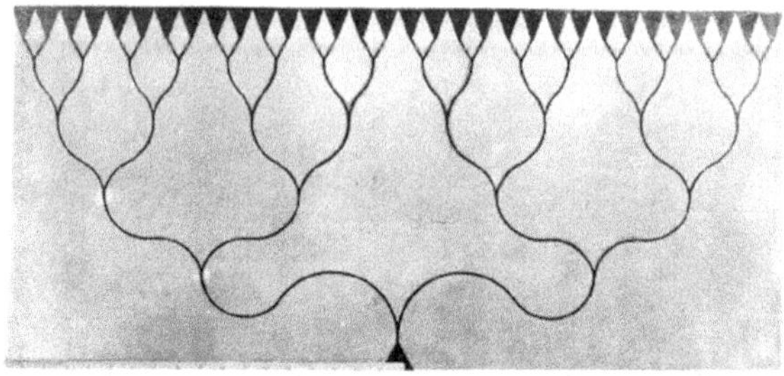

Figure 1.1 Original conformal array designed by Deshamps [3] in 1953 fed with a microstrip transmission line

and development of planar microwave structures. Stripline was also seen as an adaptation of coaxial cable, and microstrip as an adaptation of two wire transmission line. R.M. Barrett opined in 1955 that "The merits of these two systems [stripline and microstrip] are essentially the merits of their respective antecedents [coaxial cable and two wire] [7]." These viewpoints may have been part of the reason microstrip did not achieve immediate popularity in the 1950s.

The development of microstrip transmission line analysis and design methods continued in the mid to late 1960s by Wheeler [8] and Purcel *et al.* [9,10].

In 1969, Denlinger noted that rectangular and circular microstrip resonators could efficiently radiate [11]. Previous researchers had realized that in some cases, 50% of the power in a microstrip resonator would escape as radiation. Denlinger described the radiation mechanism of a rectangular microstrip resonator as arising from the discontinuities at each end of a truncated microstrip transmission line. The two discontinuities, separated by a multiple of a guide half wavelength, can be treated separately and then combined to describe the complete radiator. It was noted that the percent of radiated power to the total input power increased as the substrate thickness of the microstrip resonator increased.

These correct observations are discussed in greater detail in Chapter 2. Denlinger's results only explored increasing the substrate thickness until approximately 70% of the input power was radiated into space. Denlinger also investigated radiation from a resonant circular microstrip disc. He observed that at least 75% of the power was radiated by one circular resonator under study. In late 1969, Watkins described the fields and currents of the resonant modes of circular microstrip structures [12].

The microstrip antenna concept finally began to receive closer examination in the early 1970s when aerospace applications, such as spacecraft and missiles, produced the impetus for researchers to investigate the utility of conformal antenna designs. In 1972, Howell articulated the basic rectangular microstrip radiator fed with microstrip transmission line at a radiating edge [13]. The microstrip resonator with considerable radiation loss was now described as a microstrip antenna. A number of antenna designers received the design with considerable caution. It was difficult to believe that a resonator of this type could radiate with >90% efficiency. The narrow bandwidth

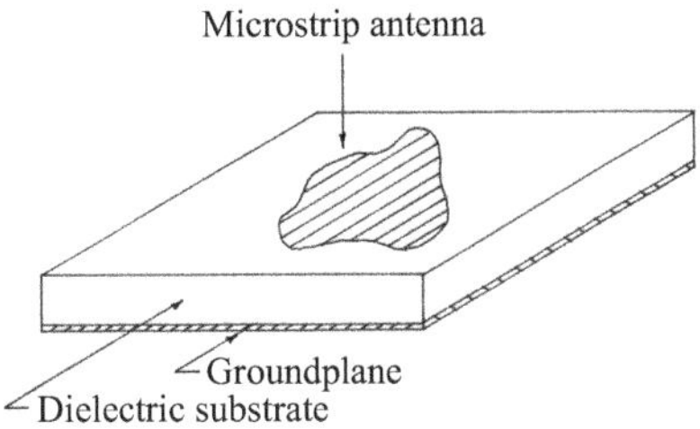

Figure 1.2 Geometry of a microstrip antenna

of microstrip antennas seemed to severely limit the number of possible applications to which the antenna could prove useful. By the late 1970s, many of these objections had not proven to derail the use of microstrip antennas in numerous aerospace applications. By 1981, microstrip antennas had become so ubiquitous and studied they were the subject of a special issue of the *IEEE Transactions on Antennas and Propagation* [14].

Today a farrago of designs have been developed which can be bewildering for designers who are new to the subject. This volume attempts to explain basic concepts and present useful designs. It will also direct the reader who wishes to research other microstrip antenna designs, which are not presented in this work, to pertinent literature.

The geometry which defines a microstrip antenna is presented in Figure 1.2. A conductive patch exists along the plane of the upper surface of a dielectric slab. This conductor area, which forms the radiating element, is generally rectangular or circular, but may be of any shape. The dielectric substrate has a groundplane on its bottom surface.

1.2 Microstrip antenna analysis methods

It was known that the resonant length of a rectangular microstrip antenna is approximately one-half wavelength, with the effective dielectric permittivity of the substrate taken into account. Following the introduction of the microstrip antenna, analysis methods were desired to determine the approximate resonant resistance of a basic rectangular microstrip radiator. The earliest useful model introduced to provide approximate values of resistance presented at the edge of a microstrip antenna is known as the transmission line model, introduced by Munson [15]. The transmission line model provides insight into the simplest microstrip antenna design, but is not complete enough to be useful when more than one resonant mode is present. In the late 1970s, Lo *et al.* developed a model of the rectangular microstrip antenna as a lossy resonant cavity [16].

Microstrip antennas, despite their simple geometry, proved to be very challenging to analyze using exact methods. In the 1980s, the method of moments became the first numerical analysis method which was computationally efficient enough so that contemporary computers could provide enough memory and CPU speed to practically analyze microstrip antennas [17–20].

The improvement of computational power and memory size of personal computers during the 1990s made numerical methods such as the finite difference time domain (FDTD) method and finite element method (FEM), which require a much more memory than moment method solutions, workable for everyday use by designers. This volume will generally use Ansoft HFSS as a full wave analysis method as well as FDTD [21,22].

1.3 Microstrip antenna advantages and disadvantages

The main advantages of microstrip antennas are

- Low-cost fabrication;
- Can easily conform to a curved surface of a vehicle or product;
- Resistant to shock and vibration (most failures are at the feed probe solder joint);
- Many designs readily produce linear or circular polarization;
- Considerable range of gain and pattern options (2.5–10.0 dBi);
- Other microwave devices realizable in microstrip may be integrated with a microstrip antenna with no extra fabrication steps (e.g. branchline hybrid to produce circular polarization or corporate feed network for an array of microstrip antennas);
- Antenna thickness (profile) is small.

The main disadvantages of microstrip antennas are

- Narrow bandwidth (5%–10% [2:1 voltage standing wave ratio] is typical without special techniques);
- Dielectric, conductor, and surface wave losses can be large for thin patches, resulting in poor antenna efficiency;
- Sensitivity to environmental factors such as temperature and humidity.

1.4 Microstrip antenna applications

A large number of commercial needs are met using microstrip and printed antennas. These include the ubiquitous global positioning system (GPS), Zigbee, Bluetooth, WiMax, Wi-Fi applications, 802.11a,b,g, and others. The most popular microstrip antenna is certainly the rectangular patch (Chapter 2). GPS applications, such as asset tracking of vehicles, and marine uses, have created a large demand for antennas. The majority of these are rectangular patches, which have been modified to produce right-hand circular polarization (RHCP), and operate at 1.575 GHz. Numerous vendors offer patches designed using ceramics with a high relative dielectric permittivity ($\epsilon_r = 6, 20, 36$) to reduce the rectangular microstrip antenna to as small a footprint as possible for a given application. Some manufactured patches are provided ready for circuit board integration with low-noise amplifiers. Rectangular patch antennas are also used for Bluetooth automotive applications (2.4 GHz) with RHCP.

In recent years, Satellite Digital Audio Radio Services have become a viable alternative to AM and FM commercial broadcasts in automobiles. The system has strict radiation pattern requirements which have been met with a combination of a printed monopole and TM_{21} mode annular microstrip antenna, which has been altered with notches to produce left-hand circular polarization at 2.338 GHz [23]. The annular microstrip antenna is addressed in Chapter 3.

Wireless local area networks (WLAN) provide short-range high-speed data connections between mobile devices (such as a laptop computer) and wireless access points. The range for wireless data links is typically around 30–90 m (100–300 feet) indoors and 600 m (2000 feet) outdoors. Wireless data links use the IEEE Standards 802.11a,b,g. The majority of WLANs use the unlicensed 2.4 GHz band (802.11b and 802.11g). The 802.11a standard uses the 5 GHz unlicensed frequency band. Multiband-printed antennas that are integrated into ceiling tiles use a microstrip diplexer (Chapter 5) to combine the signal from Global System for Mobile communication cell phones (860 MHz band), personal communications services cell phones (1.92 GHz band), and 802.11a WLAN service (2.4 GHz band) provided by two integrated microstrip dipoles [24].

WLAN systems sometimes require links between buildings that have wireless access points. This is sometimes accomplished using microstrip-phased arrays at 5 GHz (Chapter 6).

In other applications, such as warehouse inventory control, a printed antenna with an omnidirectional pattern is desired (Chapter 7). Omnidirectional microstrip antennas are also of utility for many WiMax applications (2.3, 2.5, 3.5, and 5.8 GHz are some of the frequencies currently of interest for WiMax applications) and for access points. Microstrip-fed printed slot antennas have proven useful to provide vertical polarization and integrate well into laptop computers (Chapter 7) for WLAN.

The advantages of using antennas in communication systems will continue to generate new applications which require their use. Antennas of course have the advantage of communication mobility without any required physical connection. They are the device which has enabled all the "wireless" systems that have become ubiquitous in our society. The use of transmission line, such as coaxial cable or waveguide, may have an advantage in transmission loss for short lengths, but as the distance increases the transmission loss between antennas becomes lower than any transmission line and, in some applications, outperform cables for shorter distances [25]. The amount of material cost for wired infrastructure also encourages the use of antennas in many modern communication systems.

References

[1] H. Schlesinger, *The Battery*, Harper, 2010, pp. 228–229.

[2] Grieg, D.D., and Englemann, H.F., "Microstrip—A Transmission Technique for the Kilomegacycle Range," *Proceedings of The IRE*, 1952, Vol. 40, No. 10, pp. 1644–1650.

[3] Deschamps, G.A., "Microstrip Microwave Antennas," *The Third Symposium on The USAF Antenna Research and Development Program*, University of Illinois, Monticello, Illinois, October 18–22, 1953.

[4] Bernhard, J.T., Mayes, P.E., Schaubert, D., and Mailoux, R.J., "A Commemoration of Deschamps' and Sichak's 'Microstrip Microwave Antennas': 50 Years of Development, Divergence, and New Directions," *Proceedings of the 2003 Antenna Applications Symposium*, Monticello, Illinois, September 2003, pp. 189–230.

[5] Gutton, H., and Baissinot, G., "Flat Aerial for Ultra High Frequencies," French Patent No. 703113, 1955.

[6] Wu, T.T., "Theory of the Microstrip," *Journal of Applied Physics*, March 1957, Vol. 28, No. 3, pp. 299–302.

[7] Barrett, R.M., "Microwave Printed Circuits—A Historical Survey," *IEEE Transactions on Microwave Theory and Techniques*, Vol. 3, No. 2, March 1955, pp. 1–9.

[8] Wheeler, H.A., "Transmission Line Properties of Parallel Strips Separated by a Dielectric Sheet," *IEEE Transactions on Microwave Theory of Techniques*, Vol. MTT-13, March 1965, pp. 172–185.

[9] Purcel, R.A., Massé, D.J., and Hartwig, C.P., "Losses in Microstrip," *IEEE Transactions on Microwave Theory and Techniques*, Vol. 16, No. 6, June 1968, pp. 342–350.

[10] Purcel, R.A., Massé, D.J., and Hartwig, C.P., Errata: "Losses in Microstrip," *IEEE Transactions on Microwave Theory and Techniques*, Vol. 16, No. 12, December 1968, p. 1064.

[11] Denlinger, E.J., "Radiation from Microstrip Radiators," *IEEE Transactions on Microwave Theory of Techniques*, Vol. 17, April 1969, pp. 235–236.

[12] Watkins, J., "Circular Resonant Structures in Microstrip," *Electronics Letters*, October 1969, Vol. 5, No. 21, pp. 524–525.

[13] Howell, J.Q., "Microstrip Antennas," *IEEE International Symposium Digest on Antennas and Propagation*, Williamsburg, Virginia, December 11–14, 1972, pp. 177–180.

[14] *IEEE Transactions on Antennas and Propagation*, January 1981.

[15] Munson, R.E., "Conformal Microstrip Antennas and Microstrip Phased Arrays," *IEEE Transactions on Antennas and Propagation*, January 1974, Vol. 22, No. 1, pp. 235–236.

[16] Lo, Y.T., Solomon, D., and Richards, W.F., "Theory and Experiment on Microstrip Antennas," *IEEE Transactions on Antennas and Propagations*, 1979, Vol. AP-27, pp. 137–149.

[17] Hildebrand, L.T., and McNamara, D.A., "A Guide to Implementational Aspects of the Spatial–Domain Integral Equation Analysis of Microstrip Antennas," *Applied Computational Electromagnetics Journal*, March 1995, Vol. 10, No. 1, pp. 40–51, ISSN 1054–4887.

[18] Mosig, J.R., and Gardiol, F.E., "Analytical and Numerical Techniques in the Green's Function Treatment of Microstrip Antennas and Scatterers," *IEE Proceedings*, March 1983, Vol. 130, Pt. H, No. 2, pp. 175–182.

[19] Mosig, J.R., and Gardiol, F.E., "General Integral Equation Formulation for Microstrip Antennas and Scatterers," *IEE Proceedings*, Vol. 132, Pt. H, No. 7, December 1985, pp. 424–432.

[20] Mosig, J.R., "Arbitrarily Shaped Microstrip Structures and Their Analysis with a Mixed Potential Integral Equation," *IEEE Transactions on Microwave Theory and Techniques*, February 1988, Vol. 36, No. 2, pp. 314–323.

[21] Tavlov, A., and Hagness, S.C., *Computational Electrodynamics: The Finite-Difference Time-Domain Method*, Second Edition, Artech House, 2000.

[22] Tavlov, A., Ed., *Advances in Computational Electrodynamics: The Finite-Difference Time-Domain Method*, Artech House, 1998.

[23] Licul, S., Petros, A., and Zafar, I., "Reviewing SDARS Antenna Requirements," *Microwaves & RF*, September 2003.

[24] US Patent No. US 6,307,525 B1 2017-04-07.

[25] Milligan, T., *Modern Antenna Design*, McGraw Hill, 1985, pp. 8–9.

Chapter 2
Rectangular microstrip antennas

2.1 The transmission line model

The rectangular patch antenna is very probably the most popular microstrip antenna geometry implemented by designers. Figure 2.1 shows the geometry of this antenna type. A rectangular metal patch of width $W = a$ and length $L = b$ is separated by a dielectric material from a groundplane by a distance h. The two ends of the antenna (located at 0 and b) can be viewed as radiating due to fringing fields along each edge of the antenna width W $(= a)$. The two *radiating edges* are separated by a distance L $(= b)$. The two edges along the sides of length L are often referred to as *non-radiating edges*.

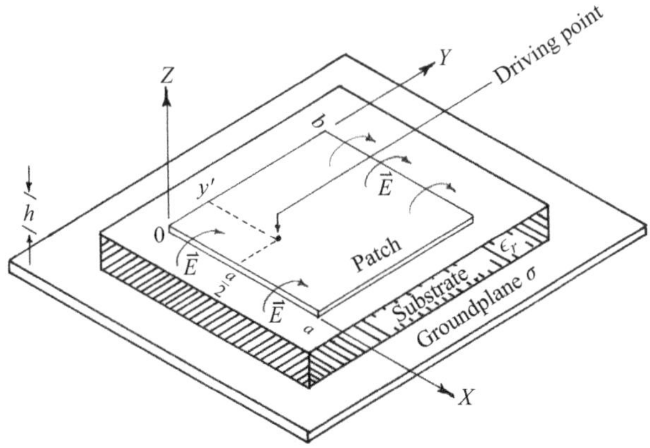

Figure 2.1 Rectangular microstrip patch geometry used to describe the transmission line model. The patch antenna is fed along the centerline of the antenna's dimension along the $\hat{x}$-axis (i.e. $x = a/2$). The feed point is located at $\acute{y}$ which is chosen to match the antenna with a desired impedance. The radiation originates from the fringing electric field at either end of the antenna. These edges are called radiating edges, *and the other two sides (parallel to the $\hat{y}$-axis) are* non-radiating edges

Numerous full wave analysis methods have been devised for the rectangular microstrip antenna [1–4]. Often these advanced methods require a considerable investment of time and effort to implement and are thus not convenient for computer-aided design (CAD) implementation.

The two analysis methods for rectangular microstrip antennas which are most popular for CAD implementation are the transmission line model and the cavity model. In this section, we will address the least complex version of the transmission line model. The popularity of the transmission line model may be gauged by the number of extensions to this model which have been developed [5–7].

The transmission line model provides a very lucid conceptual picture of the simplest implementation of a rectangular microstrip antenna. In this model, the rectangular microstrip antenna consists of a microstrip transmission line terminated with a pair of loads at either end [8,9].

As presented in Figure 2.2(a), the resistive loads at each end of the transmission line represent loss due to radiation. At resonance, the imaginary components of the input impedance seen at the driving point cancel, and therefore, the driving point impedance becomes exclusively real.

The *driving point* or *feed point* of an antenna is the location on an antenna where a transmission line is attached to provide the antenna with a source of microwave power. The impedance measured at the point where the antenna is connected to the transmission line is called the *driving point impedance* or *input impedance.* The driving point impedance Z_{drv} at any point along the centerline of a rectangular microstrip antenna can be computed using the transmission line model. The transmission line model is most easily represented mathematically using the transmission line equation written in terms of admittances, as presented in the following equation:

$$Y_{in} = Y_0 \frac{Y_L + jY_0 \tan(\beta L)}{Y_0 + jY_L \tan(\beta L)} \tag{2.1}$$

where Y_{in} is the input admittance at the end of a transmission line of length $L(=b)$, which has a *characteristic admittance* of Y_0 and a phase constant of β terminated with a complex load admittance Y_L. In other words, the microstrip antenna is modeled as a microstrip transmission line of width $W(=a)$, which determines the characteristic admittance, and is of physical length $L(=b)$ and loaded at both ends by an edge admittance Y_e, which models the radiation loss. This is shown in Figure 2.2(a).

Using 2.1, the driving point admittance $Y_{drv} = 1/Z_{drv}$ at a driving point located between the two radiating edges is expressed as

$$Y_{drv} = Y_0 \left[\frac{Y_e + jY_0 \tan(\beta L_1)}{Y_0 + jY_e \tan(\beta L_1)} + \frac{Y_e + jY_0 \tan(\beta L_2)}{Y_0 + jY_e \tan(\beta L_2)} \right] \tag{2.2}$$

where Y_e is the complex admittance at each radiating edge, which consists of an edge conductance G_e and edge susceptance B_e, as related in (2.3). The two loads are separated by a microstrip transmission line of characteristic admittance Y_0:

$$Y_e = G_e + jB_e \tag{2.3}$$

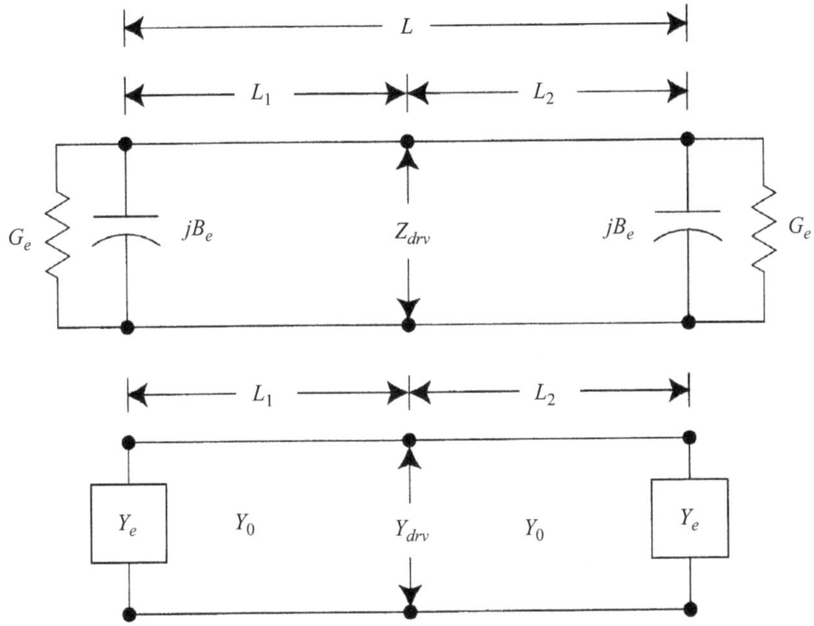

(a) Feed point between radiating edges

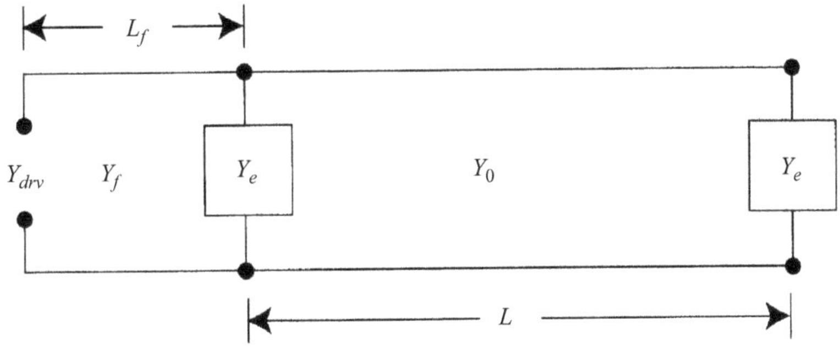

(b) Transmission line feed at radiating edge

Figure 2.2 *(a) The transmission line model of a rectangular microstrip antenna is a transmission line with two terminating loads. A driving point is chosen along the antenna length L which can be represented as a sum of* L_1 *and* L_2*. The two transmission line sections each contribute to the driving point impedance. The antenna is readily analyzed using a pair of edge admittances* Y_e *separated by two sections of a transmission line of characteristic admittance* Y_0*. (b) The microstrip antenna may be fed at one of its radiating edges using a transmission line. In this case, the transmission line model is augmented with a feed line of characteristic admittance* Y_f *of length* L_f *connected to a radiating edge. The driving point admittance* Y_{drv} *is then computed at the end of this feed line*

Approximate values of G_e and B_e may be computed using 2.4 and 2.5 [10]:

$$G_e = 0.00836\frac{W}{\lambda_0} \tag{2.4}$$

$$B_e = 0.01668\frac{\Delta l}{h}\frac{W}{\lambda_0}\epsilon_e \tag{2.5}$$

The effective relative dielectric permittivity ($W/h \geq 1$) is given as

$$\epsilon_e = \frac{\epsilon_r + 1}{2} + \frac{\epsilon_r - 1}{2}\left(1 + 12\frac{h}{W}\right)^{-\frac{1}{2}} \tag{2.6}$$

The fringing field extension, normalized to the substrate thickness h, is

$$\frac{\Delta l}{h} = 0.412\frac{(\epsilon_e + 0.3)(W/h + 0.264)}{(\epsilon_e - 0.258)(W/h + 0.8)} \tag{2.7}$$

The value Δl is the line extension due to the electric field fringing at the edge of the patch antenna. The physical size of a resonant microstrip patch antenna would be $\lambda_{\epsilon_e}/2$ were it not for the effect of fringing at the end of the rectangular microstrip antenna.[1] We may use equation (2.7) to correct for this effect and compute the physical length of a rectangular microstrip antenna, which will resonate at a desired design frequency f_r.

The first feed method is called a coaxial probe feed,[2] as shown in Figure 2.3(a). The outer shield of a coaxial transmission line is connected to the groundplane of the microstrip antenna. Metal is removed from the groundplane which is generally the same radius as the inside of the outer coaxial shield. The coaxial center conductor then passes through the dielectric substrate of the patch antenna and connects to the patch. Feeding the antenna in the center (i.e. at $a/2$) suppresses the excitation of a mode along the width of the antenna. This feed symmetry enforces the purest linear polarization along the length of the patch which can be achieved with a single direct feed.

The second feed method, shown in Figure 2.3(b), drives the antenna with a microstrip transmission line along a non-radiating edge. This feed method is modeled in an identical manner to the coaxial probe feed when using the transmission line model; in practice, it can often excite a mode along the width of the patch when $a \approx b$ and cause the antenna to radiate with an elliptical polarization. The advantage of this feed method is that it allows one to use a 50 Ω microstrip transmission line connected directly to a 50 Ω driving point impedance, which eliminates the need for impedance matching.

[1]This fringing is similar to the fringing at the end of a dipole antenna. The extra electrical length causes a dipole antenna to resonate at a length which is closer to 0.48λ rather than the 0.50λ expected if no end capacitance were present.

[2]SMA or other probe printed circuit board (PCB) connectors can have a significant effect on the measured driving point impedance of a microstrip or printed antenna. The connector itself should be included in a full-wave simulation (e.g. HFSS, CST, etc.) to obtain a measured driving point impedance which is consistent with analysis. Model what you build, and build what you model.

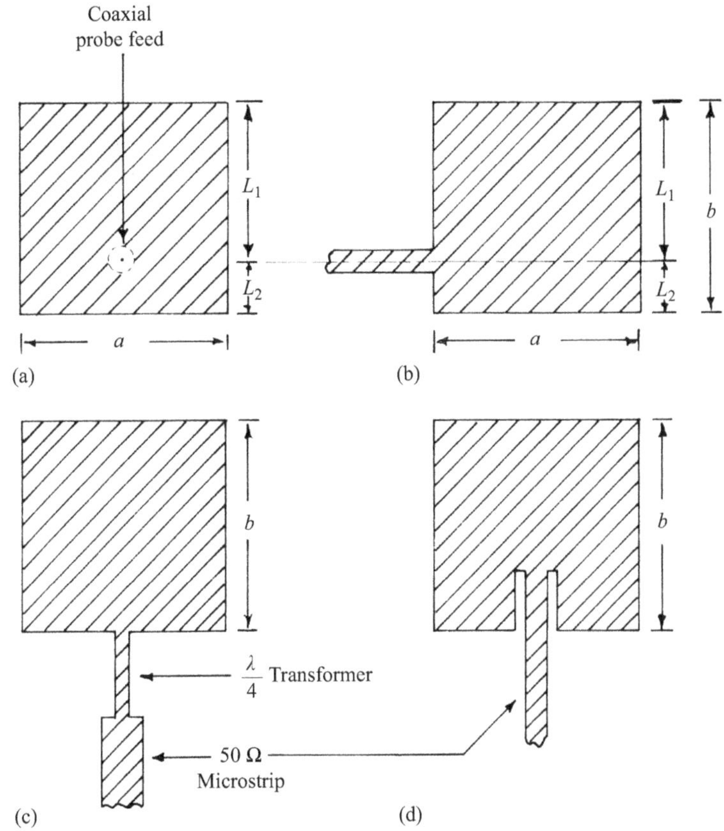

Figure 2.3 *Common methods used to feed a rectangular microstrip antenna. (a) Coaxial feed probe. (b) Microstrip transmission line feed along a nonradiating edge. (c) Microstrip transmission feed along a radiating edge. (d) Microstrip feed line into a cutout in a radiating edge which is inset to a 50 Ω driving point*

The third feed method, Figure 2.3(c), is to drive the antenna at one of its radiating edges with a microstrip transmission line. This disturbs the field distribution along one radiating edge which causes slight changes in the radiation pattern. The impedance of a typical resonant rectangular ($a < 2b$) microstrip antenna at a radiating edge is around 200 Ω. This edge resistance R_{in} is $1/(2G_e)$ at resonance. In general, one must provide an impedance transformation to 50 Ω for this feed method. This is often accomplished using a quarter-wave impedance transformer between the radiating edge impedance and a 50 Ω microstrip feed line. A quarter-wave transformer has a larger bandwidth than the antenna element and therefore does not limit it. It is possible to widen a rectangular microstrip antenna ($a > b$) so the edge resistance at resonance is 50 Ω.

In this special case, no impedance transformer is required to feed the antenna with a 50 Ω microstrip transmission line at a radiating edge.

A fourth feed method, illustrated in Figure 2.3(d), is to cut a narrow notch out of a radiating edge far enough into the patch to locate a 50 Ω driving point impedance. The removal of the notch perturbs the patch fields. A study by Basilio *et al.* indicates that a probe-fed patch antenna has a driving point resistance which follows an $R_{in}\cos^2(\pi L_2/L)$, a patch with an inset feed is measured to follow an $R_{in}\cos^4(\pi L_2/L)$ function, where $0 < L_2 < L/2$ [11]. More complex relationships have also been developed [12]. One can increase the patch width, which increases the edge conductance, until at resonance, and the edge impedance is 50 Ω. The inset distance into the patch goes to zero which allows one to directly feed a patch for this special case using a 50 Ω microstrip line at a radiating edge. The patch width is large enough in this case to increase the antenna gain considerably.

Equation (2.8) may be used to compute the resonant length (L) of a rectangular microstrip antenna. The cutback value Δl is given by (2.7) and the effective relative permittivity, ϵ_e, is given by (2.6):

$$\begin{aligned} L &= \frac{c}{2f_r\sqrt{\epsilon_e}} - 2\Delta l \\ &= \frac{\lambda_{\epsilon_e}}{2} - 2\Delta l \end{aligned} \tag{2.8}$$

$$\lambda_{\epsilon_e} = \frac{\lambda_0}{\sqrt{\epsilon_e}} \tag{2.9}$$

Equation (2.2) provides a predicted input impedance at the desired design frequency f_r. Numerical methods for obtaining the roots of an equation such as the Bisection method (Appendix B) may be used with (2.2) to determine the value of L_1 and L_2 which correspond to a desired input resistance value. The initial guesses are along b at $b_1 = 0$ ($R_{in} = 1/2G_e$) and $b_2 = b/2$ ($R_{in} \approx 0$). The predicted position of a desired driving point impedance to feed the antenna is generally close to measurement as long as the substrate height is not larger than about $0.1\lambda_0$. A good rule of thumb for an initial guess to the location of a 50 Ω feed point, when determining the position in an empirical manner, is 1/3 of the distance from the center of the antenna to a radiating edge, inward from a radiating edge.

Early investigation of the rectangular microstrip antenna, viewed as a linear transmission line resonator, was undertaken by Derneryd [13]. The input impedance characteristics of the transmission line model were altered by Derneryd in a manner which allows for the influence of mutual conductance between the radiating edges of the patch antenna. This model further allows for the inclusion of higher order linear transmission line modes.

In 1968, an experimental method to investigate the electric fields surrounding objects was developed, which used a liquid crystal sheet (LCS) backed with a resistive thin film material [14,15]. Derneryd used a liquid crystal field detector of this type to map the electric field of a narrow microstrip antenna. Derneryd's results are reproduced in Figure 2.4 along with thermal (electric field magnitude) plots produced using

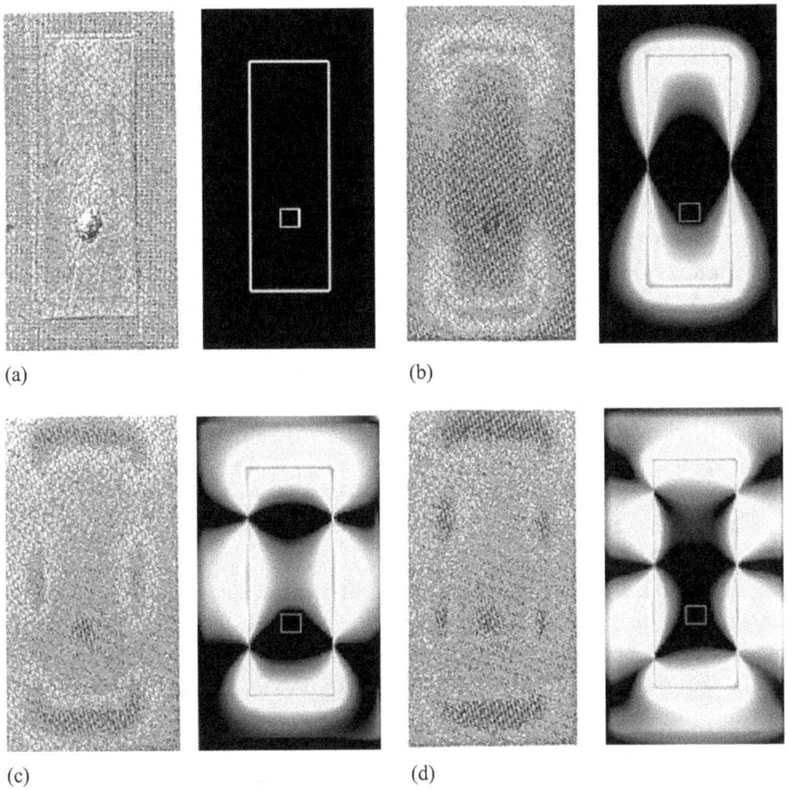

Figure 2.4 Electric field distribution surrounding a narrow patch antenna as computed using FDTD and measured using a liquid crystal sheet: (a) patch without fields, (b) 3.10 GHz, (c) 6.15 GHz, and (d) 9.15 GHz. After Dernerryd [13]

the finite difference time domain (FDTD) method. The FDTD patch analysis used $a = 10.0$ mm, $b = 30.5$ mm, $\epsilon_r = 2.55$, $h = 1.5875$ mm, and $\tan\delta = 0.001$. The feed-point location is 5.58 mm from the center of the patch antenna along the centerline. The groundplane is 20 mm × 42 mm.

Figure 2.4(a) is the antenna without an electric field present. Figure 2.4(b), just to the right of (a), is Derneryd's element analyzed with a thermal LCS, which shows the first (lowest order) mode of this antenna. The frequency for this first mode is reported to be 3.10 GHz. A sinusoidal source at 3.10 GHz with FDTD was used to model this antenna. The FDTD plot is of the total magnitude of the electric field in the plane of the antenna. The FDTD simulation thermal plot is very similar to the shape of the measured LCS thermal pattern. We see two radiating edges at either end of the antenna in the lowest mode with two non-radiating edges on the sides.

Figure 2.4(c) has Derneryd's measured LCS results with the antenna driven at 6.15 GHz. The LCS visualization shows the next higher order mode one would expect from transmission line theory. The electric field seen at either side of the center of the patch antenna along the non-radiating edges still contribute little to the antenna's radiation. In the far field[3], the radiation contribution from each side of the non-radiating edges cancel. The FDTD thermal plot result in Figure 2.4(c) is once again very similar in appearance to Derneryd's LCS thermal measurement at 6.15 GHz.

The next mode is reported by Derneryd to exist at 9.15 GHz. The measured LCS result in Figure 2.4(d) and the theoretical FDTD thermal plot once again have good correlation. As before, the radiation from the non-radiating edges will cancel in the far field.

The LCS method of measuring the near fields of microstrip antennas is still used, but other photographic and probe measurement methods have been developed as an aid to the visualization of the fields around microstrip antennas [16–19].

2.2 The cavity model

The transmission line model is conceptually simple, but has a number of drawbacks. The transmission line model is often inaccurate when used to predict the impedance bandwidth of a rectangular microstrip antenna for thin substrates. The transmission line model also does not take into consideration the possible excitation of modes which are not along the linear transmission line. The transmission line model assumes that the currents flow in only one direction along the transmission line. In reality, currents transverse to these assumed currents can exist in a rectangular microstrip antenna. The development of the cavity model addressed these difficulties.

The cavity model, originated in the late 1970s by Lo *et al.*, views the rectangular microstrip antenna as an electromagnetic cavity, with electric walls at the groundplane and the patch, and magnetic walls at each edge [20,21]. The fields under the patch are the superposition of the resonant modes of this two-dimensional radiator.[4] Equation (2.10) expresses the $(\vec{E}_z)$ electric field under the patch at a location (x, y) in terms of these modes. This model has undergone a considerable number of refinements since its introduction [22,23].

The fields in the lossy cavity are assumed to be the same as those which will exist in a thin cavity of this type. It is assumed that in this configuration where $(h \ll \lambda_0)$ only a vertical electric field $(\vec{E}_z)$ will exist, which is assumed to be constant along $\hat{z}$, and only horizontal magnetic field components ($\vec{H}_x$ and $\vec{H}_y$) exist. The magnetic

[3]The *far field* of an antenna is at a distance from the antenna where a transmitted (spherical) electromagnetic wave may be considered to be planar at the receive antenna. This distance R is generally accepted for most practical purposes to be $R \geq 2d^2/\lambda$. The value d is the largest linear dimension of transmit or receive antenna, and λ is the free space wavelength. The *near field* is a distance very close to an antenna where the reactive (non-radiating) fields are very large.

[4]The cavity model is the dual of a very short piece of rectangular waveguide which is terminated on either end with magnetic walls.

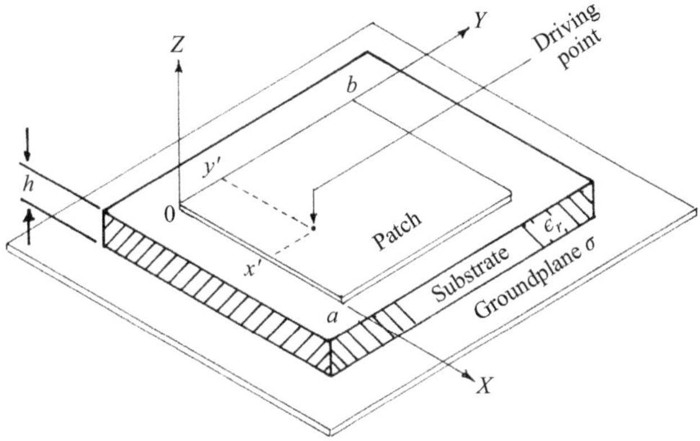

Figure 2.5 Rectangular microstrip patch geometry used for cavity model analysis

field is transverse to the $\hat{z}$-axis (Figure 2.5), and the modes are described as TM_{mn} modes (m and n are integers). The electric current on the rectangular patch antenna is further assumed to equal zero normal to each edge. Because the electric field is assumed to be constant along the $\hat{z}$ direction, one can multiply (2.10) by h to obtain the voltage difference from the patch to the groundplane. The driving point current can be mathematically manipulated to produce the ratio of voltage to current on the left side of (2.10). This creates an expression which can be used to compute the driving point impedance (2.15) at an arbitrary point $(\acute{x}, \acute{y})$ as illustrated in Figure 2.5:

$$E_z = \sum_{m=0}^{\infty}\sum_{n=0}^{\infty} A_{mn}\Phi_{mn}(x,y) \tag{2.10}$$

$$A_{mn} = j\omega\mu\frac{\langle J_z, \Phi_{mn}\rangle}{\langle \Phi_{mn}, \Phi_{mn}\rangle}\left(\frac{1}{k_c^2 - k_{mn}^2}\right) \tag{2.11}$$

$$\Phi_{mn}(x,y) = \cos\left(\frac{m\pi x}{a_{eff}}\right)\cos\left(\frac{n\pi y}{b_{eff}}\right) \tag{2.12}$$

The cavity walls are slightly larger electrically than they are physically due to the fringing field at the edges; therefore, we extend the patch boundary outward and the new dimensions become $a_{eff} = a + 2\Delta$ and $b_{eff} = b + 2\Delta$ which are used in the mode expansion. The effect of radiation and other losses is represented by lumping them into an effective dielectric loss tangent as defined in (2.19):

$$k_c^2 = \epsilon_r(1 - j\delta_{eff})k_0^2 \tag{2.13}$$

$$k_{mn}^2 = \left(\frac{m\pi}{a_{eff}}\right) + \left(\frac{n\pi}{b_{eff}}\right) \tag{2.14}$$

The driving point impedance at $(\acute{x}, \acute{y})$ may be calculated using

$$Z_{drv} = \sum_{m=0}^{\infty}\sum_{n=0}^{\infty} \frac{j\omega\alpha_{mn}}{\omega_{mn}^2 - (1 - j\delta_{eff})\omega^2} \tag{2.15}$$

$$\omega_{mn} = \frac{c_0 k_{mn}}{\sqrt{\epsilon_r}} \tag{2.16}$$

$$\alpha_{mn} = \frac{h\delta_m\delta_n}{a_{eff}b_{eff}\epsilon_0\epsilon_r}\cos^2\left(\frac{m\pi\acute{x}}{a_{eff}}\right)\cos^2\left(\frac{n\pi\acute{y}}{b_{eff}}\right)\text{sinc}^2\left(\frac{m\pi w_p}{2a_{eff}}\right) \tag{2.17}$$

where w_p is the width of the feed probe.

$$\delta_i = \begin{cases} 1 & \text{if } i = 0 \\ 2 & \text{if } i \neq 0 \end{cases} \tag{2.18}$$

The effective loss tangent for the cavity is computed from the total Q of the cavity:

$$\delta_{eff} = \frac{1}{Q_T} = \frac{1}{Q_d} + \frac{1}{Q_c} + \frac{1}{Q_r} + \frac{1}{Q_{sw}} \tag{2.19}$$

The total quality factor of the cavity Q_T consists of four components: Q_d, the dielectric loss; Q_c, the conductor loss; Q_r, the radiation loss; and Q_{sw}, the surface wave loss:

$$Q_d = \frac{1}{\tan\delta} \tag{2.20}$$

$$Q_c = \frac{1}{2}\eta_0\mu_r\left(\frac{k_0 h}{R_s}\right) \tag{2.21}$$

$$R_s = \sqrt{\frac{\omega\mu_0}{2\sigma}} \tag{2.22}$$

$$Q_r = \frac{2\omega W_{es}}{P_r} \tag{2.23}$$

where W_{es} is the energy stored:

$$W_{es} = \frac{\epsilon_0\epsilon_r ab V_0^2}{8h} \tag{2.24}$$

The power radiated into space is P_r [24]:

$$P_r = \frac{V_0^2 A\pi^4}{23{,}040}\left[(1-B)\left(1-\frac{A}{15}+\frac{A^2}{420}\right)+\frac{B^2}{5}\left(2-\frac{A}{7}+\frac{A^2}{189}\right)\right] \tag{2.25}$$

$$A = \left(\frac{\pi a}{\lambda_0}\right)^2 \tag{2.26}$$

$$B = \left(\frac{2b}{\lambda_0}\right)^2 \tag{2.27}$$

where V_0 is the input (driving point) voltage.

The Q of the surface wave loss Q_{sw} is related to the radiation quality factor Q_r [25]:

$$Q_{sw} = Q_r\left(\frac{e_r^{hed}}{1-e_r^{hed}}\right) \tag{2.28}$$

$$e_r^{hed} = \frac{P_r^{hed}}{P_r^{hed}+P_{sw}^{hed}} \tag{2.29}$$

$$P_r^{hed} = \frac{(k_0h)^2(160\pi^2\mu_r^2c_1)}{\lambda_0^2} \tag{2.30}$$

$$c_1 = 1-\frac{1}{n_1^2}+\frac{2}{5n_1^4} \tag{2.31}$$

$$n_1 = \sqrt{\epsilon_r\mu_r} \tag{2.32}$$

$$P_{sw}^{hed} = \frac{\eta_0k_0^2}{8}\,\frac{\epsilon_r(x_0^2-1)^{3/2}}{\epsilon_r(1+x_1)+k_0h\sqrt{x_0^2-1}(1+\epsilon_r^2x_1)} \tag{2.33}$$

$$x_1 = \frac{x_0^2-1}{\epsilon_r-x_0^2} \tag{2.34}$$

$$x_0 = 1+\frac{-\epsilon_r^2+\alpha_0\alpha_1+\epsilon_r\sqrt{\epsilon_r^2-2\alpha_0\alpha_1+\alpha_0^2}}{(\epsilon_r^2-\alpha_1^2)} \tag{2.35}$$

$$\alpha_0 = \sqrt{\epsilon_r-1}\tan(k_0h\sqrt{\epsilon_r-1}) \tag{2.36}$$

$$\alpha_1 = -\left[\frac{\tan(k_0h\sqrt{\epsilon_r-1})+\frac{k_0h\sqrt{\epsilon_r-1}}{\cos^2(k_0h\sqrt{\epsilon_r-1})}}{\sqrt{\epsilon_r-1}}\right] \tag{2.37}$$

The cavity model is conceptually accessible and readily implemented, but its accuracy is limited by assumptions and approximations, which are only valid for

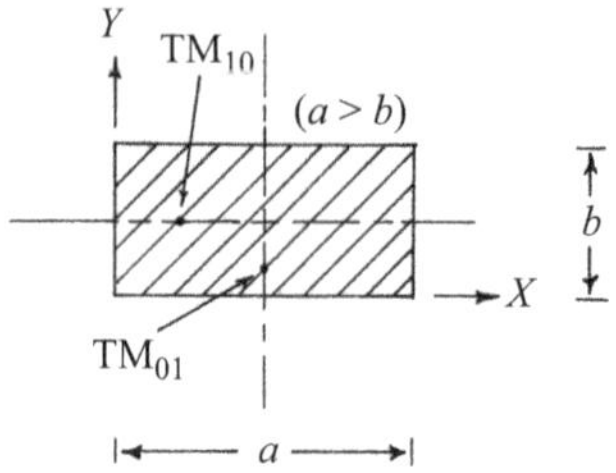

Figure 2.6 When $(a > b)$, the TM_{10} mode is the lowest order mode (lowest resonant frequency) for a rectangular microstrip antenna. The TM_{01} mode has the next highest resonant frequency

electrically thin substrates. The self-inductance of a coaxial probe used to feed the rectangular microstrip antenna is not included in this model. The cavity model is generally accurate in its impedance prediction and is within 3% of measured resonant frequency for a substrate thickness of $0.02\lambda_0$ or less. When it is thicker than this, anomalous results *may* occur [26].

2.2.1 The TM_{10} and TM_{01} modes

When a rectangular microstrip antenna has its dimension a wider than dimension b and is fed along the centerline of dimension b, only the TM_{10} mode is driven. When it is fed along the centerline of dimension a, only the TM_{01} mode is driven.

When the geometric condition $a > b$ is met, the TM_{10} mode is the lowest order mode and possesses the lowest resonant frequency of all the time harmonic modes. The TM_{01} mode is the next highest order mode and has the next lowest resonant frequency (Figure 2.6).

When $b > a$ the situation is reversed, TM_{01} becomes the mode with the lowest resonant frequency and TM_{10} has the next lowest resonant frequency.

If $a = b$, the two modes TM_{10} and TM_{01} maintain their orthogonal nature, but have identical resonant frequencies.

The integer mode index m of TM_{mn} is related to half-cycle variations of the electric field under the rectangular patch along a. Mode index n is related to the number of half-cycle electric field variations along b. In the case of the TM_{10} mode, the electric field is constant across any slice through b (i.e. the $\hat{y}$ direction), and a single half-cycle variation exists in any cut along a (i.e. the $\hat{x}$ direction). Figure 2.4 shows a narrow patch driven in the TM_{01}, TM_{02}, and TM_{03} mode according to cavity model convention.

One notes that the electric field is equal to zero at the center of a rectangular patch for both the TM_{10} and TM_{01} modes. This allows a designer the option of placing a shorting pin in the center of the rectangular patch without affecting the generation of either of the two lowest order modes. This added shorting pin or via forces the groundplane and rectangular patch to maintain an equivalent DC electrostatic potential. In many cases, the build-up of static charge on a patch antenna is undesirable

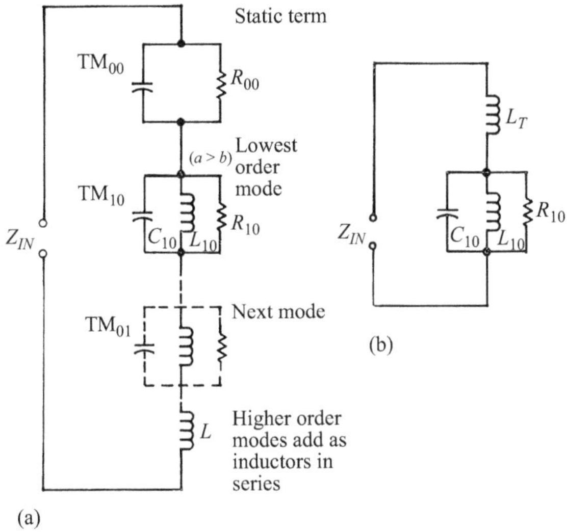

Figure 2.7 Network models used to represent a rectangular microstrip antenna: (a) general model and (b) narrow band model which is valid for the TM_{10} mode

from an electrostatic discharge (ESD) point of view, and a via may be placed in the center of the rectangular patch to address the problem.

Figure 2.7(a) shows the general network model used to represent a rectangular microstrip antenna. The TM_{00} mode is the static (DC) term of the series [27]. As described previously, the TM_{10} and/or TM_{01} are two lowest order modes which are generally driven in most applications. When this is the case, the other higher order modes are below cut-off and manifest their presence as an infinite number of small inductors which add in series with the driving point impedance. The convergent sum of these inductances may be lumped into a single series inductor which represents the contribution of the higher order modes to the driving point impedance. As the substrate thickness h of a microstrip patch increases, the contribution of the equivalent series inductance of the higher order modes to the driving point impedance becomes larger and larger, which produces a larger and larger mismatch, until the patch antenna can no longer be matched by simply choosing an appropriate feedpoint location. The cavity model does not include the small amount of intrinsic self-inductance introduced by a coaxial feed probe [26]. Increasing the thickness of the substrate also increases the impedance bandwidth of the element. These two properties (impedance bandwidth and match) may need to be traded off in a design.

The cavity model is accurate enough to use for many engineering designs. Its advantage is that it is expressed with closed-form equations, which allow efficient computation and ease of implementation. Its disadvantage is its accuracy when compared with more rigorous methods.

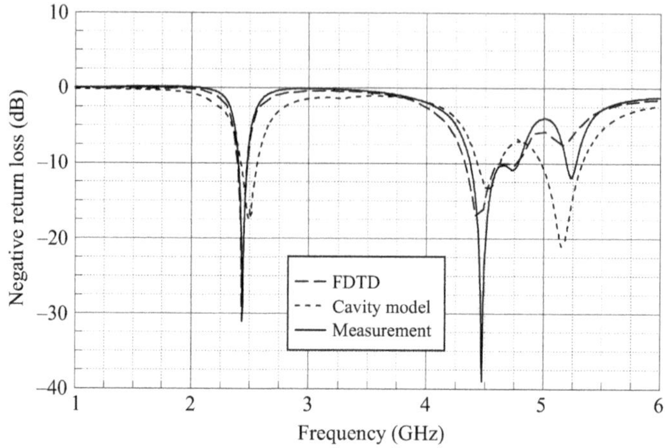

Figure 2.8 Comparison of predicted negative return loss of a rectangular microstrip patch (of parameters in Table 2.1) by the cavity model and FDTD to measurement

Table 2.1 A 2.45 GHz linear microstrip antenna (groundplane dimensions = 63.5 mm × 63.5 mm)

a	b	h	ϵ_r	tan δ	$\acute{x}$	$\acute{y}$
34.29 mm	30.658 mm	3.048 mm	3.38	0.0027	$a/2$	7.733 mm

The cavity model equations presented previously were implemented for a rectangular patch antenna with $a = 34.29$ mm and a resonant length of $b = 30.658$ mm (TM_{01}). The feed point is 7.595 mm from the center of the patch, $\acute{x} = a/2$ and $\acute{y} = 7.734$ mm. The dielectric thickness is $h = 3.048$ mm (0.120 inch) with $\epsilon_r = 3.38$ and tan $\delta = 0.0027$ (these values are in Table 2.1). The measured maximum return loss of a patch fabricated using these dimensions is 30.99 dB at 2.442 GHz. The FDTD method was also used to analyze this patch antenna. The impedance results for the cavity model, FDTD, and measurement is presented in Figure 2.8. The cavity model predicts a maximum return loss at 2.492 GHz, which is about a 2% error versus measurement. FDTD predicts 2.434 GHz which is a 0.33% error. These resonance values are presented in Table 2.2. The cavity model predicts a larger bandwidth for the first resonance than is actually measured, and it is fairly good at predicting the next higher resonance but then deviates significantly. The groundplane size of the fabricated antenna, also used in the FDTD analysis, is 63.5 mm × 63.5 mm with the dielectric flush to each groundplane edge.

Table 2.2 Microstrip antenna resonance values

Analysis method	**Resonance frequency**
Cavity model	2.492 GHz
Measurement	2.442 GHz
FDTD	2.434 GHz

2.3 Radiation pattern of a linear rectangular patch

The transmission line model, combined with the measured and computed thermal plots, suggests a model for the computation of radiation patterns of a rectangular microstrip patch antenna in the TM_{01} mode. The fringing fields at the edge of a microstrip antenna which radiate are centered about each edge of the antenna. This implies that the radiation pattern would be comparable to a pair of radiating slots centered about each radiating edge of the patch driven in phase. These slots can be viewed as equivalent to slots in a groundplane with a uniform electric field across them. This is illustrated in Figure 2.9(a). In Figure 2.9(b) is an FDTD thermal plot of the magnitude of the electric field distribution of a microstrip antenna cut through the $\hat{x}$–$\hat{y}$ plane. We can see that the two radiating edges, and the fields which radiate, form a semi-circle about each edge. The electric field extends outward from each edge along the dielectric substrate about the same amount as the dielectric thickness.

The radiating slots have a length b and are estimated to be of h (the substrate thickness) across. The two slots form an array. When the dielectric substrate is air $\epsilon_r \approx 1.0$, the resonant length a is nearly $\lambda_0/2$. When a pair of radiation sources have this spacing in free space, the array produces a maximum directivity. As the dielectric permittivity increases, the resonant length of the patch along a decreases, which decreases the spacing between the radiating slots. The slots no longer optimally add broadside to the rectangular microstrip antenna, which decreases the directivity and hence increases the pattern beamwidth.

The electric field from a single slot with a voltage across the slot of V_0 is given as [28]:

$$E_\phi = -j2V_0bk_0\frac{e^{-jk_0r}}{4\pi r}F(\theta,\phi) \tag{2.38}$$

$$E_\theta = 0 \tag{2.39}$$

$$F(\theta,\phi) = \frac{\sin(k_0(h/2)\sin\theta\cos\phi)}{k_0(h/2)\sin\theta\cos\phi}\cdot\frac{\sin(k_0(b/2)\cos\theta)}{k_0(b/2)\cos\theta}\sin\theta \tag{2.40}$$

$$k_0 = \frac{2\pi}{\lambda_0} \tag{2.41}$$

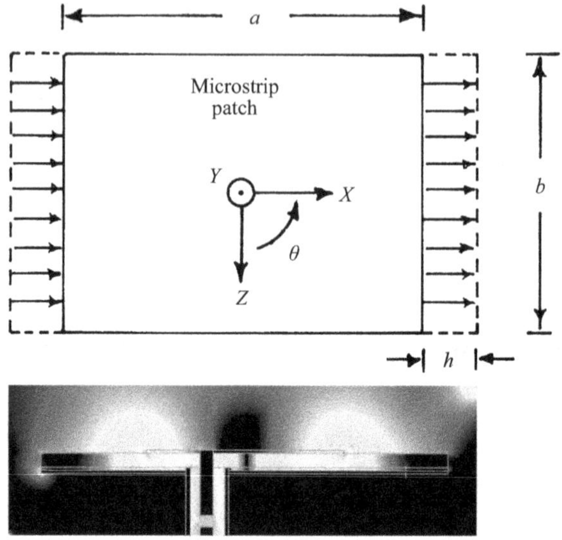

(a)

(b)

Figure 2.9 (a) Top view of a rectangular microstrip patch with a pair of equivalent slots located at a distance a *apart. The electric fields across the slots radiate in phase. (b) Side view FDTD thermal plot of the electric field for the patch analyzed in Figure 2.8 fed with a square coaxial cable. This plot demonstrates that the radiating electric fields are approximately constant at each radiating edge of the patch and extend for a distance which is nearly the thickness of the substrate. (Note that the virtual short circuit at the center of the patch, under the antenna, is clearly visible)*

For two slots spaced at a distance a apart the E-plane radiation pattern is

$$F_E(\phi) = \frac{\sin(k_0(h/2)\cos\phi)}{k_0(h/2)\cos\phi}\cos(k_0 b/2\cos\phi) \tag{2.42}$$

The H-plane pattern is independent of the slot spacing a and is given by

$$F_H(\theta) = \frac{\sin(k_0 b\cos\theta)}{k_0 b\cos\theta}\sin\theta \tag{2.43}$$

The angle θ is measured from the $\hat{z}$-axis and ϕ is measured from the $\hat{x}$-axis.

The directivity of a microstrip antenna can be approximated by starting with the directivity of a single slot [10]:

$$D = \frac{4b^2\pi^2}{I_1\lambda_0^2} \tag{2.44}$$

$$I_1 = \int_0^{\pi} \sin^2\left(\frac{k_0 b\cos\theta}{2}\right)\tan^2\theta\sin\theta d\theta \tag{2.45}$$

In the case of a microstrip antenna with a pair of radiating slots, the directivity D_s is

$$D_s = \frac{2D}{1 + g_{12}} \tag{2.46}$$

$$g_{12} = \frac{1}{120\pi^2} \int_0^{\pi} \frac{\sin^2\left(\frac{\pi b \cos\theta}{\lambda_0}\right) \tan^2\theta \sin\theta J_0\left(\frac{2\pi a}{\lambda_0} \sin\theta\right)}{G} d\theta \tag{2.47}$$

where $J_0(x)$ is the zeroth-order Bessel function with argument x.

$$G = \frac{1}{R_r} \tag{2.48}$$

where R_r is the radiation resistance,

$$R_r = \frac{120\pi^2}{I_1} \tag{2.49}$$

The integrals of (2.45) and (2.47) may be accurately evaluated numerically with Gaussian quadrature (Appendix B). The directivity estimates and pattern functions do not take groundplane effects into account and are often lower than measured. These equations are very useful for estimating the directivity and radiation pattern of a rectangular microstrip antenna. It is always best to use a more powerful technique of analysis such as FDTD or finite element method (FEM) to refine the pattern prediction of a given design.

Figure 2.10 shows measured E- and H-plane patterns of the 2.45 GHz microstrip antenna of Table 2.1 plotted with results from the slot pair model, and results using FDTD. The FDTD method results were obtained using a single frequency square coaxial source and the patterns calculated using the surface equivalence theorem [29,30]. One can see the measured and FDTD results are very similar for the upper hemisphere in both E- and H-plane patterns. Equation (2.46) was used to compute the directivity for the slot model. The E-plane slot model pattern results are close for $\pm 45°$ but begin to deviate at low angles. The H-plane slot model is close up to about $\pm 60°$. The slot model does not take groundplane affects into account, but is clearly very accurate considering the simplicity of the model used.

The important parameter which determines the directivity of a microstrip antenna is the relative dielectric permittivity, ϵ_r, of the substrate. When the substrate is air ($\epsilon_r \approx 1.0$), the two antenna edges are approximately half of a free space wavelength apart ($\lambda_0/2$). This spacing produces an array spacing for the slot model which produces maximum directivity. It is possible to achieve a directivity of almost 10 dB with an air loaded rectangular microstrip patch antenna. As the dielectric permittivity of the substrate is increased, the slots become closer to each other in terms of free space wavelengths, and no longer array to produce as high a directivity as in the free space case. As the substrate dielectric permittivity of a rectangular microstrip

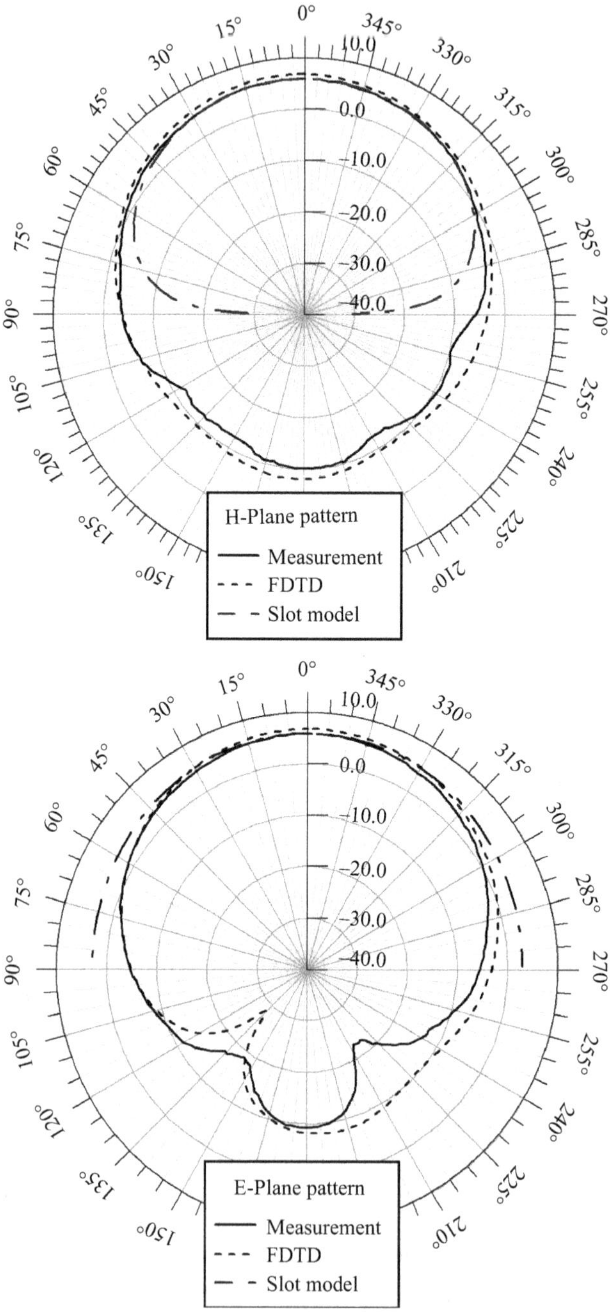

Figure 2.10 *Comparison of the measured and predicted radiation pattern of the 2.45 GHz linear microstrip antenna of Table 2.1 using FDTD analysis, and the slot model for the TM_{01} mode*

Table 2.3 Directivity (dB) of a square linear microstrip antenna versus ϵ_r (2.45 GHz, $h = 3.048$ mm, tan $\delta = 0.0005$) [square groundplane dimensions for FDTD = 63.5 mm × 63.5 mm (antenna centered)]

Slot directivity estimate vs. FDTD		
ϵ_r	**Slot model**	**FDTD**
1.0	8.83	8.00
2.6	6.56	7.11
4.1	5.93	6.82
10.2	5.24	6.54
20.0	5.01	6.45

antenna increases, the directivity of a patch antenna decreases. Table 2.3 presents a comparison of the directivity predicted by the slot model and FDTD for a square microstrip antenna. For low values of relative dielectric permittivity ($\epsilon_r < 4.0$), the slot model is within approximately 1 dB. For ($\epsilon_r > 4.0$), the directivity of the slot model is still within about 1.5 dB. The slot model can be useful for estimating directivity.

2.4 Quarter-wave microstrip antenna

Understanding the electric field distribution under a rectangular microstrip antenna allows us to develop useful variations of the original $\lambda/2$ rectangular microstrip antenna design. In the case where a microstrip antenna is fed to excite the TM_{01} mode exclusively, a virtual short circuit plane exists in the center of the antenna, parallel to the x-axis, centered between the two radiating edges. This virtual shorting plane can be replaced with a physical metal shorting plane to create a rectangular microstrip antenna which is half its original length (approximately $\lambda_{eff}/4$), as illustrated in Figure 2.11. Only a single radiating edge remains with this design, which reduces the radiation pattern directivity compared with a half-wavelength patch. This rectangular microstrip antenna design is known as a *quarter-wave microstrip patch* or *half patch* antenna. The use of a single shorting plane to create a quarter-wave patch antenna was first described by Sanford and Klein in 1978 [31]. Later, Post and Stephenson [32] described a transmission line model to predict the driving point impedance of a $\lambda/4$ microstrip antenna.

The length of a quarter-wavelength patch antenna for a given operating frequency f_r is

$$L = \frac{c}{4f_r\sqrt{\epsilon_e}} - \Delta l \tag{2.50}$$

$$= \frac{\lambda_{\epsilon_e}}{4} - \Delta l \tag{2.51}$$

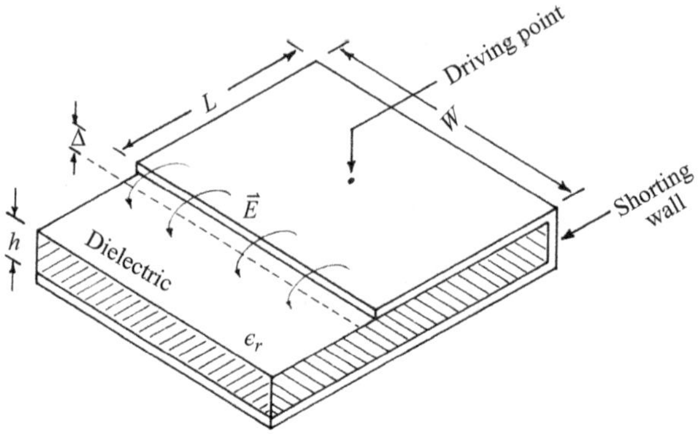

Figure 2.11 A quarter-wave microstrip antenna has a shorting wall which replaces the virtual short found in a half-wave microstrip antenna

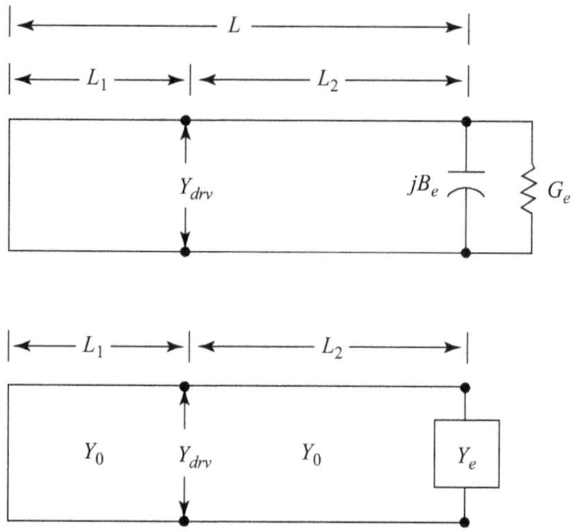

Figure 2.12 Transmission line model of a quarter-wave microstrip antenna

The cutback value Δl is given by (2.7) and the effective relative permittivity ϵ_e is given by (2.6):

$$Y_{drv} = Y_0 \frac{Y_e + jY_0 \tan(\beta L_2)}{Y_0 + jY_e \tan(\beta L_2)} - jY_0 \cot(\beta L_1) \tag{2.52}$$

The transmission line model of a quarter-wave microstrip antenna is presented in Figure 2.12. Equation (2.52) represents the driving point admittance at a point along L represented by $L = L_1 + L_2$. The final term in (2.52) is a pure susceptance at the driving point, which is due to the shorted transmission line stub. The admittance at the driving point, from the section of transmission line that translates the edge admittance Y_e along a transmission line of length L_2, resonates when its susceptance cancels the susceptance of the shorted stub. The 50 Ω input resistance location may be found from equation (2.52) with an appropriate root finding method such as the bisection method (Appendix B). The 50 Ω driving point impedance location is not exactly at the same position relative to the center short as the 50 Ω driving point location of a half-wavelength patch is to its virtual shorting plane. This is because for the case of the half-wavelength patch, two radiators exist, that have a mutual coupling term which disappears in the quarter-wavelength case. Equation (2.52) does not take this difference into account, but provides a good engineering starting point.

Computations and experimental verification reveal the impedance bandwidth of a quarter-wave microstrip antenna designed using structural foam (see A.3.7 of Appendix A) with a relative permittivity near unity have significantly larger bandwidths than their half-wave counterparts. As the value of ϵ_r becomes greater than about 2.3, the bandwidth of a quarter-wave patch has less bandwidth than its half-wavelength equivalent [33]. The bandwidth difference between quarter- and half-wave antennas appears consistent with the change in radiating edge coupling as described by James *et al.* [34].

The short circuit of the quarter-wave patch antenna is critical. To maintain the central short, considerable current must exist within it. Deviation from a low impedance short circuit will result in a significant change in the resonant frequency of the antenna and modify the radiation characteristics [35]. A design of this type often uses a single piece of sheet metal with uniform width, which is stamped into shape, and utilizes air as a dielectric substrate.

When $a = b$, the TM_{01} and TM_{10} modes have the same resonant frequency (square microstrip patch). If the patch is fed along the diagonal, both modes can be excited with equal amplitude, and in phase. This causes all four edges to become radiating edges. The two modes are orthogonal and therefore independent. Because they are in phase, the resultant of the electric field radiation from the patch is slant linear along the diagonal of the patch.

When a square microstrip patch is operating with identical TM_{01} and TM_{10} modes, a pair of shorting planes exist centered between each of the pairs of radiating slots (Figure 2.13). We can replace the virtual shorting planes, which divide the patch into four sections, with physical shorting planes. We may remove one section (i.e. quadrant) and drive it separately due to the symmetry of the modes (Figure 2.14). This produces an antenna which has one-fourth the area of a square patch antenna [36]. This provides a design option for applications in which volume is restricted.

2.5 Circular-polarized design

2.5.1 Single feed CP design

There are essentially two methods used to create rectangular circularly polarized microstrip antennas. The first is to feed the patch at a single point and perturb its boundary, or interior, so that two orthogonal modes exist at a single frequency, which have identical magnitudes, and differ in phase by 90°. The second is to directly

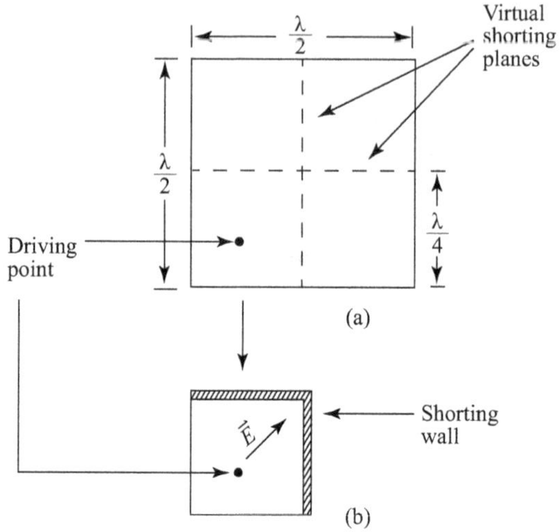

Figure 2.13 *Development of a λ/4 × λ/4 microstrip antenna from a square microstrip antenna. When a square microstrip antenna is driven along the diagonal, two virtual shorting planes appear. (a) Replacing the virtual shorting planes with physical shorting planes allows one to remove a quarter section of the original antenna and drive it independently (b)*

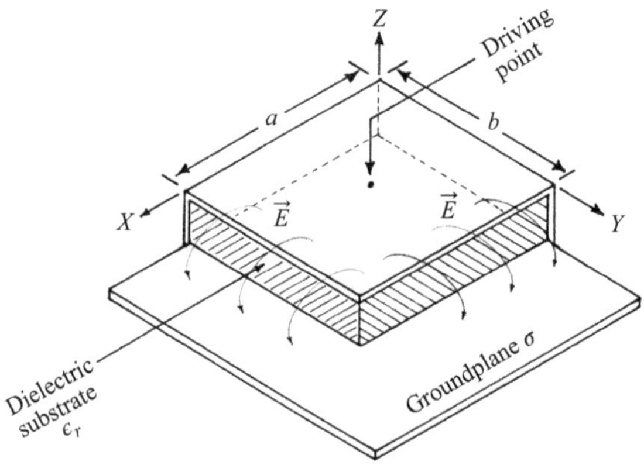

Figure 2.14 *λ/4 × λ/4 microstrip antenna*

feed each of two orthogonal modes with a microwave device which provides equal amplitudes and a 90° phase difference (e.g. 90° branchline hybrid). This section addresses the first type of design.

In Figure 2.15, we see four common methods used to create circularly polarized radiation from a rectangular microstrip antenna with a single driving point. The first

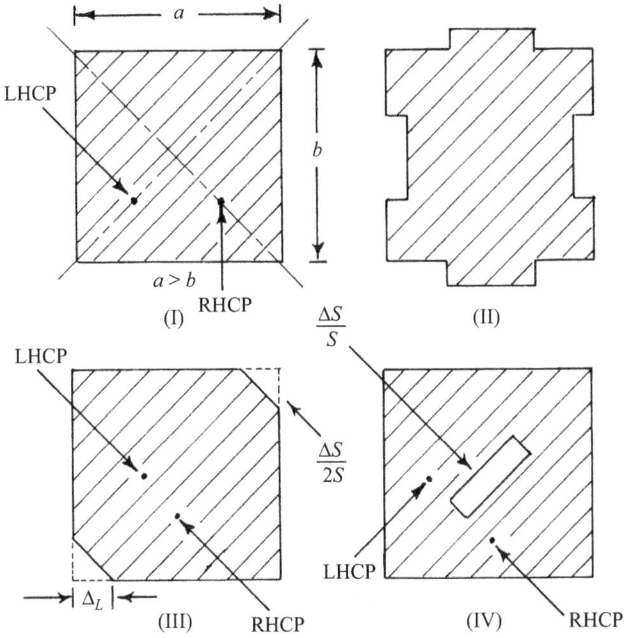

Figure 2.15 Four methods for generating circular polarization from a rectangular microstrip antenna using a single feed: (I) Using the aspect ratio of a patch to generate two orthogonal modes with equal amplitude and 90° out of phase; (II) use of indentations and/or tabs; (III) cutting off corners to create orthogonal modes; and (IV) introduction of a diagonal slot

method (I) is to choose an aspect ratio a/b such that the TM_{10} and TM_{01} modes both exist at single frequency where their magnitudes are identical and their phases differ by 90°. The two orthogonal modes radiate independently and sum in the far field to produce circular polarization.

The second method presented in Figure 2.15(II) is essentially the same as (I), but uses two rectangular tabs and two rectangular indentations to perturb the modes to produce a 90° phase difference. This situation is the most general geometry which describes a case of this type of circularly polarized patch. One could use a single tab, a single indent, a pair of tabs or a pair of indents to perturb a rectangular microstrip antenna and produce circular polarization.

The third method, illustrated in Figure 2.15(III), is to remove a pair of corners from the microstrip antenna. This creates a pair of diagonal modes (no longer TM_{10} and TM_{01} as the rectangular shape of the patch has been altered) which can be adjusted to have identical magnitudes, and a 90° phase difference between these modes.

The fourth method, illustrated in Figure 2.15(IV), is to place a slot diagonally across the patch. The slot does not disturb the currents flowing along it, but electrically lengthens the path across it. The dimensions of the slot can be adjusted to produce

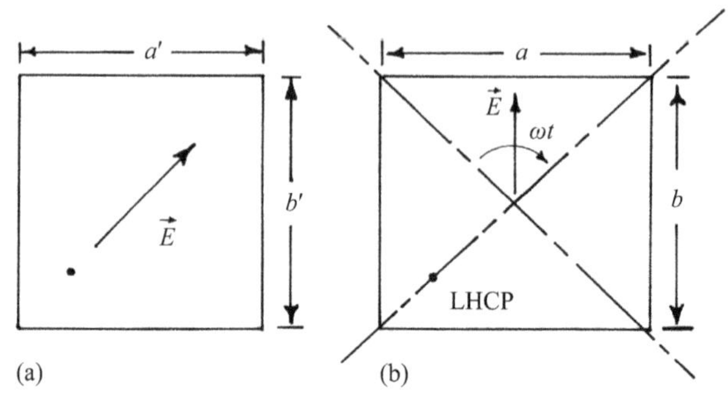

Figure 2.16 *Development of a rectangular patch with circular polarization from a square patch. (a) Square patch fed along a diagonal produces TM_{10} and TM_{01} modes which are equal in magnitude and identical in phase. These two modes add together and produce linear polarization along the diagonal of the patch antenna. (b) The ratio of a/b may be adjusted to detune each mode slightly, so that at a single frequency, the amplitudes of each mode are equal, but their phase differs by 90°, producing a rotating electric field phasor*

circular polarization. It is important to keep the slot narrow so that radiation from the slot will be minimal. One only wishes to produce a phase shift between modes and not create a secondary slot radiator. Alternatively, one can place the slot across the patch and feed along the diagonal [37].

Figure 2.16 illustrates how one designs a patch of Type (I). The figure on the left (a) shows a perfectly square patch antenna, probe fed in the lower left, along the diagonal. This patch will excite the TM_{10} and TM_{01} modes with identical amplitudes and in phase. The two radiating edges which correspond to each of the two modes have a phase center which is located at the center of the patch. Therefore, the phase center of the radiation from the TM_{10} and TM_{01} modes coincide and are located in the center of the patch. When ($a = b$), i.e. a square patch, the two modes will add in the far field to produce slant linear polarization along the diagonal. If the aspect ratio of the patch is changed so that $a > b$, the resonant frequency of each mode shifts. The TM_{10} mode shifts down in frequency, and the TM_{01} mode shifts up, compared with the original resonant frequency of the slant linear patch. Neither mode is exactly at resonance. This slightly non-resonant condition causes the edge impedance of each mode to possess a phase shift. When the phase angle of one edge impedance is +45°, and the other is −45°, the total difference of phase between the modes is 90°. This impedance relationship clearly reveals itself when the impedance versus frequency of the patch is plotted on a Smith chart. The frequency of optimum circular polarization is a point on a Smith chart which is the vertex of a vee-shaped kink.

Figure 2.17 presents the results of a cavity model analysis of a patch radiating left-hand circular polarization (LHCP) using a rectangular microstrip antenna with an appropriate a/b ratio. The antenna operates at 2.2 GHz, its substrate thickness is

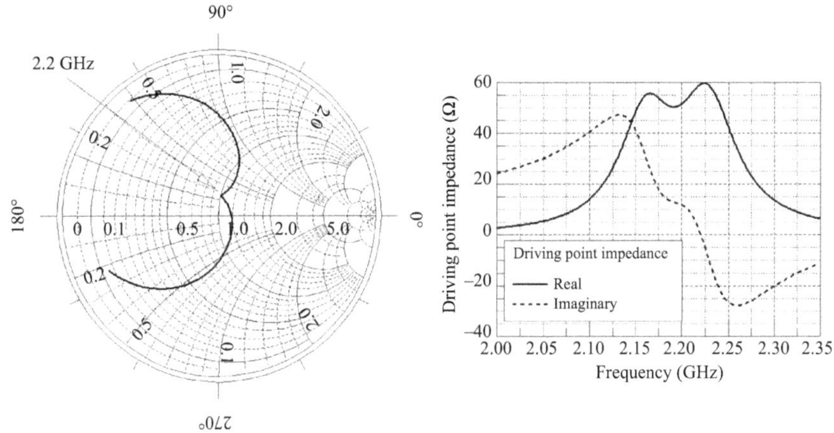

Figure 2.17 The Smith chart shows the impedance kink formed when the aspect ratio a/b has been adjusted to properly produce circular polarization. The rectangular plot shows the impedance as real and imaginary. The TM_{10} and TM_{01} mode resonant peaks, which combine to produce circular polarization, are clearly identifiable

1.5748 mm, with $\epsilon_r = 2.5$ and $\tan\delta = 0.0019$, $a = 40.945$ mm, $b = 42.25$ mm. The patch is fed at $\acute{x} = 13.5$ mm, $\acute{y} = 14.5$ mm, and $W_p = 1.3$ mm. The approximate a/b ratio was arrived upon using trial and error.

The design of a rectangular circularly polarized patch is difficult to realize due to the sensitivity of the patch to physical dimensions and dielectric permittivity. One method is to start with the case of the slant linear patch. The slant linear patch has $a = b$ and is therefore square and has its dimensions chosen to produce resonance at a desired design frequency. The ratio of a/b when the square patch aspect ratio has been adjusted to produce circular polarization has been derived using a perturbation technique [21]:

$$\frac{a}{b} = 1 + \frac{1}{Q_0} \tag{2.53}$$

The Q of the unperturbed slant linear patch (Q_0) is given by[5]

$$\frac{1}{Q_0} = \frac{1}{Q_d} + \frac{1}{Q_c} + \frac{1}{Q_r} + \frac{1}{Q_{sw}} \tag{2.54}$$

[5]The Q of a square rectangular microstrip antenna driven as a slant linear patch or as a linear patch is essentially identical. When a patch is square, the TM_{10} and TM_{10} modes are degenerate, the energy storage in the TM_{10} and the TM_{01} mode are identical as is the amount of energy loss in each for the slant linear case. If all the energy is stored in a single TM_{10} or TM_{01}, as occurs when the patch is driven in the linear case, the same total amount of energy will be lost as in the slant linear case. In both situations, the energy stored per cycle versus energy lost is the same and therefore so is the Q.

Table 2.4 2.2 GHz LHCP microstrip antenna trial-and-error design, $W_p = 1.3$ mm

Trial and error LHCP design						
a	b	h	ϵ_r	$\tan\delta$	$\acute{x}$	$\acute{y}$
40.945 mm	42.250 mm	1.5748 mm	2.5	0.0019	13.5 mm	14.5 mm

If the slant linear patch has the dimension $\acute{a}$ ($=\acute{b}$), the new dimensions of the circularly polarized patch will be

$$a = \acute{a} + \Delta_L \tag{2.55}$$

$$b = \acute{a} - \Delta_L \tag{2.56}$$

We can write

$$\Delta_L = \frac{\acute{a}}{2Q_0 + 1} \tag{2.57}$$

The use of (2.57) is illustrated by using the circularly polarized patch of Table 2.4 which has the proper impedance relationship to produce LHCP. The design values for that example were developed by adjusting the patch aspect ratio by trial and error until a circular polarization kink appeared. The center frequency of LHCP operation is 2.2 GHz.

We arrive at a slant linear patch design by taking the average of the values used to create the CP patch of Table 2.4: $(a + b)/2 = (42.250\text{ mm} + 40.945\text{ mm})/2 \approx 41.6\text{ mm}$. This average gives us a value of a slant linear patch to which we can apply (2.57) to compute an aspect ratio which should produce CP. The new patch has a resonance at 2.2 GHz, with a resistance of 88 Ω. The total Q (i.e. Q_0) from the cavity model is computed as 29.3 at 2.2 GHz. Equation (2.57) allows us to compute the length change required to produce circular polarization:

$$\Delta_L = \frac{41.6\text{ mm}}{(2 \cdot 29.3) + 1} = 0.698\text{ mm} \tag{2.58}$$

We can now find the values of a and b:

$$a = 41.6\text{ mm} + 0.698\text{ mm} = 42.298\text{ mm} \tag{2.59}$$

$$b = 41.6\text{ mm} - 0.698\text{ mm} = 40.902\text{ mm} \tag{2.60}$$

The driving point impedance of the slant linear patch and the patch modified to produce circular polarization using the a and b values computed with (2.55) and (2.56) are plotted in Figure 2.18. Again the cavity model has been used to compute the driving point impedance. It can be seen that in this case, the computation has the

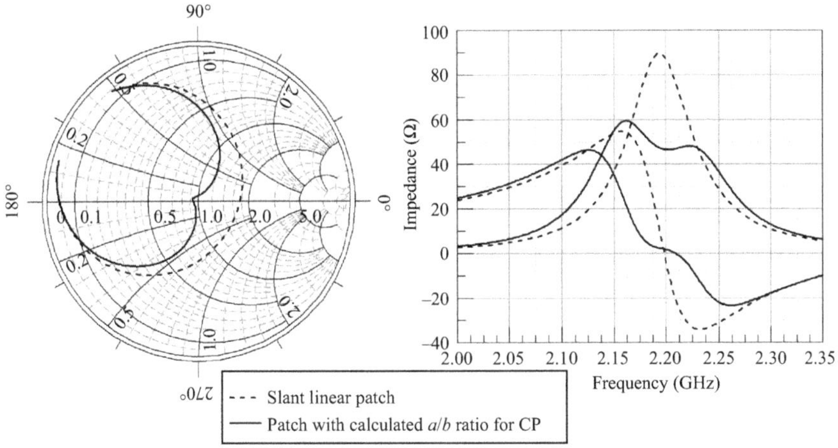

Figure 2.18 The rectangular patch antenna of Figure 2.17 has its dimensions averaged to create a slant linear patch which resonates at 2.2 GHz using cavity model analysis (dashed lines). Next, (2.55) and (2.56) are used to compute the values of a and b required to produce circular polarization at 2.2 GHz which is then analyzed using the cavity model (solid lines)

advantage that it produces a better match for the circularly polarized patch, which has been modified to produce circular polarization, then does the trial-and-error method of the original patch. The input impedance at 2.2 GHz for the patch modified to produce CP is $46.6 + j1.75\ \Omega$. This is about half of the input resistance value of the slant linear patch. This calculation provides some insight into the sensitivity of the driving point impedance location of the design to physical parameters of the patch.

The cavity model can be used to compute the axial ratio of a circularly polarized rectangular patch [38]. The relationship between electric field and axial ratio is [39]:

$$AR = \sqrt{\frac{1 + |E_x/E_y|^2 + T}{1 + |E_x/E_y|^2 - T}} \tag{2.61}$$

where AR is the axial ratio.

$$T = \sqrt{1 + \left|\frac{E_x}{E_y}\right|^4 + 2\left|\frac{E_x}{E_y}\right|^2 \cos(2\psi)} \tag{2.62}$$

where ψ is the phase of E_x/E_y.

The ratio of electric field components E_x and E_y from the cavity model may be approximated by

$$\frac{E_x}{E_y} \approx \frac{b(k_c^2 - k_{01}^2)}{a(k_c^2 - k_{10}^2)} \frac{\cos(\pi \acute{x}/a)}{\cos(\pi \acute{y}/b)} \operatorname{sinc}\left(\frac{\pi w_p}{2a}\right) \tag{2.63}$$

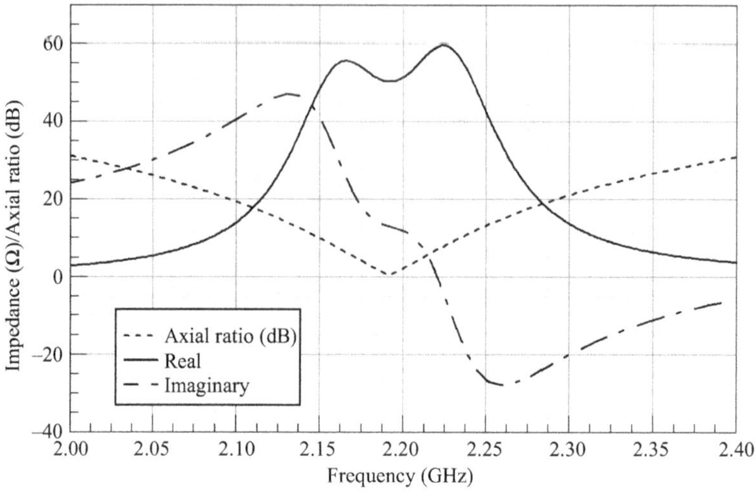

Figure 2.19 The rectangular CP patch antenna of Table 2.4 analyzed using the cavity model. The real and imaginary components of the driving point impedance are plotted with the computed axial ratio in dB

Table 2.5 2.2 GHz LHCP microstrip antenna $W_p = 1.3$ mm

Antenna Q LHCP design using equation (2.57)						
a	b	h	ϵ_r	$\tan\delta$	$\acute{x}$	$\acute{y}$
42.298 mm	40.902 mm	1.5748 mm	2.5	0.0019	13.5 mm	14.5 mm

$$\frac{E_x}{E_y} = \begin{cases} j & \text{RHCP} \\ -j & \text{LHCP} \end{cases} \tag{2.64}$$

The input impedance calculated with the values in Table 2.4, using the cavity model, is plotted with the axial ratio as a function of frequency and presented in Figure 2.19. We see the minimum axial ratio occurs between the resonant peaks and where the imaginary value flattens out. Table 2.5 illustrates that often the driving point location which produces optimum axial ratio performance and driving point match is not exactly along the patch diagonal.

The sensitivity of this type of design is such that we need accurate values of Q to obtain the most accurate value of Δ_L possible. The cavity model often does not produce as accurate values for the Q of a slant linear patch as does the FDTD method or direct measurement. When the antenna is matched and driven in a single RLC-type impedance mode, the frequency of the maximum return loss f_0, divided by the bandwidth between 3 dB values, will provide a good approximation for the Q of a patch using measured or simulated data from a full-wave analysis method. Equation (2.57) works well with the cavity model, but experience indicates that

$$\Delta_L = \frac{\acute{a}}{3.395Q_0 + 1} \tag{2.65}$$

is more appropriate when using measured or computed (e.g. FDTD) S_{11} dB results of a slant linear patch to design a circularly polarized element.

Equation (2.57) also shows that as the antenna Q increases, Δ_L decreases. When a high dielectric permittivity material is used as a substrate, the Q of the antenna becomes larger, which means that the impedance bandwidth has become narrower. The high dielectric permittivity also decreases the size of the patch, which drives down the value of Δ_L, which tightens any manufacturing tolerances considerably

A more complex iterative approach, which uses the cavity model to compute single feed circularly polarized rectangular patch designs, is presented by Lumini *et al.* [38]. Another design approach is to use a genetic algorithm optimization with the cavity model to develop a circularly polarized rectangular microstrip antenna design [40]. This method has the advantage that it optimizes for driving point match and axial ratio simultaneously. This eliminates first developing a slant linear patch and then using (2.55) and (2.56) to compute the dimensions to produce circular polarization. Experience with genetic algorithms indicates that it produces a design which is no better than the more straightforward method previously described.

Figure 2.15(II) uses indentation tabs to produce circular polarization. This type of design is undertaken experimentally.

Figure 2.15(III) has a pair of corners cut off to produce circular polarization. This creates a pair of diagonal modes (no longer TM_{10} and TM_{01} as the shape of the patch has been altered) which can be adjusted to have identical magnitudes, and a 90° phase difference between these modes. The antenna is fed along the centerline in this case so it will excite each of the diagonal modes with equal amplitude. In Figure 2.15, we see that if the upper right-hand corner, and lower left-hand corner are reduced, we can view the situation as reducing the capacitance along that diagonal making it more inductive. The opposite diagonal from lower right to upper left remains unchanged and has a larger capacitance by comparison. The amount of the area removed can be adjusted so the phase of the chopped corner diagonal is 45°, and the phase of the unmodified diagonal is −45°. This situation creates right-hand circular polarization (RHCP). Leaving the feed point position unchanged and removing the opposite pair of corners reverses the phase and thus the polarization sense (Figure 2.20).

We will define the total area removed to perturb the patch so it produces circular polarization as ΔS (Figure 2.15(III)). The total area S of the unperturbed square patch prior to the corner removal to produce circular polarization is $S = a \cdot b = \acute{a}^2$ $(a = b = \acute{a})$ (Figure 2.20). It has been reported that the ratio of the change in area ΔS, to the original area of the patch S is related to the Q of the uncut antenna Q_0 computed using (2.54) by [41]:

$$\frac{\Delta S}{S} = \frac{1}{2Q_0} \tag{2.66}$$

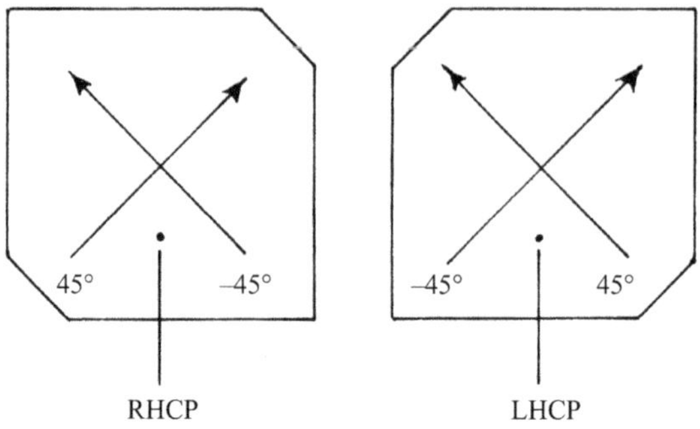

Figure 2.20 *One may cut off a pair of opposing corners of a rectangular microstrip antenna to produce circular polarization. One can view cutting off a corner as reducing the capacitance of that diagonal mode. This will produce a more inductive impedance across the two chopped corners which will cause the electric field to have a phase of 45° compared with the −45° of the electric field with the capacitive impedance across the uncropped corners. Reversing the position of the corners reverses the polarization sense*

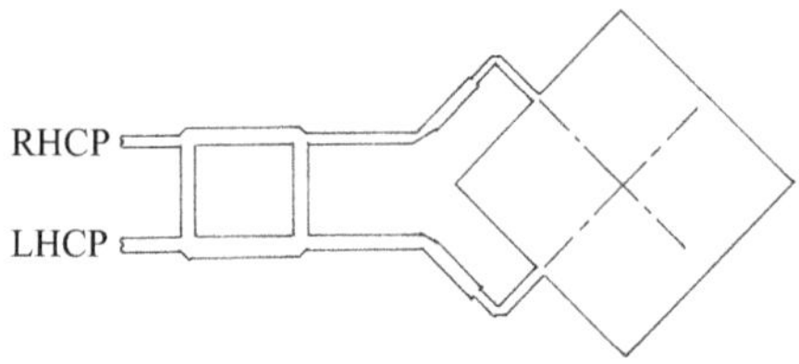

Figure 2.21 *Circular polarization may be synthesized using a 90° branchline hybrid*

The area to cut from each corner of the unperturbed patch, as shown in Figure 2.15(III), is half of the perturbation area S, calculated using (2.66) or $\Delta S/2S$. In terms of the length along each edge which is cut off, we have

$$\Delta_L = \frac{\acute{a}}{\sqrt{Q_0}} \tag{2.67}$$

Figure 2.15(IV) uses a diagonal slot to produce circular polarization. A guideline for choosing the slot area is to make it equal to $\Delta S/S$.

2.5.2 Dual-feed CP design

Figure 2.21 shows the use of a 90° branchline hybrid to feed a microstrip antenna and create circular polarization. In this case, one begins with a square microstrip

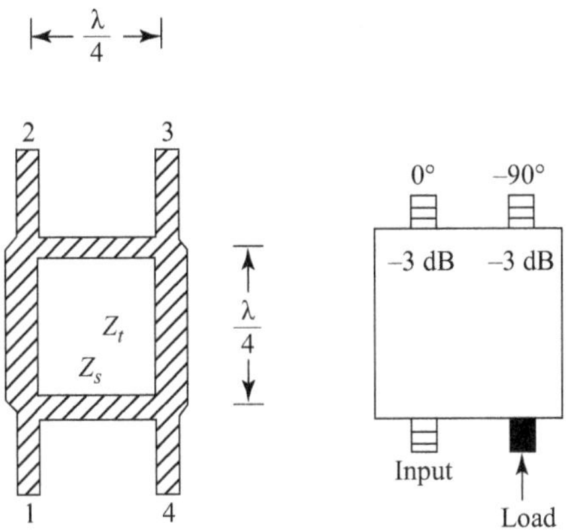

Figure 2.22 A 90° branchline hybrid realized in microstrip or stripline, and as often packaged commercially

antenna. The TM_{01} and TM_{10} modes will have the same resonant frequency and are orthogonal. Each mode is fed independently using the branchline hybrid, which provides equal amplitude, and the required 90° phase shift, at the hybrid's center frequency, to produce circular polarization. Figure 2.21 shows the branchline hybrid inputs which will produce LHCP and RHCP. In practice, if the antenna is fed RHCP, the LHCP port is terminated in a matched load and vice versa.

The branchline hybrid will enforce equal amplitudes and nearly correct phase over a wide bandwidth, but as the patch edge impedance mismatches with frequency, the rejected power will appear at the terminated port, and power is lost to maintain good circular polarization compared with a single feed design. The input impedance bandwidth and the axial ratio bandwidth are far greater than the single feed design, but when antenna efficiency is taken into account, the amount of power lost into the load of the hybrid is approximately the same as the power lost from impedance and polarization mismatch in a single feed CP antenna design. This design can also be implemented using a pair of probe feeds (one for the TM_{10} mode and one for the TM_{01} mode), and an external branchline 90° hybrid realized with coaxial cable.

2.5.3 Quadrature (90°) hybrid

The design of a rectangular patch with circular polarization described in Section 2.5.2 requires a branchline hybrid, also known as a quadrature hybrid. A branchline quadrature hybrid provides a 3 dB power split between a pair of output ports and a 90° difference between them. The left-hand illustration of Figure 2.22 shows a branchline hybrid as it would appear realized in stripline or microstrip. The shunt branches have a characteristic impedance Z_s and the through or series branch has a characteristic impedance of Z_t.

At the branch-line hybrid design frequency, the scattering parameters are [42]:

$$S_{21} = -j\frac{Z_t}{Z_0} \tag{2.68}$$

$$S_{31} = -\frac{Z_t}{Z_s} \tag{2.69}$$

$$S_{11} = 0.0 \tag{2.70}$$

$$S_{41} = 0.0 \tag{2.71}$$

The illustration on the right of Figure 2.22 shows how a commercial hybrid appears with coaxial connectors. Some hybrids have a built-in load on one port as shown, and others require the user to provide a load. This allows one to have one input which produces RHCP and the other LHCP as shown in Figure 2.21. This allows a system to switch between polarization if desired.

When a 3 dB split between ports is desired, with a reference impedance of Z_0 (generally 50 Ω), the shunt branches should have $Z_s = Z_0$, and the through branches $Z_t = Z_0/\sqrt{2}$ (35.4 Ω for a 50 Ω system). The length of the branches is $\lambda/4$. When port 1 is used as an input port, port 2 receives half of the input power and is the phase reference for port 3. Port 3 receives half of the input port power with a phase which is 90° behind port 2. The split waves cancel at port 4, which is called the isolated port. A load is generally placed on this port to absorb any imbalance, which stabilizes the phase difference between ports 2 and 3.

If port 4 is the input port, then port 1 becomes the isolation port, port 3 is the 0° phase reference port with half of the input power, and port 2 becomes the −90° port.

In practice, there is often a slight imbalance in the power split between ports 2 and 3. We note that (2.69) has Z_s in its denominator. This allows one to change the characteristic impedance of the shunt branches *slightly* and obtain a more even power split.

The bandwidth of a branchline hybrid is limited, by the quarter-wave length requirement on the branches, to 10%–20%. One must also take the discontinuities at the transmission line junctions into account to produce a design which operates as desired. One can increase the bandwidth of a branchline coupler by adding cascading sections [43]. In 2004, Qing added an extra section to produce a three-stub hybrid coupler and created a microstrip antenna design with 32.3% 2:1 voltage standing wave ratio (VSWR) bandwidth and 42.6% 3 dB axial ratio bandwidth [44].

Quadrature hybrids which have unequal power division and/or unequal characteristic impedances at each port can also be designed [45].

2.5.4 *Impedance and axial ratio bandwidth*

The impedance bandwidth of a rectangular microstrip antenna can be determined with the total Q used in the cavity model. For a linear rectangular microstrip antenna,

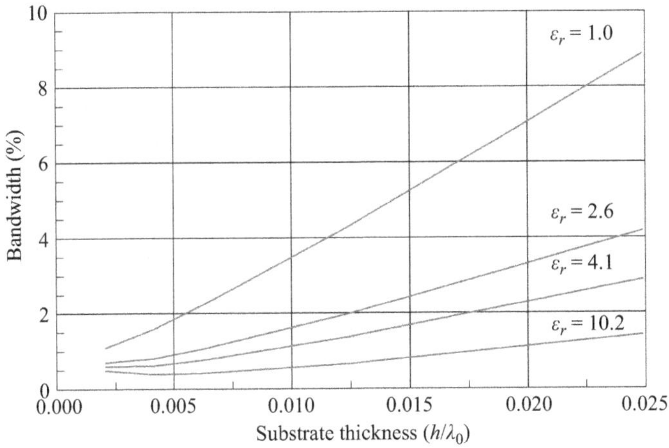

Figure 2.23 Normalized bandwidth of a square microstrip antenna as a function of substrate thickness and relative dielectric permittivity predicted by the cavity model

driven in a single mode, the normalized impedance bandwidth is related to the total Q by [46]:

$$BW_{linear} = \frac{S-1}{Q_T\sqrt{S}} \quad (S{:}1\ \text{VSWR}) \tag{2.72}$$

When a linear microstrip antenna design is very close to achieving an impedance bandwidth design goal, one can obtain a tiny amount of extra impedance bandwidth by designing the antenna to have a 65 Ω driving point resistance at resonance, rather than a perfectly matched 50 Ω input resistance. The perfect match at one frequency is traded for a slightly larger overall 2:1 VSWR bandwidth [47]. The impedance bandwidth also increases slightly when the width of the rectangular microstrip antenna is increased. The largest bandwidth increase occurs as the substrate relative dielectric permittivity ϵ_r is *decreased* and/or the substrate thickness is increased. The affect substrate thickness and dielectric permittivity have on impedance bandwidth, as computed with the cavity model, is illustrated in Figure 2.23 for a square linearly polarized microstrip antenna.

One must recall that as the substrate thickness is increased, higher order modes provide a larger and larger contribution to an equivalent series inductance, which in turn produces a larger and larger driving point mismatch. A desirable driving point impedance must be traded for impedance bandwidth.

Equations (2.73) and (2.74) have been developed to relate the impedance bandwidth of a rectangular patch antenna radiating circular polarization to total Q as well as its expected axial ratio bandwidth. We can substitute $S = 2$ in (2.72) and (2.73) forming the ratio of circular to linear bandwidth. This reveals that the impedance bandwidth of a circularly polarized microstrip antenna compared to a linear antenna is larger by a factor of two. The two detuned resonances which sum to create circular polarization increase the total impedance bandwidth:

$$BW_{circular} = \frac{\sqrt{2(S-1)}}{Q_T} \tag{2.73}$$

$$BW_{axialratio} = \frac{AR-1}{Q_T\sqrt{AR}} \tag{2.74}$$

The received power (P_R) bandwidth of a patch is independent of polarization and given by

$$BW_{P_R} = \frac{2}{Q_T}\sqrt{\frac{1-p_{min}}{p_{min}}} \tag{2.75}$$

where p is the fraction of power received by a matched load (load resistance is equal to driving point resistance at resonance) to the power received by the antenna at its resonant frequency ($0 < p < 1$). The received power reaches maximum when $p = 1$ and becomes zero when $p = 0$. In (2.75), p_{min} is the minimum acceptable receive power coefficient for a given design.

Langston and Jackson have written the above expressions in terms of a normalized frequency variable for comparison [48]. The axial ratio bandwidth is the smallest for a transmitting single feed circularly polarized patch. The receive-power bandwidth is larger than the axial ratio or impedance bandwidth.

2.6 Efficiency

The antenna efficiency e relates the gain and directivity of an antenna:

$$G = eD \tag{2.76}$$

where G is the antenna gain and D is the directivity.

The efficiency of a rectangular microstrip antenna can be calculated from the cavity model in terms of the cavity Q's [49]. The radiated efficiency is the power radiated, divided by the total power, which is the sum of the radiated, surface wave, conductor loss, and dielectric loss. The stored energy is identical for all the cavity Q's. This allows us to write

$$e = \frac{Q_T}{Q_r} \tag{2.77}$$

expanded out:

$$e = \frac{Q_d Q_c Q_{sw}}{Q_{sw}Q_cQ_d + Q_{sw}Q_cQ_r + Q_{sw}Q_rQ_d + Q_rQ_dQ_c} \tag{2.78}$$

When multiplied by 100%, (2.77) gives the antenna efficiency in percent, as predicted by the cavity model. We can readily see from (2.78) that as Q_{sw}, Q_c, and Q_d become large compared with Q_r, the antenna's efficiency approaches 100%. In

Table 2.6 Cavity model losses in a square linear microstrip antenna versus h (2.45 GHz, a = b = 56.46 mm), tan δ = 0.0025, $\epsilon_r = 1.1$ (foam dielectric)

Cavity model foam dielectric efficiency					
	h	η_r	η_{sw}	η_c	η_d
(0.030")	762 μm	83.41 %	0.01 %	6.86 %	9.71 %
(0.060")	1524 μm	92.67 %	0.03 %	1.91 %	5.39 %
(0.090")	2286 μm	95.38 %	0.05 %	0.87 %	3.70 %
(0.030")	3048 μm	96.63 %	0.06 %	0.50 %	2.81 %

Table 2.7 HFSS Q-losses—solid copper in a square linear MSA versus substrate thickness—lossless connector (2.45 GHz), tan δ = 0.0025, $\epsilon_r = 1.1$ (foam dielectric), Cu thickness = 17 μm (100 mm × 100 mm groundplane)

Full-wave model foam dielectric efficiency					
	h	η_r	η_{sw}	η_c	η_d
(0.030")	762 μm	72.98 %	12.21 %	5.69 %	9.12 %
(0.060")	1524 μm	87.68 %	5.35 %	1.69 %	5.28 %
(0.090")	2286 μm	92.37 %	3.24 %	0.82 %	3.56 %
(0.120")	3048 μm	94.95 %	2.13 %	0.42 %	2.50 %

other words, we desire to minimize the radiation Q, and maximize Q_s, Q_c, and Q_d, the surface wave Q, conductor Q, and dielectric Q, respectively, for maximum efficiency.

It is instructive to calculate the losses from each of the mechanisms separately. We can calculate $\eta_r = Q_T/Q_r$ for radiation efficiency, $\eta_{sw} = Q_T/Q_{sw}$ for the surface wave efficiency, $\eta_c = Q_T/Q_c$ for the conductor efficiency, and $\eta_d = Q_T/Q_d$ for the dielectric efficiency. When these are added together we are able to account for 100% of the power in the antenna.

In order to obtain some understanding of the contribution from each of the loss mechanisms of a rectangular microstrip antenna, with respect to substrate thickness h and relative dielectric permittivity ϵ_r, we have computed the losses for three typical dielectric values ($\epsilon_r = 1.1$, 2.6, and 10.2).

First, we will compute the predicted losses using a cavity model surface wave analysis. Table 2.6 shows the efficiency components η_r (radiation), η_{sw} (surface wave), η_c (conductor), and η_d (dielectric) for a square linear microstrip antenna operated at 2.45 GHz with a very low dielectric permittivity substrate ($\epsilon_r = 1.1$). We note the loss contribution from surface waves is insignificant in this case. The next largest loss is that due to the dielectric and then the conductors. As the substrate thickness h of the antenna is increased, the radiation efficiency increases.

Modern full-wave analysis tools now allow one to compute these losses more directly. The method used is described in Appendix B.5. When the computations are

Table 2.8 Solid copper HFSS Q-losses in a square linear MSA versus substrate thickness—lossless connector (2.45 GHz), tan $\delta = 0.0025$, $\epsilon_r = 2.6$, Cu thickness = 17 μm (100 mm × 100 mm groundplane)

Full-wave model dielectric efficiency $\epsilon_r = 2.6$					
	h	η_r	η_{sw}	η_c	η_d
(0.030")	762 μm	67.88 %	15.43 %	6.00 %	10.69 %
(0.060")	1524 μm	84.71 %	6.97 %	1.96 %	6.37 %
(0.090")	2286 μm	90.44 %	4.29 %	0.93 %	4.34 %
(0.120")	3048 μm	93.21 %	3.19 %	0.55 %	3.06 %

Table 2.9 Solid copper—HFSS Q-losses in a square linear MSA versus substrate thickness—no connector losses (2.45 GHz), tan $\delta = 0.0025$, $\epsilon_r = 10.2$, Cu thickness = 17 μm (100 mm × 100 mm groundplane)

Full-wave model dielectric efficiency $\epsilon_r = 10.2$					
	h	η_r	η_{sw}	η_c	η_d
(0.030")	762 μm	47.78 %	22.26 %	9.05 %	20.91 %
(0.060")	1524 μm	70.13 %	11.38 %	3.43 %	15.06 %
(0.090")	2286 μm	80.74 %	7.57 %	1.69 %	9.99 %
(0.120")	3048 μm	86.48 %	5.53 %	0.83 %	7.15 %

repeated using HFSS (Table 2.7), surface waves now have a much larger contribution to the antenna's overall non-radiative losses. The cavity model surface wave loss analysis assumes a TM_0 surface wave that propagates along a shorted dielectric slab, [25] like the geometry which extends beyond the edges of a rectangular microstrip antenna. This dielectric type of surface wave requires the antenna substrate possess a relative dielectric permittivity greater than one to propagate [50].

The surface wave loss for a 2450 MHz rectangular microstrip antenna with a vacuum substrate (i.e. $\epsilon_r = 1$ and tan $\delta = 0$) remains close to the values shown in Table 2.7. Clearly, the surface wave loss cannot be from a TM_0 mode. It is doubtful that the losses are due to the proximity effect. A candidate for these losses might be the generation of Zenneck surface waves along the groundplane-dielectric interface [51,52]. Zenneck surface waves are lossy and non-radiative. The subject of surface waves is complex and the losses may be from a different mechanism entirely [53].

The next analysis case is for $\epsilon_r = 2.6$ (Table 2.8). We can see that the surface wave contribution has increased slightly compared with $\epsilon_r = 1.1$ (Table 2.7). The surface wave loss decreases in proportion to the thickness of the substrate, as does the conductor and dielectric losses. If a designer wants to maximize the space wave contribution in this case, then the thickest possible substrate appears to be a good choice.

When the relative dielectric permittivity is increased to $\epsilon_r = 10.2$ (Table 2.9), we see that the amount of surface wave power increases significantly compared with the $\epsilon_r = 2.6$ case in Table 2.8. The thinnest substrate only radiates 47.78 % of the total input power into the space wave. As h increases from 762 μm to 1524 μm, the amount of power lost to surface waves and dielectric loss is almost equivalent. As before, the thickest substrate one can use, without producing excessive mismatch, will minimize the overall non-radiative losses and maximize the radiated power. This is often a good design path to use because of the difficulty involved in making experimental efficiency measurements [54].

2.7 Design of microstrip antenna with dielectric cover

Microstrip antennas are often enclosed in dielectric covers (i.e. radomes) to protect them from harsh environments. These can range from vacuum molded or injection molded plastic enclosures, which leave an air gap between the radiating patch and the radome, to bonding a plastic material directly to the antenna.

Bonding dielectric material directly to the antenna can provide a high degree of hermetic sealing. When the substrate material is Teflon based, bonding can be technically involved to produce good adhesion. In some commercial applications, the injection molding of a plastic radome which surrounds the antenna element and seals it, has been implemented. In these cases, the use of a full wave simulator such as Ansoft HFSS is best for the refinement of a design prior to prototyping, but the use of a quick quasi-static analysis can provide initial design geometry for refinement and design sensitivity prior to optimization.

A number of approaches have been forwarded to analyze a microstrip antenna with a dielectric cover [55–58]. Here, we will utilize the transmission line model to analyze a rectangular microstrip antenna with a dielectric cover. A quasi-static analysis of a microstrip transmission line with a dielectric cover forms the basis of this analysis [59]. An effective relative dielectric permittivity for the geometry shown in Figure 2.24 is defined in (2.79) and the characteristic impedance is related in (2.80):

$$\epsilon_e = \frac{C_{\epsilon_r}}{C_0} \tag{2.79}$$

$$Z_0 = \frac{Z_{air}}{\sqrt{\epsilon_e}} \tag{2.80}$$

$$Z_{air} = \frac{1}{cC_0} \tag{2.81}$$

where ϵ_e is the effective relative dielectric permittivity of microstrip line, Z_0 is the characteristic impedance of microstrip line, Z_{air} is the characteristic impedance of microstrip line with no dielectrics present, C_{ϵ_r} is the capacitance per unit length with dielectrics present, C_0 is the capacitance per unit length with only free space present, and c is the speed of light in a vacuum.

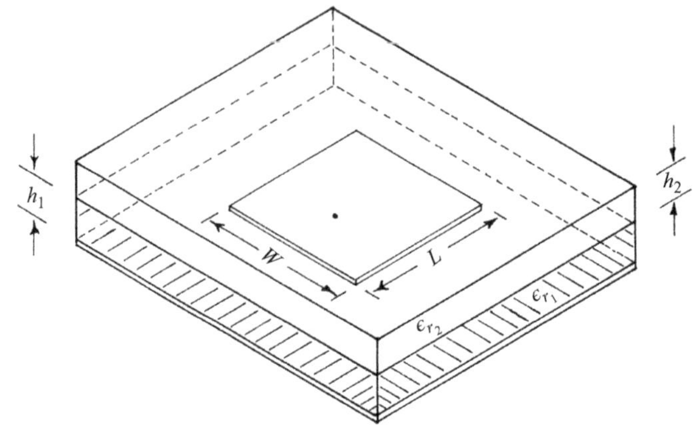

Figure 2.24 Rectangular microstrip patch geometry of a dielectric covered microstrip antenna which is analyzed with the transmission line model. The patch antenna is fed along the centerline of the antenna's width (i.e. W/2). The feedpoint is represented by the black dot

Using the substitution of $\alpha = \beta h_1$ in Bahl *et al.* [60], we are able to write the capacitance as

$$\frac{1}{C} = \frac{1}{\pi \epsilon_0} \int_0^{\infty} \left(1.6 \frac{\sin(\alpha W/2h_1)}{(\alpha W/2h_1)} + 2.4(\alpha W/2h_1)^{-2} \right.$$
$$\cdot \left. \left[\cos(\alpha W/2h_1) - \frac{2\sin(\alpha W/2h_1)}{(\alpha W/2h_1)} + \sin^2(\alpha W/4h_1) \cdot (\alpha W/4h_1)^{-2} \right] \right)^2$$
$$\cdot \left(\left[\epsilon_{r_2} \frac{\epsilon_{r_2} \tanh(\alpha h_2/h_1) + 1}{\epsilon_{r2} + \tanh(\alpha h_2/h_1)} + \epsilon_{r_1} \coth(\alpha) \right] \alpha \right)^{-1} d\alpha \tag{2.82}$$

where W is the width of microstrip transmission line (patch width), h_1 is the thickness of dielectric substrate, ϵ_{r_1} is the relative dielectric permittivity of substrate, h_2 is the thickness of dielectric superstrate (dielectric cover/radome), and ϵ_{r_2} is the relative dielectric permittivity of radome dielectric superstrate (dielectric cover/radome).

The integration of (2.82) is efficiently computed using Gaussian quadrature as presented in Appendix B. The slot admittance is modified when a dielectric cover is added to a rectangular microstrip antenna design. The modification is slight and is best computed using (2.4) when compared with the accuracy of more complicated alternatives [61,62]. The edge susceptance may be written as

$$B_e = j\omega C_{slot} \tag{2.83}$$

The capacitance of the radiating slot for a rectangular microstrip antenna with a cover layer is obtained using:

$$C_{slot} = \frac{W}{2}\left[\frac{\epsilon_e(L)}{cZ_{air}(L)} - \epsilon_0\epsilon_r\frac{L}{h_1}\right] \tag{2.84}$$

where $\epsilon_e(L)$ and $Z_{air}(L)$ is the effective dielectric permittivity with dielectrics present, and characteristic impedance of a microstrip line of width L with only air present, respectively. Equation (2.83) provides the edge capacitance of a microstrip transmission line with a dielectric superstrate of width L *rather than* W. The left-hand term inside of (2.84) is the transmission line capacitance, and the right-hand term is the capacitance per unit length of a parallel plate capacitor of width L. The fringing field capacitance is computed by the difference, which is used as an approximation of the slot capacitance.

When a microstrip antenna is covered with a dielectric substrate in practice, an air gap may exist. This air gap has a strong effect on the effective dielectric permittivity of the patch, which in turn affects characteristic impedance and resonant frequency of the antenna.

The resonant frequency of the microstrip antenna with a dielectric cover is computed using:

$$f = \frac{c}{2(L + 2\Delta l)\sqrt{\epsilon_e}} \tag{2.85}$$

$$\Delta l = 0.412h_1\frac{(\epsilon_e + 0.3)(W/h_1 + 0.264)}{(\epsilon_e - 0.258)(W/h_1 + 0.8)} \tag{2.86}$$

2.8 Design guidelines for rectangular microstrip antenna

There are a number of antenna performance trade-offs with respect to substrate dielectric permittivity and thickness to consider when designing a linear rectangular microstrip antenna [63]. Clearly, if one needs to feed a patch with a coaxial transmission line, then a probe feed is a good choice. If the design requires a microstrip feed, a non-radiating edge feed can make sense, but the patch needs to be narrow enough to decrease any excitation of a secondary mode. A narrower patch has a slightly decreased bandwidth compared with a wide patch. If the impedance bandwidth requirement is greater than a narrow patch can provide, then one can turn to a feed along a radiating edge. A quarter-wave transformer feed on a radiating edge produces the least amount of perturbation of the patch radiation, but if the design constraints do not allow for enough area to implement the transformer, an inset feed can be utilized. In either case, if the patch is fed along a centerline which drives the lowest order mode, the driving point impedance presented by the next dominant mode is along a shorting plane for that next mode and mismatched (not driven) even if the patch is square.

A useful beginning patch width for a linear microstrip antenna is

$$W = \frac{c}{2f_r}\left(\frac{\epsilon_r + 1}{2}\right)^{-1/2} \tag{2.87}$$

Patch thickness is an important parameter to consider. If the patch thickness is too thin, the efficiency and impedance bandwidth are decreased. When the patch is too thick, it can produce a series inductive mismatch at the driving point from higher order cavity modes.

At higher frequencies, Gopinath has presented an analysis which allows one to choose a substrate thickness which maximizes the Q of a microstrip line at a given frequency [64].

The resonant length of a rectangular microstrip antenna is computed with

$$L = \frac{c}{2f_r\sqrt{\epsilon_e}} - 2\Delta l \tag{2.88}$$

When designing a square microstrip patch, one can use

$$L_0 = \frac{c}{2f_r\sqrt{\epsilon_r}} \tag{2.89}$$

to provide an initial length L_0. This value can then be equated to the width of the antenna W to produce a new resonant length L_1 using (2.88), (2.6), and (2.7). This process is continued until the value becomes fixed. The solution generally becomes fixed by the fifth iteration [L_5] (Appendix B, Section B.3).

If ESD is a consideration, one can place an electrical short at the center of the patch using a via or soldered shorting pin. Gold flashing can be used to protect a copper element from many environmental hazards. Tin immersion is another alternative which is useful in some situations to prevent copper degradation.

In some designs, a higher frequency resonance of a rectangular microstrip may coincide with a band which is to be isolated for system design reasons. This problem can sometimes be resolved by using a circular microstrip patch which has resonances with different frequency spacings than those of a rectangular patch.

The desired feedpoint impedance may be located using the transmission line model with a root finding algorithm, such as the bisection method (Appendix B, Section B.4). Experience indicates the relationships used to compute the wall admittance, (2.4) and (2.5) predict accurate values for the feedpoint location when the relative dielectric permittivity is between $2.2 < \epsilon_r < 3.8$ which is often encountered in practice, and have proven more accurate than alternative expressions for uncovered rectangular microstrip antennas. In other situations, the feedpoint will require experimental determination. One can also use the cavity model to predict the location of a desired driving point impedance, but its results are somewhat sensitive to the effective probe diameter used in the computation.

The directivity of a linear rectangular microstrip antenna can be estimated using (2.46) which is generally within 1–2 dB of measurement for most groundplane sizes. A more accurate directivity can be computed with a more powerful technique such as

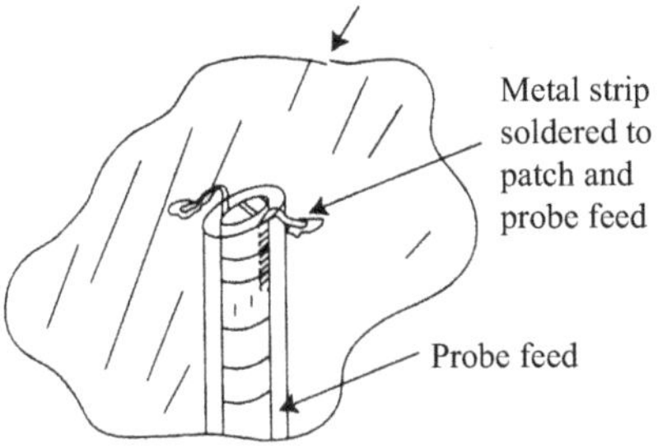

Figure 2.25 Vulnerability of a probe feed to shock and vibration can be mitigated by the use of two (or more) small metal strips. Each strip is soldered to the feed pin, extend through a minute gap along the pin which is provided by a slightly larger than required hole in the dielectric substrate. The strips are then soldered to the patch with a small radius of slack. This slack allows the feed pin to move up and down without solder failure

the FDTD or FEM. The antenna efficiency can be computed from (2.77) and used to calculate the antenna gain.

As discussed previously, the directivity of a linear rectangular microstrip antenna depends on the substrate's relative dielectric permittivity ϵ_r. Directivity increases as the dielectric permittivity decreases and will decrease, in an asymptotic manner, as the dielectric constant is increased (see Table 2.3).

In some design environments, a microstrip antenna must survive a considerable amount of shock and vibration. When a patch is probe fed, the solder joint on the top of a microstrip patch which connects the feed probe to the patch is vulnerable to failure. Under large vibrational shock, the probe can punch itself through the upper solder joint leaving a microscopic ring-shaped gap between the solder and the feed probe. Often this ring-shaped gap is too small to be seen without a microscope, but will produce antenna failure. One solution to this problem is to use a pair of thin metal strips soldered along the feed probe and whose ends are bent at right angles with a small amount of slack and soldered to the patch. The feed pin with soldered strips on either side pass through a hole which is large enough to allow the feed pin to move axially without interference. The small amount of extra slack, which is left as a small radius at the right angle bend of the strip before the end of the strip is soldered, allows for movement. This is illustrated in Figure 2.25.

Cross-polarization is produced by the existence of higher order modes on the patch. This is elaborated upon in Section 2.12 and in Chapter 3 for circular microstrip antennas.

In the case of a linearly polarized rectangular microstrip antenna, we generally feed the antenna along the centerline ($x = 0$). This will drive the TM_{01} mode and theoretically not excite the TM_{10} mode, which is mismatched.

When a square patch is used to produce circular polarization with two orthogonal (microstrip or probe) feeds, any error in centering the feeds along the y-axis and x-axis will increase cross-polarization. A square microstrip antenna has the property that both TM_{01} and TM_{10} modes have the same resonant frequency, and the undesired mode may be readily excited by error in the driving point location. One way to mitigate this cross-polarization is by using multiple probe feeds to generate circular polarization [65].

A resonant-cavity model may be used to estimate the amount of cross-polarization produced by probe placement error [66]. These computations were undertaken by Mishra and Milligan [67]. For a square patch to have no less than 25 dB cross-polarization, the feed probe must not vary more than 0.75% of the antenna width (a). A square patch designed on a substrate with a 2.32 relative dielectric permittivity and a 3.0 GHz operating frequency, with $\approx$ 30 mm width must be within 0.22 mm to achieve 25 dB cross-polarization.

When only linear polarization is desired, one can increase the 25 dB cross-polarization driving point location tolerance by using a rectangular patch ($a > b$). By shifting the TM_{10} resonance to a frequency twice that of the 2:1 VSWR band-edge location, the probe location tolerance increases from 0.75% to 2% of the patch width (a).

These examples demonstrate that when feeding a patch with a probe feed, the feed point location tolerance for low cross-polarization is *very* tight, and small location errors will quickly increase cross-polarization levels.

2.9 Design guidelines for a circularly polarized microstrip antenna

The design of a circularly polarized microstrip antenna which uses a rectangular patch with an aspect ratio begins with the design of a square microstrip antenna. Equations (2.8) and (2.7) may be iterated (Appendix B, Section B.4) to create a square patch. The resonant frequency tends to be slightly low when the iterated converged value is used. The cavity model may be used to refine the patch size and make it more accurate. The Q of a single mode, TM_{01} or TM_{10}, of a square patch can be determined with measurement, or computed using the cavity model, to determine Q_T. The cavity model can then be used to compute a driving point location which has an input resistance at a resonance of about 88 Ω. Equations (2.55) and (2.56) are then used to compute the patch dimensions that produce circular polarization.

Experimental optimization is often required to complete the design of a circularly polarized rectangular patch antenna. The Smith chart of Figure 2.18 illustrates the impedance trace one needs in order to achieve circular polarization. The frequency which exists at the vertex of a kink in the Smith chart impedance, which forms a 90° angle, is the point at which optimum circular polarization occurs. The polarization sense of the antenna may be determined by consulting Figure 2.15(I). Often the impedance at the kink of the impedance trace is not well matched and frequently has a capacitive component. In the case of a probe-fed circularly polarized rectangular

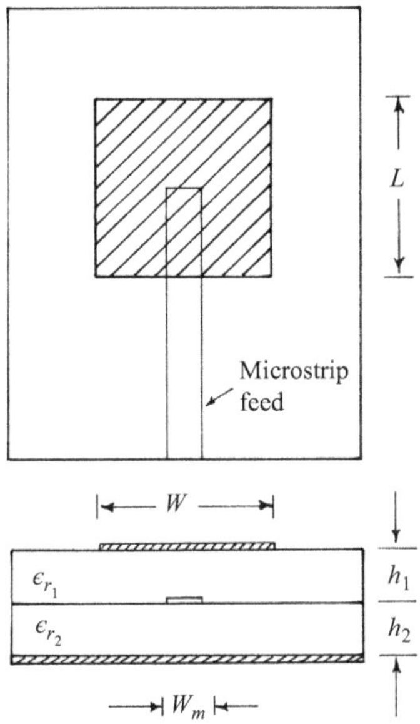

Figure 2.26 Rectangular microstrip patch with an electromagnetically coupled feed

microstrip antenna, one can move the driving point location off the patch diagonal and generally match the antenna at the CP frequency.

When a dielectric superstrate (radome) covers a microstrip antenna (Figure 2.24) which generates circular polarization, the axial ratio bandwidth will remain unchanged [68].

A variety of GPS RHCP rectangular microstrip antenna designs, which use high dielectric permittivity ceramic material as a substrate, are offered by a number of manufacturers. A common GPS antenna design has substrate dimensions of 25 mm × 25 mm × 4 mm with $\epsilon_r = 20$ which is optimized to operate on a 70 mm × 70 mm groundplane. This antenna design is electrically small at 1.575 GHz, and its performance is strongly affected by electrically small groundplane dimensions. The groundplane dimensions affect the resonant frequency and radiation patterns adversely, and these effects must be included in the design. The limitations of electrically small antennas are discussed in Section 7.6.1 in Chapter 7.

2.10 Electromagnetically coupled rectangular microstrip antenna

One may use a rectangular microstrip patch which is fed with a microstrip transmission line electromagnetically coupled to the patch. The geometry of this design is defined

Table 2.10 Single-element gain of the UWMSA examples

ϵ_r	Gain (dBi)
1.00	12.84
2.60	10.29

in Figure 2.26. Microstrip antennas of this configuration are very difficult to directly analyze, and most designs are empirical, or designed by trial and error with a full wave simulator [69]. One can use (2.79), (2.80), (2.81), and (2.82) to determine the width which corresponds to a 50 Ω microstrip line embedded between two dielectrics ϵ_{r_1} and ϵ_{r_2}. Alternatively, many full-wave electromagnetic simulation programs allow one to compute the characteristic impedance of an embedded microstrip transmission line. Altering the width of a patch generally allows one to match the antenna to the transmission line, and patch length determines the resonant frequency.

An example EMC patch, designed to operate at 2.45 GHz, has substrate heights of $h_1 = h_2 = 1.524$ mm. The dielectric permittivity of both layers ϵ_{r_1} and $\epsilon_{r_2} = 2.6$ with a 0.0025 tan δ. The patch width is $W = 44.0$ mm and the resonant length is $L = 34.0$ mm The width of the 50 Ω microstrip feedline is $W_m = 4.0$ mm. The groundplane width and length are 128 mm and 130 mm, respectively. The patch is centered on the substrate and the feedline extends under the patch to the patch center. The antenna has about 3.2% impedance bandwidth, with 7.3 dBi gain.

We can see from this design example that another advantage of this geometry is the ability to feed a rectangular microstrip antenna directly with a 50 Ω microstrip transmission line.

2.11 Ultra-wide rectangular microstrip antennas

In Chapter 4, we will investigate the use of a matching network to increase the bandwidth of a microstrip antenna. The example used requires a microstrip antenna with an edge resistance of 92.5 Ω. This antenna is obtained by increasing the width of the patch beyond that generally suggested [70]. The symmetry of the feed is such that driving a mode along the length of the patch does not occur when it is driven at the frequency of the TM_{01} mode. Rectangular microstrip antennas which are very wide compared with their resonant length will be referred to as *Ultra-Wide* Rectangular Microstrip Antennas (UWMSA).

The geometry of a UWMSA 50 Ω patch is presented in Figure 2.27. The antenna has a width W_{50} and length L. The patch is fed with a 50 Ω microstrip transmission line whose width is designated as W_M. As we have discussed previously, the

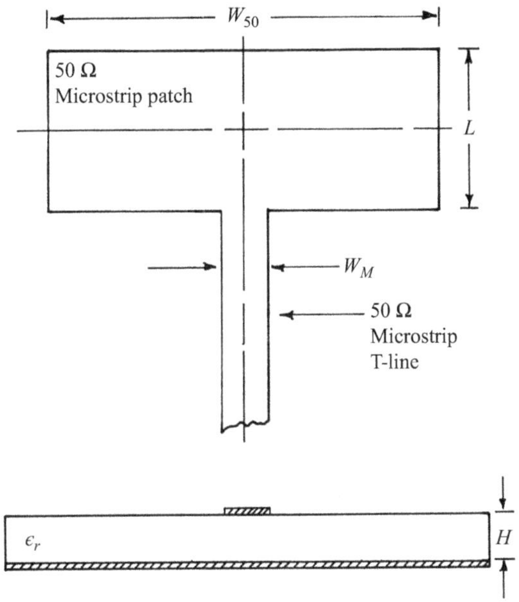

Figure 2.27 Ultra-wide microstrip antenna geometry

gain of a microstrip antenna is dependent upon the relative dielectric permittivity ϵ_r of the substrate. We will examine the radiation patterns and bandwidth of the UWMSA for $\epsilon_r = 1.0$ and 2.6 and the special case of a 50 Ω patch. The substrate thickness is 2.286 mm (0.090 inch). The operating frequency is 5.25 GHz. For $\epsilon_r = 1.0$, $W_{50} = 68.0$ mm, $L = 24.76$ mm, and $W_M = 11.0$ mm and when $\epsilon_r = 2.6$, $W_{50} = 54.0$ mm, $L = 15.10$ mm, and $W_M = 6.2$ mm.

The computed radiation patterns of the ultra-wide microstrip antenna are presented in Figure 2.28. The top polar plot is of a patch with air dielectric ($\epsilon_r = 1.0$). The bottom patch has a dielectric substrate with $\epsilon_r = 2.6$. Table 2.10 presents the single element gain of the UWMSA examples. The air loaded UWMSA has as much gain as a typical 2 × 2 rectangular patch array on a dielectric substrate. A two-patch array, with $\epsilon_r = 2.08$, fed on their non-radiating edges has been reported to possess the same gain as a single ultra-wide microstrip antenna [71].

2.12 Rectangular microstrip antenna cross-polarization

The cross-polarization characteristics of a rectangular microstrip antenna have been studied in detail [72]. One way to understand the cross-polarization mechanism of a half-wave rectangular microstrip antenna is by analyzing it in terms of equivalent magnetic current around its periphery. The magnetic current around a microstrip

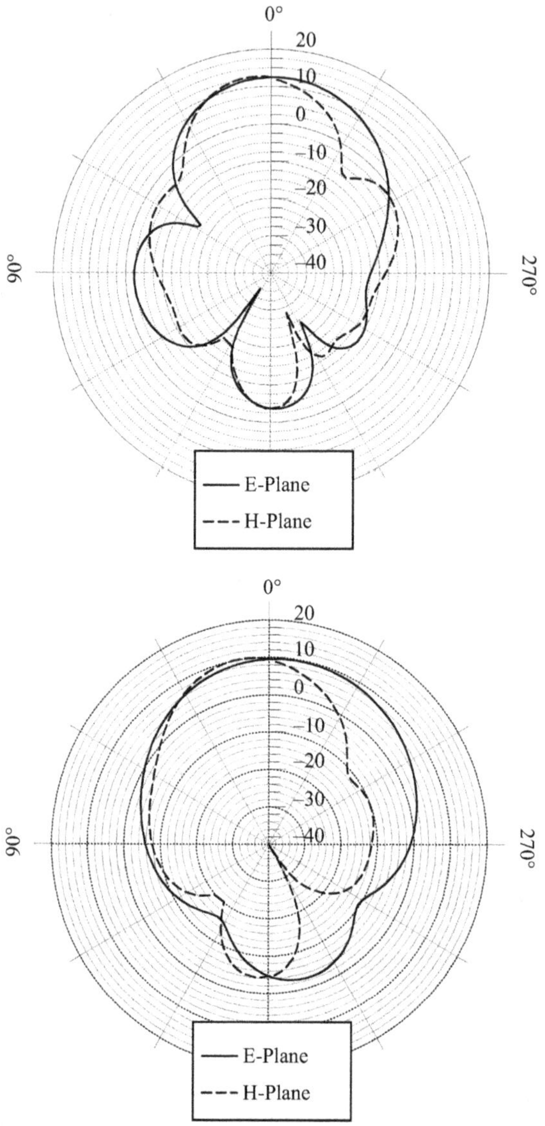

Figure 2.28 Predicted radiation pattern of the 2.45 GHz 50 Ω UWMSA of Table 2.10 using FDTD analysis for $\epsilon_r = 1.0$ (top) and $\epsilon_r = 2.6$ (bottom)

antenna, in its dominant mode, is given in Figure 2.29. The cavity model is used to determine the electric field around the periphery, and the equivalent magnetic current is

$$\vec{M} = 2HE_z(x, y)\hat{z} \times \hat{n} \tag{2.90}$$

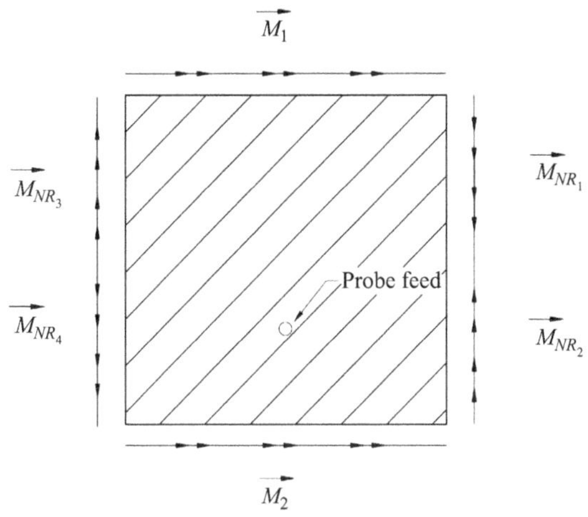

Figure 2.29 Equivalent magnetic current around a rectangular microstrip antenna driven in its lowest order mode

where H is the substrate thickness and E_z is the $\hat{z}$ directed electric field along the edges of the rectangular patch antenna. One can see that the two magnetic currents at either radiating edge of the patch, $\vec{M}_1$ and $\vec{M}_2$, are in phase, and radiate to produce broadside radiation. The right-hand side of the patch has a pair of opposing currents, $\vec{M}_{NR_1}$ and $\vec{M}_{NR_2}$, which by themselves would cancel in the far field. The left-hand edge also has a pair of opposing currents, $\vec{M}_{NR_3}$ and $\vec{M}_{NR_4}$, that are opposite in phase with those on the right-hand edge.

This assumes that the magnitude of the currents are all equivalent. They are not identical because the probe feed is asymmetric on the patch and drives the fields asymmetrically. The magnitudes of $\vec{M}_{NR_1}$ and $\vec{M}_{NR_3}$ tend to be smaller than $\vec{M}_{NR_2}$ and $\vec{M}_{NR_4}$. The cross-polarization is largest along the diagonals of the patch antenna.

One can also view co- and cross-polarization in terms of electric field components $\vec{E}_x$ and $\vec{E}_y$, which are in the same directions as the radiating electric currents. In this case, the desired electric field is along the $\hat{y}$-axis, and the undesired is along $\hat{x}$-axis [73,74]. The ideal is to reduce the $\vec{E}_x$ component to zero with $\vec{E}_y$ remaining as the principle polarization radiated.

A common method used to reduce cross-polarization in a rectangular microstrip antenna is to use differential feeding [75]. A second probe, inset by the same amount as the first, but inset from the opposite edge, is introduced. The second probe feed has a current of the same amplitude as the first feed probe, but is of opposite phase. The superposition of the two non-symmetric modes produces a symmetric mode, with equal amplitude magnetic currents along the non-radiating edges, and appropriate phase that reduces the cross-polarization significantly. GPS accuracy is increased as one changes from a single probe feed, to dual probe feed, to a four location probe feed [65].

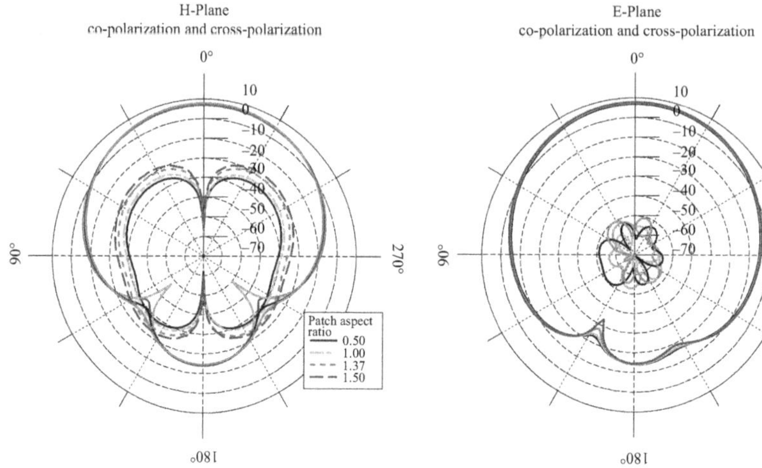

Figure 2.30 Co- and cross-polarization computed by HFSS for an SMA probe-fed rectangular microstrip antenna, at 2.45 GHz. Patch aspect ratios of 0.5, 1.0, 1.37, and 1.5 are presented

The aspect ratio of the patch has a significant effect on the cross-polarization. In one study, it was found that when a rectangular patch is fed along a non-radiating edge, with a microstrip transmission line, that the maximum ratio of co- to cross-polarization occurs for a width-to-length ratio of 1.5 [76]. For the case of a probe-fed patch, driven along the centerline between non-radiating edges, the optimum ratio has been reported as 1.37 [77]. In Figure 2.30, the computed E- and H-plane co- and cross-polarization patterns for a 2.45 GHz SMA probe-fed microstrip antenna are reported. The substrate thickness is 1.542 mm, $\epsilon_r = 2.33$ with $\tan\delta = 0.0012$. We note that the H-plane cross-polarization increases with an increasing patch aspect ratio. For the narrowest patch $a/b = 0.50$, the cross-polarization is lower. This is expected as the magnetic current along the non-radiating edges is closer and cancels more effectively in the far field. The worst-case cross-polarization occurs for the largest aspect ratio $a/b = 1.5$ when the currents are farthest apart. The E-plane cross-polarization is low as this plane is along the centerline between the magnetic currents $\vec{M}_{NR_1}$, $\vec{M}_{NR_2}$, $\vec{M}_{NR_3}$, and $\vec{M}_{NR_4}$, which are responsible for the cross-polarized radiation. They maximally cancel in the far field along the E-plane.

It has been recently reported that matching a probe-fed microstrip antenna to operate below its natural resonance allows for a decrease in cross-polarization of up to about 7 dB over its operating bandwidth [78]. Cross polarization is shown to decrease with increased relative dielectric permittivity of the substrate and decrease as the substrate thickness decreases.

The cross-polarization of an inset-fed rectangular patch increases as the length of the inset into the patch increases [79]. It also appears to be the case for a probe-fed patch. Cavity model analysis indicates that cross-polarization increases with increased thickness for low permittivity substrates. Cross polarization decreased with increasing

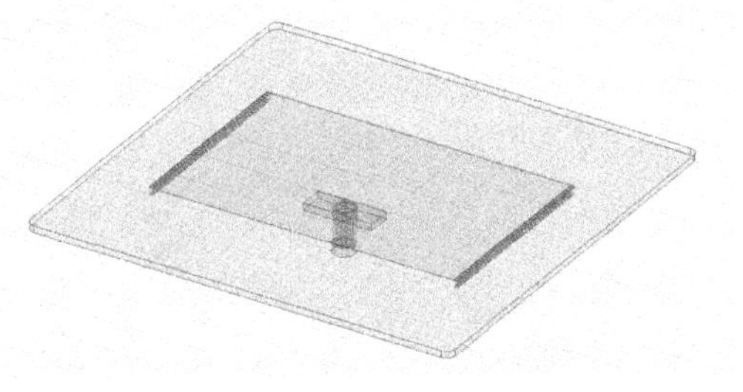

Figure 2.31 Shorted non-radiating edge (SNRE) rectangular patch antenna geometry. The shorting walls are realized with vias

thickness for high dielectric permittivity substrates [80]. Cross polarization is also dependent on frequency.

2.12.1 MSA with shorted non-radiating edges

It has been demonstrated that a rectangular microstrip antenna can be shorted along its non-radiating edges and produce an antenna with significantly decreased cross-polarization [81] (Figure 2.31). A modified cavity model (MCM) with shorted edges was used to analyze the geometry. This model does not allow a TM_{10} or TM_{01} mode to exist. The lowest order mode is TM_{11}. The resonant frequency f_r for this mode is

$$f_r = \frac{c}{2\sqrt{\epsilon_e}}\sqrt{\left(\frac{1}{L+2\Delta L}\right)^2 + \left(\frac{1}{W}\right)^2} \tag{2.91}$$

where c is the speed of light, ϵ_e is the effective dielectric permittivity as given by (2.6), L is the resonant length of the patch (b), W is the width (a), and ΔL is the edge extension from the fringing field at the radiating edges of the patch expressed by (2.7).

The shorted edges produce a minimum amount of electric field at the non-radiating edges, which are the source of the magnetic currents that contribute to the antenna's cross-polarization. Despite the differences in the modes between a conventional rectangular microstrip antenna and one with shorted edges, radiation from the dominant mode is still maximum broadside to the antenna.

For comparison, the same dielectric substrate used in Section 2.12 is used in an HFSS analysis. The resonant frequency is 2.45 GHz. An aspect ratio of 1.5 is chosen for the SNRE patch as it had the worst-case cross-polarization in the group of rectangular microstrip antennas analyzed in Section 2.12. The antenna interior width is 70.275 mm and its resonant length is 46.85 mm. Vias with a radius of 0.5 mm were used to create shorting walls along the non-radiating edge.

The co- and cross-polarization patterns of the conventional rectangular patch, and the SNRE rectangular patch, are given in Figure 2.32. The decrease in cross-polarization for the SNRE rectangular patch is at least 20 dB in the H-plane and similar in magnitude for the E-plane.

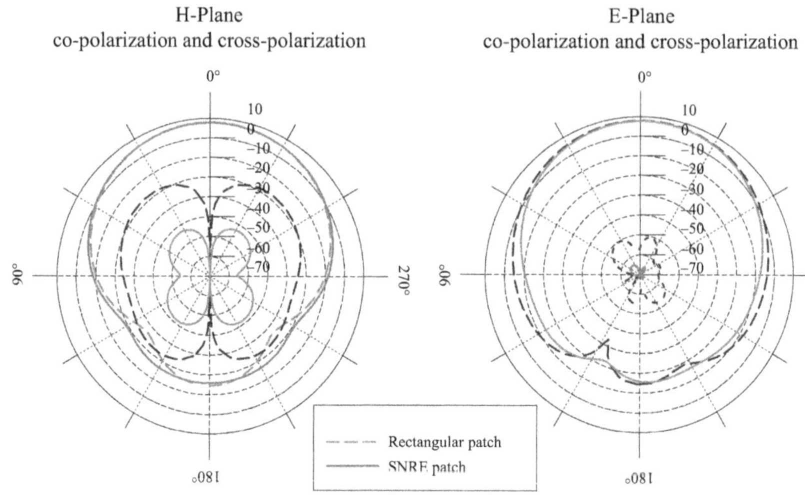

Figure 2.32 Comparison of a conventional rectangular microstrip patch and a shorted non-radiating edge rectangular antenna co- and cross-polarization. Both antennas have an aspect ratio of 1.5

The 2:1 VSWR impedance bandwidth for the conventional patch is 39.6 MHz (1.62%) and 32.3 MHz (1.32%) for the SNRE patch, so there is little bandwidth penalty for the increase in cross-polarization performance. The predicted efficiency for the conventional patch is 91.64% and for the SNRE patch 87.94%. The gain of the rectangular patch is 7.81 dBi, and the SNRE patch is 7.56 dBi.

Given $\epsilon_e = 2.23$ and $\Delta L = 0.796$ mm, (2.91) predicts a resonant frequency of 2.518 GHz using conventional microstrip parameters, whereas HFSS predicts 2.450 GHz.

References

[1] Hildebrand, L.T., and McNamara, D.A., "A Guide to Implementational Aspects of the Spatial–Domain Integral Equation Analysis of Microstrip Antennas," *Applied Computational Electromagnetics Journal*, March 1995, Vol. 10, No. 1, pp. 40–51, ISSN 1054–4887.

[2] *IEEE Transactions on Antennas and Propagation*, January 1981.

[3] Mosig, J.R., and Gardiol, F.E., "General Integral Equation Formulation for Microstrip Antennas and Scatterers," *IEE Proceedings*, Vol. 132, Pt. H, No. 7, December 1985, pp. 424–432.

[4] Mosig, J.R., "Arbitrarily Shaped Microstrip Structures and Their Analysis with a Mixed Potential Integral Equation," *IEEE Transactions on Microwave Theory and Techniques*, February 1988, Vol. 36, No. 2, pp. 314–323.

[5] Pues, H., and Van de Capelle, A., "Accurate Transmission-Line Model for the Rectangular Microstrip Antenna," *IEE Proceedings*, December 1984, Vol. 131, Pt. H, No. 6.

[6] Dearnley, R.W., and Barel, A.R.F., "A Broad-Band Transmission Line Model for a Rectangular Microstrip Antenna," *IEEE Transactions on Antennas and Propagations*, January 1989, Vol. 37, No. 1.

[7] Bhattacharyya, A.K., and Garg, R., "Generalized Transmission Line Model for Microstrip Patches," *IEE Proceedings*, 1985, Vol. 132, Pt. H, pp. 93–98.

[8] Derneryd, A.G., "Linearly Polarized Microstrip Antennas," *IEEE Transactions on Antennas and Propagations*, November 1976, Vol. AP-24, pp. 846–851.

[9] Munsen, R.E., "Conformal Microstrip Antennas and Microstrip Phased Arrays," *IEEE Transactions on Antennas and Propagations*, January 1974, Vol. AP-22, pp. 74–78.

[10] Bahl, I.J., and Bhartia, P., *Microstrip Antennas*, Artech House, 1980, 2nd Printing 1982, p. 46.

[11] Basilio, L., Khayat, M., Williams, J., and Long S., "The Dependence of the Input Impedance on Feed Position of Probe and Microstrip Line-Fed Patch Antennas," *IEEE Transactions on Antennas and Propagation*, January 2001, Vol. 49, No. 1, pp. 45–47.

[12] Hu, Y., Jackson, D.R., Williams, J.T., Long, S.A., and Komanduri, V.R., "Characterization of the Input Impedance of the Inset-Fed Rectangular Microstrip Antenna," *IEEE Transactions on Antennas and Propagation*, October 2008, Vol. 56, No. 10, pp. 3314–3317.

[13] Derneryd, A.G., "A Theoretical Investigation of the Rectangular Microstrip Antenna Element," *IEEE Transactions on Antennas and Propagations*, July 1978, Vol. AP-26, pp. 532–535.

[14] Augustine, C.F., "Field Detector Works in Real Time," *Electronics*, June 1968, pp. 118–122.

[15] Fergason, J.L., "Liquid Crystals," *Scientific American*, 1964, pp. 77–85.

[16] Yang, K., Gerhard, D., Yook, J.G., Papapolymerou, I., Katehi, L.P.B., and Whitaker, J.F., "Electrooptic Mapping and Finite-Element Modeling of the Near-Field Pattern of a Microstrip Patch Antenna," *IEEE Transactions on Microwave Theory and Techniques*, February 2000, pp. 288–294.

[17] Bokhari, S.A., Zücher, J.F., Mosig, J.R., and Gardiol, F.E., "Near Fields of Microstrip Antennas," *IEEE Transactions on Antennas and Propagation*, 1995, Vol. 43, No. 2, pp. 188–197.

[18] Levine, E., "Special Measurement Techniques for Printed Antennas," *Handbook of Microstrip Antennas*, James, J.R., and Hall, P.S. (eds.), 1989, Chapter 16.

[19] Frayne, P.G., "Microstrip Field Diagnostics," *Handbook of Microstrip Antennas*, James, J.R., and Hall, P.S. (eds.), 1989, Chapter 21.

[20] Lo, Y.T., Solomon, D., and Richards, W.F., "Theory and Experiment on Microstrip Antennas," *IEEE Transactions on Antennas and Propagations*, 1979, Vol. AP-27, No. 2, pp. 137–149.

[21] Richards, W.F., Lo, Y.T., and Harrison, D.D., "An Improved Theory for Microstrip Antennas and Applications," *IEEE Transactions on Antennas and Propagations*, 1981, Vol. AP-29, pp. 38–46.
[22] Lee, H.F., and Chen, W. (eds.), *Advances in Microstrip and Printed Antennas*, John Wiley and Sons, 1997, pp. 223–242.
[23] Gan, Y.-B., Chua, C.-P., and Li, L.-W., "An Enhanced Cavity Model for Microstrip Antennas," *Microwave and Optical Technology Letters*, March 2004, Vol. 40, No. 6, pp. 520–523.
[24] Thouroude, D., Himdi, M., and Daniel, J.P., "CAD-Oriented Cavity Model for Rectangular Patches," *Electronics Letters*, June 1990, Vol. 26, No. 13, pp. 842–844.
[25] Pozar, D.M., "Rigorous Closed-Form Expressions for the Surface Wave Loss," *Electronics Letters*, June 1990, Vol. 26, No. 13, pp. 954–956.
[26] Schaubert, D.H., Pozar, D.M., and Adrian, A., "Effect of Microstrip Antenna Substrate Thickness and Permittivity: Comparison of Theories with Experiment," *IEEE Transactions on Antennas and Propagation*, June 1989, Vol. 37, No. 6, pp. 677–682.
[27] Cabedo-Fabres, M., Antonino-Daviu, E., Valero-Nogueira, A., and Bataller, M.F., "The Theory of Characteristic Modes Revisited: A Contribution to the Design of Antennas for Modern Applications," *IEEE Antennas and Propagation Magazine*, October 2007, Vol. 49, No. 5, pp. 52–67.
[28] Bahl, I.J., and Bhartia, P., *Microstrip Antennas*, Artech House, 1980, 2nd Printing 1982, p. 48–50.
[29] Tavlov, A., *Computational Electrodynamics The Finite-Difference Time-Domain Method*, Artech House, 1995, pp. 213–218.
[30] Kunz, K.S., and Luebbers, R.J., *The Finite Difference Time Domain Method for Electromagnetics*, CRC, 1993, p. 109.
[31] Sanford, G.E., and Klein, L., "Recent Developments in the Design of Conformal Microstrip Phased Arrays," *IEE Conference on Maritime and Aeronautical Satellites for Communication and Navigation*, IEE Conference Publication 160, London, pp. 105–108.
[32] Post, R.E., and Stephenson, D.T., "The Design of Microstrip Antenna Array for A UHF Space Telemetry Link," *IEEE Transactions on Antennas and Propagations*, January 1981, Vol. AP-29, pp. 129–134.
[33] Lee, K.F., Guo, Y.X., Hawkins, J.A., and Luk K.M., "Theory and Experiment on Microstrip Patch Antennas with Shorting Walls," *IEEE Transactions on Antennas and Propagation*, December 2000, Vol. 147, No. 6, pp. 521–525.
[34] James, J.R., Hall, P.S., and Wood, C., *Microstrip Antenna Theory and Design*, Peter Peregrinus Ltd., 1981, p. 106.
[35] Milligan, T., *Modern Antenna Design*, McGraw Hill, 1985, p. 118.
[36] Bancroft, R., Unpublished Witnessed/Notarized Engineering Notebook, October 23, 1998.
[37] Lu, J.-H., and Yang, K.-P., "A Simple Design for Single-Feed Circularly-Polarized Microstrip Antennas," *Proceedings of National Science Council ROC(A)*, 2000, Vol. 24, No. 2, pp. 130–133.

[38] Lumini, F., Cividanes, L., and Lacava, J.C. da S., "Computer Aided Design Algorithm For Singly Fed Circularly Polarized Rectangular Microstrip Patch Antennas," *International Journal of RF & Microwave Computer Aided Engineering*, Vol. 9, 1999, pp. 32–41.

[39] Balanis, C.A., *Antenna Theory Analysis and Design*, Harper & Row, New York, 1982, p. 51.

[40] Lee, D., and Lee, S., "Design of a Coaxially Fed Circularly Polarized Rectangular Microstrip Antenna Using a Genetic Algorithm," *Microwave and Optical Technology Letters*, September 2000, Vol. 26, No. 5, pp. 288–291.

[41] Suzuki, Y., *Analysis, Design, and Measurement of Small and Low-Profile Antennas*, Hirasawa, K., and Haneishi, M. (eds.), Artech House, 1992, pp. 144–145.

[42] Mongia, R., Bahl, I., and Bhartia, P., *RF and Microwave Coupled-Line Circuits*, Artech House, 1999, pp. 244–247.

[43] Malherbe, J.A.G., *Microwave Transmission Line Coupler*, Artech House, 1988.

[44] Qing, X.M., "Broadband Aperture-Coupled Circularly Polarized Microstrip Antenna Fed By A Three-Stub Hybrid Coupler," *Microwave and Optical Technology Letters*, January 2004, Vol. 40, No. 1, pp. 38–41.

[45] Sainati, R.A., *CAD of Microstrip Antennas for Wireless Applications*, Artech House, 1996, pp. 124–129.

[46] Langston, W.L., and Jackson, D.R., "Impedance, Axial Ratio, and Receive-Power Bandwidths of Microstrip Antennas," *IEEE International Symposium*, June 16–21, 2002, San Antonio, TX, pp. 882–885.

[47] Milligan, T., *Modern Antenna Design*, McGraw Hill, 1985, p. 104.

[48] Langston, W.L., and Jackson, D.R., "Impedance, Axial Ratio, and Receive-Power Bandwidths of Microstrip Antennas," *IEEE Transactions on Antennas and Propagation*, October 2004, Vol. 52, No. 10, pp. 2769–2773.

[49] Sainati, R.A., *CAD of Microstrip Antennas for Wireless Applications*, Artech House, 1996, p. 54.

[50] Schelkunoff, S.A., "Anatomy of Surface Waves," *IRE Transactions on Antennas and Propagation*, December 1959, pp. S133–S139.

[51] Hill, D.A., and Wait, J.R., "Excitation of the Zenneck Surface Wave by a Vertical Aperture," *Radio Science*, November–December 1978, Vol. 13, No. 6, pp. 969–977.

[52] Wait, J.R., "The Ancient and Modern History of EM Ground-Wave Propagation," *IEEE Antennas and Propagation Magazine*, October 1998, Vol. 40, No. 5, pp. 7–24.

[53] Sarkar, T.K., Abdallah, M.N., Salazar-Palma, M., and Dyab, W.M., "Surface Plasmons/Polaritons, Surface Waves, and Zenneck Waves," *IEEE Antennas & Propagation Magazine*, June 2017, pp. 77–93.

[54] Pozar, D., "Comparison of Three Methods for the Measurement of Printed Antenna Efficiency," *IEEE Transactions on Antennas and Propagation*, January 1988, Vol. 36, No. 1, pp. 136–139.

[55] Roudot, B., Mosig, J.R., and Gardiol, F.E., "Radome Effects on Microstrip Antenna Parameters," *17th European Microwave Conference*, Rome, Italy, September 1987, pp. 771–777.

[56] Verma, A.K., Bhupal, A., Rostamy, Z., and Srivastava, G.P., "Analysis of Rectangular Patch Antenna with Dielectric Cover," *IEICE Transactions*, May 1991, Vol. E 74, No. 5, pp. 1270–1276.

[57] Bernhard, J.T., and Tousignant, C.J., "Resonant Frequencies of Rectangular Microstrip Antennas with Flush and Spaced Dielectric Substrates," *IEEE Transactions on Antennas and Propagation*, February 1999, Vol. 47, No. 2, pp. 302–308.

[58] Zhong, S.Z., Liu, G., and Qasim, G., "Closed Form Expressions for Resonant Frequency of Rectangular Patch Antennas With Multidielectric Layers," *IEEE Transactions on Antennas and Propagation*, September 1994, Vol. 42, No. 9, pp. 1360–1363.

[59] Bahl, I.J., and Stuchly, S.S., "Variational Method for the Analysis of Microstrip-Like Transmission Lines," *IEEE Transactions on Microwave Theory and Techniques*, August 1968, Vol. MTT-16, No. 8, pp. 529–535.

[60] Bahl, I.J., and Stuchly, S.S., "Analysis of a Microstrip Covered with a Lossy Dielectric," *IEEE Transactions on Microwave Theory and Techniques*, February 1980, Vol. MTT-28, No. 2, pp. 104–109.

[61] Shavit, R., "Dielectric Cover Effect on Rectangular Microstrip Antenna Array," *IEEE Transactions on Antennas and Propagation*, August 1994, Vol. 42, No. 8, pp. 1180–1184.

[62] Sainati, R.A., *CAD of Microstrip Antennas for Wireless Applications*, Artech House, 1996, pp. 70–71.

[63] James, J.R., Henderson, A., and Hall, P.S., "Microstrip Antenna Performance is determined by Substrate Constraints," *Microwave System News (MSN)*, August 1982, pp. 73–84.

[64] Gopinath, A., "Maximum Q-Factor of Microstrip Resonators," *IEEE Transactions on Microwave Theory and Techniques*, February 1981, Vol. MTT-29, No. 2, pp. 128–131.

[65] Caizzone, S., Circiu, M.-S., Elmarissi, W., Enneking, C., Felux, M., and Yinusa, K.A., "Antenna Pattern Uniformity—Effects on Pseudorange Tracking Error," *GPS World*, February 2018, pp. 18–32.

[66] Benalia, A., and Gupta, K.C., "Faster Computation of Z-matrices for Rectangular Segments in Planar Microstrip Circuits," *IEEE Transactions on Microwave Theory and Techniques*, June 1986, Vol. MTT-34, pp. 733–736.

[67] Mishra, R.K., and Milligan, T., "Cross-Polarization Tolerance Requirements of Square Microstrip Patches," *IEEE Antennas and Propagation Magazine*, April 1996, Vol. 38, No. 2, pp. 56–58.

[68] Chen, W., Wong, K., and Row, J., "Superstrate Loading Effects on the Circular Polarization and Crosspolarization Characteristics of a Rectangular Microstrip Antenna," *IEEE Transactions on Antennas and Propagation*, February 1994, Vol. 42, No. 2, pp. 260–264.

[69] Waterhouse, R.B. (ed.), *Microstrip Antennas a Designer's Guide*, Kluwer Academic Publishers, 2003, pp. 52–54.

[70] Bahl, I.J., and Bhartia, P., *Microstrip Antennas*, Artech House, 1980, 1982 printing, p. 57.

[71] MahabubAlam, M., RifatAhmmedAoni, M., and Islam, T., "Gain Improvement of Micro Strip Antenna Using Dual Patch Array Micro Strip Antenna," *Journal of Emerging Trends in Computing and Information Sciences*, December 2012, Vol. 3, No. 12, pp. 1642–1648.

[72] Bhardwaj, S., and Rahmat-Samii, Y., "Revisiting the Generation of Cross-Polarization in Rectangular Patch Antennas: A Near-Field Approach," *IEEE Antennas and Propagation Magazine*, February 2014, Vol. 56, No. 1, pp. 14–38.

[73] Ludwig, A.C., "The Definition of Cross Polarization," *IEEE Transactions on Antennas and Propagation*, January 1973, Vol. AP-21, No. 1, pp. 116–119.

[74] Knittel, G.H., "Comments on The Definition of Cross Polarization" (and authors reply), *IEEE Transactions on Antennas and Propagation*, November 1973, pp. 917–918.

[75] Petosa, A., Ittipiboon, A., and Gagnon, N., "Suppression of Unwanted Probe Radiation in Wideband Probe-Fed Microstrip Patches," *Electronics Letters*, March 1999, Vol. 35, No. 5, pp. 355–357.

[76] Yang, S.L.S., Lee, K.F., Kishk, A.K., and Luk, K.M., "Cross Polarization Studies of Rectangular Patch Antenna," *Microwave and Optical Technology Letters*, August 2008, Vol. 50, No. 8, pp. 2099–2103.

[77] Abbas, D.M.S., Paul, S., Sen, J., *et al.*, "Aspect Ratio: A Major Controlling Factor of Radiation Characteristics of Microstrip Antenna," *Journal of Electromagnetic Analysis and Applications*, 2011, Vol. 3, pp. 452–457.

[78] Sarkar, C., Guha, D., Kumar, C., and Antar, Y.M.M., "New Insight and Design Strategy to Optimize Cross-Polarized Radiations of Microstrip Patch Over Full Bandwidth by Probe Current Control," *IEEE Transactions on Antennas and Propagation*, August 2018, Vol. 66, No. 8, pp. 3902–3909.

[79] Hu, Y., Jackson, J.T., Williams, S.A., and Komanduri, V.R., "Characterization of the Input Impedance of the Inset-Fed Rectangular Microstrip Antenna," *IEEE Transactions on Antennas and Propagation*, October 2008, Vol. AP-56, No. 10, pp. 3314–3318.

[80] Lee, R.Q., Huynh, T., and Lee, K.F., "Cross Polarization Characteristics of Rectangular Patch Antennas," *IEEE International Symposium on Antennas and Propagation*, 2, San Jose, June 1989, pp. 636–639.

[81] Ghosh, D., Ghosh, S.K., Chattopadhyay, S., *et al.*, "Physical and Quantitative Analysis of Compact Rectangular Microstrip Antenna with Shorted Non-Radiating Edges for Reduced Cross-Polarized Radiation Using Modified Cavity Model," *IEEE Antennas and Propagation Magazine*, August 2014, Vol. 56, No. 4, pp. 61–72.

Chapter 3

Circular microstrip antennas

3.1 Circular microstrip antenna properties

In Chapter 2, we have seen that the rectangular microstrip antenna has a number of useful designs. The circular microstrip antenna offers a number of radiation pattern options not readily implemented using a rectangular patch. The fundamental mode of the circular microstrip patch antenna is the TM_{11}. This mode produces a radiation pattern that is very similar to the lowest order mode of a rectangular microstrip antenna. The next higher order mode is the TM_{21} which can be driven to produce circularly polarized radiation with a monopole-type pattern. This is followed in frequency by the TM_{02} mode which radiates a monopole pattern with linear polarization. In the late 1970s, liquid crystals were used to experimentally map the electric field of the driven modes surrounding a circular microstrip antenna and optimize them [1].

In Figure 3.1, the geometry of a circular microstrip antenna is defined. The circular metallic patch has a radius a, and a driving point located at r, at an angle ϕ measured from the $\hat{x}$-axis. As with the rectangular microstrip antenna, the patch is spaced a distance h from a groundplane. A substrate of ϵ_r separates the patch and the groundplane.

An analysis of the circular microstrip antenna, which is very useful for engineering purposes, was been undertaken by Derneryd and will be utilized here [2]. The electric field between the patch and groundplane, $\vec{E}_z$, is given by

$$E_z = E_0 J_n(kr)\cos(n\phi) \tag{3.1}$$

The magnetic field components are described with

$$H_r = -\frac{j\omega\epsilon n}{k^2 r} E_0 J_n(kr)\sin n\phi \tag{3.2}$$

$$H_\phi = -\frac{j\omega\epsilon}{k} E_0 \acute{J}_n(kr)\cos n\phi \tag{3.3}$$

where k is the propagation constant in the dielectric which has a dielectric permittivity $\epsilon = \epsilon_0\epsilon_r$. J_n is the Bessel function of the first kind of order n. $\acute{J}_n$ is the derivative of the Bessel function with respect to its argument, ω is the angular frequency ($\omega = 2\pi f$). The open circuited edge condition requires that $\acute{J}_n(ka) = 0$. For each mode of a circular microstrip antenna, there is an associated radius, which is dependent on the zeros of the derivative of the Bessel function. Bessel functions in this analysis are analogous to sine and cosine functions in rectangular coordinates. E_0 is the value of the electric field at the edge of the patch across the gap.

The resonant frequency, f_{nm}, for each TM mode of a circular microstrip antenna is given by*

$$f_{nm} = \frac{A_{nm} \cdot c}{2\pi a_{eff} \sqrt{\epsilon_r}} \tag{3.4}$$

where A_{nm} is the mth zero of the derivative of the Bessel function of order n. The constant c is the speed of light in free space, and a_{eff} is the effective radius of the patch. A list of the first four Bessel function zeros used with (3.4) is presented in Table 3.1.

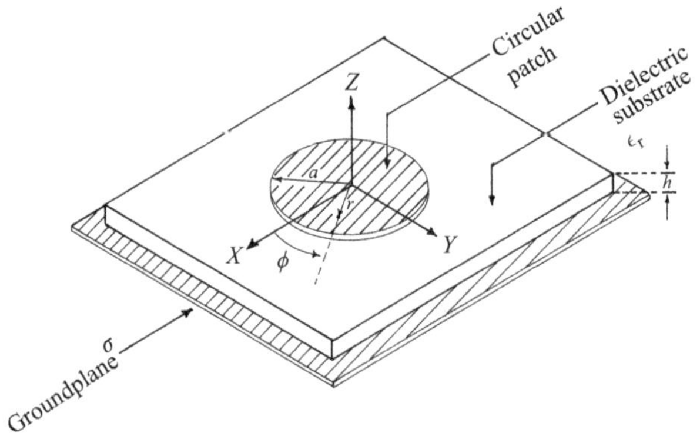

Figure 3.1 Circular microstrip antenna geometry. The circular microstrip antenna is a metal disk of radius a, has a driving point location at r, which makes an angle ϕ with the $\hat{x}$-axis. The thickness of the substrate is h, where $h \ll \lambda_0$, which has a relative dielectric permittivity of ϵ_r

Table 3.1 First four Bessel function zeros used with equation (3.4)

Bessel function zeros and corresponding modes	
A_{nm}	TM_{nm}
1.84118	1,1
3.05424	2,1
3.83171	0,2
4.20119	3,1

*In the case of a rectangular microstrip antenna, the modes are designated by TM_{mn} where m is related to x and n is related to y. The modes for a circular microstrip antenna were introduced as TM_{nm}, where n is related to ϕ and m is related to r (often designated ρ). The reversal of indices can be a source of confusion.

The effective radius of the circular patch, a_{eff}, is given by

$$a_{eff} = a \cdot \left[1 + \frac{2h}{\pi a \epsilon_r}\left(\ln\left\{\frac{\pi a}{2h}\right\} + 1.7726\right)\right]^{1/2} \tag{3.5}$$

where $a/h \gg 1$, a is the physical radius of the antenna.

Equations (3.4) and (3.5) may be combined to produce

$$a = \frac{A_{nm} \cdot c}{2\pi f_{nm}\sqrt{\epsilon_r}}\left[1 + \frac{2h}{\pi a \epsilon_r}\left(\ln\left\{\frac{\pi a}{2h}\right\} + 1.7726\right)\right]^{-1/2} \tag{3.6}$$

The form of (3.6) is

$$a = f(a) \tag{3.7}$$

which can be solved using fixed point iteration (Appendix B, Section B.3) to compute a design radius, given a desired value of A_{nm} from Table 3.1, which determines the mode TM_{nm}, and given the desired resonant frequency f_{nm}, at which the antenna is to operate [3].

An initial approximation for the radius a_0 to begin the iteration is

$$a_0 = \frac{A_{nm} \cdot c}{2\pi f_{nm}\sqrt{\epsilon_r}} \tag{3.8}$$

The initial value a_0 is placed into the right-hand side of (3.6) to produce a value for a, this value is designated a_1, which is then placed into the right-hand side to produce a second more refined value for a, designated a_2, and so on. Experience indicates that no more than five iterations are generally required to produce a stable solution.

The lowest order mode, TM_{11}, is the bipolar mode which is analogous to the lowest order mode of a rectangular microstrip antenna. In Figure 3.2, we can see the electric field concentrated at each end of the antenna for the $n = 1$ mode with a sign reversal. The mode number n corresponds to the number of sign reversals in π radians of ϕ.

The next resonant mode is the TM_{21} mode, which is called the quadrapolar mode. Figure 3.2 shows the electric field distribution for the $n = 2$ mode. We note the four concentrations of electric field with alternating signs. This mode is the first of a family of modes which may be used to create a circularly polarized monopole type pattern.

The third mode is the TM_{02} unipolar mode. In this situation, the mode index n is zero, which implies that no sign reversals occur because the cosine in (3.1) becomes unity for all values of ϕ, and is therefore independent of the angle ϕ. Figure 3.2 shows the $n = 0$ mode, and illustrates the uniform electric field around the edge of the circular antenna. This mode radiates a monopole type pattern. Following the introduction of the mathematical analysis equations for directivity, driving point impedance, and efficiency, we will examine these modes in more detail.

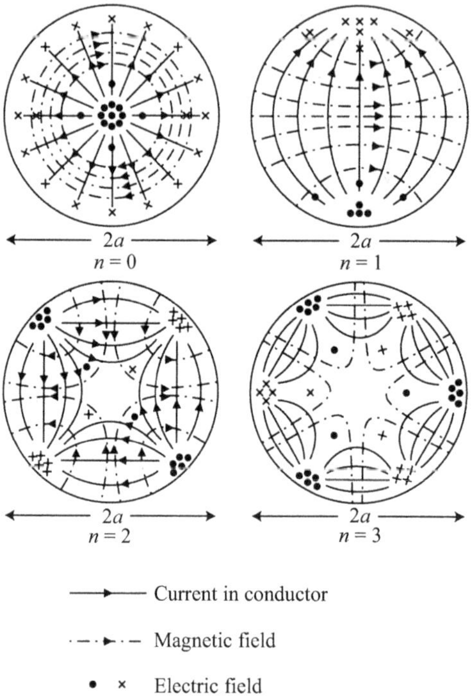

Figure 3.2 Electric and magnetic field patterns of a circular microstrip antenna at resonance

3.2 Directivity

A very useful formulation for the directivity of the fundamental mode of a circular microstrip antenna was presented by Derneryd [2].

The radiation conductance of a circular microstrip antenna is given by

$$G_{rad} = \epsilon_{n0} \frac{(k_0 a)^2}{480} \int_0^{\frac{\pi}{2}} \left[B_M^2(k_0 a \sin\theta) + B_P^2(k_0 a \sin\theta)\cos^2\theta \right] \sin\theta \, d\theta \tag{3.9}$$

where

$$\epsilon_{n0} = \begin{cases} 2 \ (n = 0) \\ 1 \ (n \neq 0) \end{cases} \tag{3.10}$$

and

$$B_P(X) = J_{n-1}(X) + J_{n+1}(X) \tag{3.11}$$

$$B_M(X) = J_{n-1}(X) - J_{n+1}(X) \tag{3.12}$$

The directivity of a circular patch for the $n = 1$ mode is expressed as

$$D = \frac{(k_0 a)^2}{120\, G_{rad}} \tag{3.13}$$

The losses associated with the dielectric may be expressed as

$$G_{\epsilon_r} = \frac{\epsilon_{n0} \tan\delta}{4\mu_0 h f_{mn}} \left[(ka)^2 - n^2\right] \tag{3.14}$$

The ohmic loss associated with the conductors is

$$G_{cu} = \frac{\epsilon_{n0} \cdot \pi (\pi f_{mn} \mu_0)^{(-3/2)}}{4h^2 \sqrt{\sigma}} \left[(ka)^2 - n^2\right] \tag{3.15}$$

The total conductance is

$$G = G_{rad} + G_{\epsilon_r} + G_{cu} \tag{3.16}$$

3.3 Input resistance and impedance bandwidth

The input resistance at resonance as a function of radius r is

$$Z_{in}(r) = \frac{1}{G} \frac{J_n^2(kr)}{J_n^2(ka)} \tag{3.17}$$

The cavity Q's, which allow us to compute the impedance bandwidth of a circular microstrip antenna, may be defined as

The radiation Q is

$$Q_R = \frac{240\left[(ka)^2 - n^2\right]}{h\mu f_r (k_0 a)^2 I_1} \tag{3.18}$$

where

$$\begin{aligned} I_1 = \int_0^{\pi} & \left[\{J_{n+1}(k_0 a \sin\theta) - J_{n-1}(k_0 a \sin\theta)\}^2 \right. \\ & \left. + \cos^2\theta \{J_{n+1}(k_0 a \sin\theta) - J_{n-1}(k_0 a \sin\theta)\}^2\right] \sin\theta \, d\theta \end{aligned} \tag{3.19}$$

The dielectric Q is

$$Q_D = \frac{1}{\tan\delta} \tag{3.20}$$

The conductor Q is

$$Q_C = h\sqrt{\mu_0 \pi f_r \sigma_c} \tag{3.21}$$

as related previously:

$$\frac{1}{Q_T} = \frac{1}{Q_R} + \frac{1}{Q_D} + \frac{1}{Q_C} \tag{3.22}$$

The impedance bandwidth (S:1 VSWR) of a circular microstrip antenna is given by

$$BW = \frac{100(S-1)}{Q_T\sqrt{S}} \tag{3.23}$$

3.3.1 TM_{11}, TM_{21}, and TM_{02} impedance bandwidth

The cavity model equations of Section 3.3 allow us to compare the bandwidth of the bipolar TM_{11}, quadrapole TM_{21}, and monopole TM_{02} mode of a circular patch antenna. The results are given in Table 3.2 for a foam dielectric. The bandwidth of the TM_{11} mode has a slightly larger impedance bandwidth than the TM_{02} mode, which has a slightly larger bandwidth than the TM_{21} mode.

The same computations are carried out for a dielectric substrate with a relative permittivity of 2.6. The results are presented in Table 3.3. Again, the fundamental TM_{11} mode has a larger impedance bandwidth than does the TM_{02}, but they are closer than they were in the foam dielectric case. The quadrapole mode quickly lags the TM_{11} and TM_{02} bandwidths as the thickness increases. For the thickest substrate computed, 3048 μm, the percentage bandwidth of the fundamental mode patch is about twice that of the quadrapole mode.

One must be mindful of the bandwidth decrease associated with the TM_{21} quadrapole mode.

When a high dielectric permittivity material is used, $\epsilon_r = 10.2$, only the TM_{11} mode has a bandwidth greater than 1% with the thickest substrate (Table 3.4). The TM_{21} mode essentially remains constant, at around 0.30%. The TM_{02} only makes it above 0.5% for the thickest of substrates.

Table 3.2 Cavity model bandwidth comparison (2.45 GHz), tan $\delta = 0.0025$, $\epsilon_r = 1.1$ (foam dielectric)

TM_{11}, TM_{21}, and TM_{02} mode
Foam dielectric bandwidth comparison

	h	**TM_{11}**	**TM_{21}**	**TM_{02}**
(0.030")	762 μm	1.17 %	0.83 %	0.95 %
(0.060")	1524 μm	2.11 %	1.36 %	1.60 %
(0.090")	2286 μm	3.22 %	1.99 %	2.36 %
(0.120")	3048 μm	4.46 %	2.70 %	3.18 %

Table 3.3 Cavity model bandwidth comparison (2.45 GHz), tan $\delta = 0.0025$, $\epsilon_r = 2.6$

TM_{11}, TM_{21}, and TM_{02} Mode
$\varepsilon_r = 2.6$ Bandwidth Comparison

	h	**TM_{11}**	**TM_{21}**	**TM_{02}**
(0.030")	762 μm	0.93 %	0.58 %	0.91 %
(0.060")	1524 μm	1.53 %	0.81 %	1.47 %
(0.090")	2286 μm	2.22 %	1.09 %	2.05 %
(0.120")	3048 μm	2.94 %	1.39 %	2.66 %

Table 3.4 Cavity model bandwidth comparison (2.45 GHz), tan $\delta = 0.0025$, $\epsilon_r = 10.2$

TM_{11}, TM_{21}, and TM_{02} mode $\varepsilon_r = 10.2$ bandwidth comparison				
	h	**TM_{11}**	**TM_{21}**	**TM_{02}**
(0.030")	762 μm	0.53 %	0.34 %	0.40 %
(0.060")	1524 μm	0.70 %	0.31 %	0.43 %
(0.090")	2286 μm	0.91 %	0.32 %	0.50 %
(0.120")	3048 μm	1.14 %	0.35 %	0.59 %

3.4 Gain, radiation pattern, and efficiency

The antenna efficiency for a microstrip antenna is

$$e = \frac{Q_C Q_D}{Q_C Q_D + Q_C Q_R + Q_D Q_R} \tag{3.24}$$

The radiation patterns may be calculated using:

$$E_\theta = j^n \frac{Vak_0}{2} \frac{e^{-jk_0 r}}{r} \cos n\phi [J_{n+1}(k_0 a \sin\theta) - J_{n-1}(k_0 a \sin\theta)] \tag{3.25}$$

$$E_\phi = j^n \frac{Vak_0}{2} \frac{e^{-jk_0 r}}{r} \cos\theta \sin n\phi [J_{n+1}(k_0 a \sin\theta) + J_{n-1}(k_0 a \sin\theta)] \tag{3.26}$$

where V is the edge voltage at $\phi = 0$:

$$V = hE_0 J_n(ka) \tag{3.27}$$

When $n = 1$, (3.13) may be used to compute the antenna directivity. One must numerically integrate (3.25) and (3.26) to obtain directivity estimates of a circular patch when $n \neq 1$. The efficiency obtained from (3.24) allows one to compute the gain of a circular microstrip antenna.

3.4.1 TM_{11} efficiency

The cavity model for a circular patch antenna in (3.4) does not include surface wave losses. Using the method of Section B.5 in Appendix B, we can separate out the losses for a circular patch as was done for a rectangular microstrip antenna in Section 2.6.

The TM_{11} mode corresponds to the TM_{01} mode of a square microstrip antenna, which allows us to compare the two design options directly. The amount of energy lost in surface waves, η_{sw}, for a circular patch with foam substrate is shown in Table 3.5. It is lower than the corresponding rectangular patch antenna (Table 2.7). The surface wave loss for a 762 μm thick rectangular patch is 12.21%, and 9.52% for a circular patch. The copper and dielectric losses are comparable. A 3048 μm thick rectangular patch has surface wave losses of 2.13%, whereas the circular patch is 0.61%. The overall efficiency of a circular patch is slightly better than its rectangular counterpart, but at the expense of slightly narrower bandwidth (Table 3.7).

The loss attributed to surface waves cannot be a TM_0 surface wave trapped within the substrate dielectric, as the foam case should have very little trapping, and when reduced to vacuum, the computed losses persist. The TM_0 surface wave has no low frequency cut-off and is generally assumed to couple to the fields of a circular patch [4].

When the relative dielectric permittivity increases to 2.6, the surface wave losses are also slightly lower than for a corresponding rectangular patch (Table 2.8). In

Table 3.5 TM_{11} Mode Foam Dielectric Efficiency HFSS Q-Losses—solid copper in a circular linear MSA versus substrate thickness—lossless connector (2.45 GHz), tan $\delta = 0.0025$, $\epsilon_r = 1.1$ (foam dielectric), Cu thickness = 17 μm (100 mm × 100 mm groundplane)

TM_{11} mode foam dielectric efficiency					
	h	η_r	η_{sw}	η_c	η_d
(0.030")	762 μm	76.67 %	9.52 %	5.42 %	8.38 %
(0.060")	1524 μm	91.55 %	1.84 %	1.74 %	4.87 %
(0.090")	2286 μm	94.99 %	1.17 %	0.61 %	3.23 %
(0.120")	3048 μm	96.55 %	0.61 %	0.47 %	2.36 %

Table 3.6 HFSS Q-Losses—solid copper in a circular linear MSA versus substrate thickness—lossless connector (2.45 GHz), tan $\delta = 0.0025$, $\epsilon_r = 2.6$, Cu thickness = 17 μm (100 mm × 100 mm groundplane)

TM_{11} mode efficiency $\varepsilon_r = 2.6$					
	h	η_r	η_{sw}	η_c	η_d
(0.030")	762 μm	70.96 %	12.24 %	6.40 %	10.41 %
(0.060")	1524 μm	88.42 %	3.16 %	2.00 %	6.42 %
(0.090")	2286 μm	92.90 %	2.07 %	0.79 %	4.25 %
(0.120")	3048 μm	95.31 %	0.98 %	0.56 %	3.15 %

Table 3.7 HFSS Q-Losses—solid copper in a circular linear MSA versus substrate thickness—lossless connector (2.45 GHz), tan $\delta = 0.0025$, $\epsilon_r = 10.2$, Cu thickness = 17 μm (100 mm × 100 mm groundplane)

TM_{11} mode efficiency $\varepsilon_r = 10.2$					
	h	η_r	η_{sw}	η_c	η_d
(0.030")	762 μm	50.03 %	18.84 %	9.98 %	21.15 %
(0.060")	1524 μm	73.70 %	11.19 %	3.04 %	12.08 %
(0.090")	2286 μm	83.32 %	6.93 %	1.41 %	8.34 %
(0.120")	3048 μm	88.89 %	4.38 %	0.64 %	6.10 %

Table 3.6, a circular patch with a 762 μm thick substrate has surface wave losses of 12.24%, but are 15.43% for the rectangular case. The thickest substrate computed is 3048 μm. The surface wave loss is 0.98% for a circular patch, and in the case of a rectangular patch it is 3.19%. The conductor and dielectric losses track very closely between the two geometries.

The TM_{11} mode efficiency for 10.2 relative permittivity has a slightly lower surface wave loss at 762 μm than its rectangular counterpart of Table 2.9, but for thicker substrate values, is comparable to the square patch case. The conductor and dielectric losses for the circular high dielectric case are also similar to the square patch.

3.4.2 TM_{21} efficiency

The efficiency of the TM_{21} mode versus thickness for a foam dielectric is given in Table 3.8. The TM_{21} mode with a foam dielectric has larger losses than the fundamental TM_{11} mode. The overall efficiency for a TM_{21} patch, with 762 μm thick substrate at 2.45 GHz, is 66.57%, whereas the TM_{11} patch has 76.67% efficiency, which is about 10% more efficient. The difference quickly decreases with thickness until there is only a 2.82% difference for the thickest (3048 μm) substrate analyzed.

When the relative dielectric permittivity is increased to 2.6 (Table 3.9), the surface wave losses for the thinnest substrate, 762 μm, are about 16% worse for the TM_{21}

Table 3.8 HFSS Q-Losses—solid copper in a circular linear MSA versus substrate thickness—lossless connector (2.45 GHz), $\tan\delta = 0.0025$, $\epsilon_r = 1.1$ (foam dielectric), Cu thickness = 17 μm (200 mm × 200 mm) groundplane

TM_{21} mode foam dielectric efficiency					
	h	η_r	η_{sw}	η_c	η_d
(0.030")	762 μm	66.57 %	13.71 %	7.42 %	12.30 %
(0.060")	1524 μm	84.61 %	5.53 %	2.27 %	7.58 %
(0.090")	2286 μm	91.07 %	2.80 %	0.86 %	5.26 %
(0.120")	3048 μm	93.73 %	1.77 %	0.77 %	3.74 %

Table 3.9 HFSS Q-Losses—solid copper in a circular linear MSA versus substrate thickness—lossless connector (2.45 GHz), $\tan\delta = 0.0025$, $\epsilon_r = 2.6$, Cu thickness = 17 μm (200 mm × 200 mm groundplane)

TM_{21} mode $\varepsilon_r = 2.6$ dielectric efficiency					
	h	η_r	η_{sw}	η_c	η_d
(0.030")	762 μm	55.14 %	27.72 %	6.24 %	10.19 %
(0.060")	1524 μm	85.15 %	5.73 %	2.02 %	7.09 %
(0.090")	2286 μm	91.76 %	2.37 %	0.78 %	5.08 %
(0.120")	3048 μm	94.15 %	1.62 %	0.24 %	3.99 %

Table 3.10 HFSS Q-Losses—solid copper in a circular linear MSA versus substrate thickness—lossless connector (2.45 GHz), tan δ = 0.0025, $\epsilon_r = 10.2$, Cu thickness = 17 μm (200 mm × 200 mm groundplane)

TM_{21} mode $\varepsilon_r = 10.2$ dielectric efficiency					
	h	η_r	η_{sw}	η_c	η_d
(0.030")	762 μm	13.27 %	30.56 %	19.06 %	37.11 %
(0.060")	1524 μm	31.22 %	20.60 %	9.38 %	38.80 %
(0.090")	2286 μm	48.77 %	13.69 %	5.46 %	32.08 %
(0.120")	3048 μm	62.08 %	10.20 %	3.20 %	24.52 %

Table 3.11 HFSS Q-Losses—solid copper in a circular linear MSA versus substrate thickness—lossless connector (2.45 GHz), tan δ = 0.0025, $\epsilon_r = 1.1$ (foam dielectric), Cu thickness = 17μm (200 mm × 200 mm groundplane)

TM_{02} mode foam dielectric efficiency					
	h	η_r	η_{sw}	η_c	η_d
(0.030")	762 μm	72.62 %	9.59 %	6.74 %	11.05 %
(0.060")	1524 μm	86.26 %	4.48 %	2.05 %	7.20 %
(0.090")	2286 μm	91.85 %	2.39 %	0.72 %	5.04 %
(0.120")	3048 μm	94.11 %	1.83 %	0.23 %	3.82 %

patch than for the TM_{11}. However, for the thickest substrate value, 3048 μm, the efficiency difference is only about 1.2%. For thicker substrates, as was the case for foam dielectric, the difference in a practical design is negligible.

When the relative permittivity is increased to 10.2 (Table 3.10), the efficiencies of a TM_{21} antenna compared with the TM_{11} are much lower for the same thickness of the substrate. The TM_{11} values are found in Table 3.7, and the TM_{21} numbers are in Table 3.10. For the thinnest substrate 762 μm, we have an overall efficiency of 13.27% for the TM_{21} patch versus 50.03% for the TM_{11}. In the case of the thickest patch, 3048 μm, the best overall efficiency is 62.08% for the TM_{21} patch and 88.89% for the TM_{11}. The surface wave losses are very large for the TM_{21}, but the dielectric losses are much worse. The conductor losses are also larger for the TM_{21} patch compared with the TM_{11}. Clearly the choice of antenna mode affects the efficiency expected.

3.4.3 TM_{02} efficiency

In Table 3.11, the losses for a TM_{02} mode antenna on dielectric foam at 2.45 GHz are presented. When we compare these values with a TM_{11} mode patch at the same frequency, given in Table 3.5, we see the surface wave losses for the TM_{02} mode

Table 3.12 HFSS Q-Losses—solid copper in a circular linear MSA versus substrate thickness—lossless connector (2.45 GHz), tan $\delta = 0.0025$, $\epsilon_r = 2.6$, Cu thickness = 17 μm (200 mm × 200 mm groundplane)

TM_{02} mode efficiency $\varepsilon_r = 2.6$					
	h	η_r	η_{sw}	η_c	η_d
(0.030")	762 μm	69.69 %	11.33 %	7.00 %	11.99 %
(0.060")	1524 μm	85.98 %	4.85 %	1.83 %	7.34 %
(0.090")	2286 μm	91.29 %	2.50 %	0.84 %	5.37 %
(0.120")	3048 μm	93.60 %	2.15 %	0.23 %	4.02 %

Table 3.13 HFSS Q-Losses—solid copper in a circular linear MSA versus substrate thickness—lossless connector (2.45 GHz), tan $\delta = 0.0025$, $\epsilon_r = 10.2$, Cu thickness = 17 μm (200 mm × 200 mm groundplane)

TM_{02} mode efficiency $\varepsilon_r = 10.2$					
	h	η_r	η_{sw}	η_c	η_d
(0.030")	762 μm	25.38 %	29.95 %	15.26 %	29.41 %
(0.060")	1524 μm	47.30 %	24.23 %	4.70 %	23.76 %
(0.090")	2286 μm	60.17 %	16.86 %	3.48 %	19.49 %
(0.120")	3048 μm	70.70 %	11.40 %	2.40 %	15.50 %

is essentially the same for the thinnest, 762 μm, substrate. But as the thickness is increased, the rate of decrease in surface wave loss is slower than the rate of the TM_{11} mode. The dielectric losses also follow the surface wave trend. They start out about 2.5% larger than the TM_{11} mode, and end about 1.5% larger at the 3048 μm thickness. The conductor losses are comparable in magnitude with thickness for both modes, with the TM_{02} loss slightly lower than the TM_{11} for the thickest substrate examined.

The computed results from increasing the relative permittivity to 2.6 are presented in Table 3.12. When compared with the TM_{11} mode results in Table 3.6, we note that the conductor and dielectric losses are similar in magnitude and have similar decreases with increasing thickness. The surface wave decrease is similar, except for the largest thickness, where it appears to plateau.

When the relative permittivity of the substrate is increased to 10.2, we see in Table 3.13 that the surface wave loss and dielectric loss are both approximately 30% each for the thinnest substrate (762 μm). The overall radiation efficiency is only 25.38%. For the case of the TM_{02} mode, the conductor losses are the smallest contribution. As in all previous cases, the surface wave contribution decreases as the substrate thickness increases, as does the copper and dielectric losses. The overall efficiency for the TM_{02} high dielectric case is slightly better than the TM_{21} mode.

Clearly, the thickest patch possible is best if one wishes to produce reasonable efficiency for higher order mode patch designs. The TM_{11} mode has the highest efficiency for a given substrate thickness.

3.5 Circular microstrip antenna radiation modes

3.5.1 The TM_{11} bipolar mode

The TM_{11} mode of a circular microstrip antenna is analogous to the lowest order mode of a rectangular patch antenna. This field distribution can be seen in the Figure 3.2 for the $n = 1$ mode. This mode is essentially similar in design utility to a rectangular microstrip antenna driven in the TM_{10} mode. The impedance bandwidth is slightly smaller for a circular patch than its rectangular counterpart. The center of a circular patch driven in the TM_{11} mode may be shorted at its center, if a DC short is required.

We will use a circular microstrip antenna with a radius of 21.21 mm on a dielectric substrate which is 1.524 mm thick and has a relative dielectric permittivity of $\epsilon_r = 2.6$, and $\tan\delta = 0.0025$, to illustrate the properties of the TM_{11} mode. A finite difference time domain (FDTD) analysis of this antenna placed on a circular groundplane, which has a 33.43 mm radius, produces a resonant frequency of 2.435 GHz. Equation (3.4) predicts the resonant frequency to be 2.467 GHz for the TM_{11} mode. The antenna is fed 7 mm from the center of the antenna at $\phi = 0$. Figure 3.3 shows the E-plane and H-plane radiation patterns. The directivity of the antenna is computed as 7.12 dB by FDTD. Using (3.24), the antenna efficiency is computed as 78.37%, which reduces the directivity peak by 1.06 dB, for a gain of 6.06 dBi.

As is the case with the rectangular patch, the pattern directivity of a TM_{11} circular patch antenna decreases as the relative dielectric permittivity of the substrate increases.

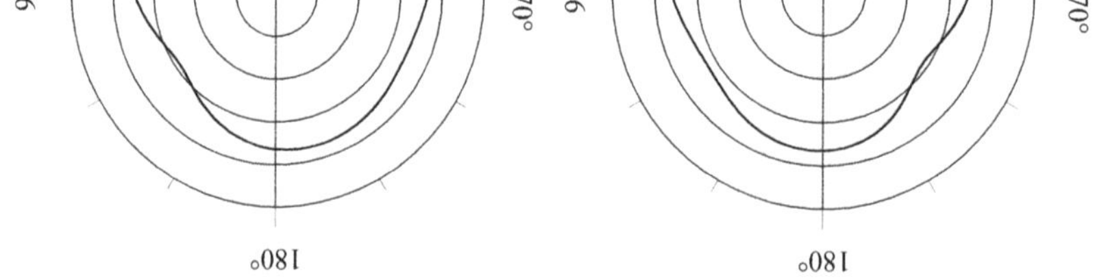

Figure 3.3 E-plane and H-plane patterns of a circular microstrip antenna driven in the TM_{11} mode

3.5.2 TM_{11} bipolar mode circular polarized design

Lo and Richards developed a perturbation relationship to design circularly polarized rectangular and circular microstrip antennas using the TM_{11} mode [5]. They extended their work on rectangular microstrip antennas and demonstrated that a circular microstrip antenna may be stretched into an ellipse, which will produce circular polarization from the superposition of the radiation of a pair of orthogonal modes, when it is fed at $\phi = \pm 45°$. The ratio of the semi-major to semi-minor axes which will produce circular polarization is given by (3.28). As with the design of a circularly polarized rectangular microstrip antenna, the Q of the unperturbed circular patch is first obtained to compute the a/b ratio of a patch, which will produce circular polarization:

$$\frac{a}{b} = 1 + \frac{1.0887}{Q} \tag{3.28}$$

The value of antenna Q can be computed using the cavity model equations (3.18), (3.20), and (3.21) with (3.29):

$$\frac{1}{Q} = \frac{1}{Q_R} + \frac{1}{Q_D} + \frac{1}{Q_C} \tag{3.29}$$

One can also measure the Q of the antenna experimentally, or use results from a full-wave analysis such as FDTD with (3.30) to estimate Q [6]:

$$Q \approx \frac{f_0}{\Delta f} = \frac{(\text{resonance frequency})}{(\text{bandwidth})_{3\text{dB}}} \tag{3.30}$$

where f_0 is the resonant frequency of the patch antenna and Δf is the bandwidth between 3 dB return loss points.

The antenna must have a single apparent resonance with reasonable symmetry for this equation to apply.

If the radius of the unperturbed circular patch, which operates at the desired design frequency f_0, is designated as $\acute{a}$, the semi-major axis a, and semi minor axis b, of the ellipse, which produce circular polarization (Figure 3.4) may be written as

$$a = \acute{a} + \Delta_L \tag{3.31}$$

$$b = \acute{a} - \Delta_L \tag{3.32}$$

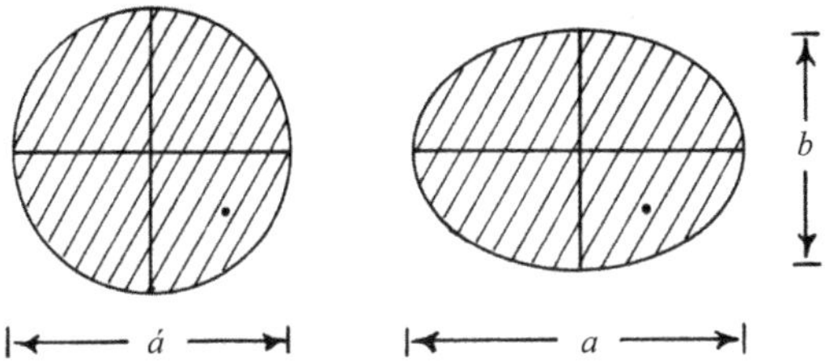

Figure 3.4 Circular microstrip antenna, and the antenna perturbed into an ellipse to produce circular polarization (heavy dot is RHCP feed)

using (3.31) and (3.32) with (3.28), we can write

$$\Delta_L = \frac{\acute{a}}{\left(\frac{2Q}{1.0887}\right) + 1} \tag{3.33}$$

The FDTD analysis of the circular patch example of Section 3.5.1 produced a negative return loss plot from which we use (3.30) to obtain a value of 13.08 for Q. We can compute Δ_L as

$$\Delta_L = \frac{21.21 \text{ mm}}{\left(\frac{2 \cdot (13.08)}{1.0887} + 1\right)} = 0.84742 \text{ mm}$$

and from (3.31) and (3.32) because we used the radius ($\acute{a} = 2a$) we obtain the semi-major and semi-minor axis values:

$$a/2 = 21.21 \text{ mm} + 0.84742 \text{ mm} = 22.057 \text{ mm}$$

$$b/2 = 21.21 \text{ mm} - 0.84742 \text{ mm} = 20.363 \text{ mm}$$

An FDTD analysis was undertaken to evaluate the circular polarization produced using (3.28). The patch feedpoint location is $x = 15.0$ mm, $y = -15.0$ mm, with the ellipse centered in the x–y plane, with a 33.43 mm radius circular groundplane.

In Figure 3.5, are synthesized rotating linear principle plane patterns from an FDTD analysis, driven with a sinusoidal source, with a square coaxial probe at 2.45 GHz [7]. The performance of the antenna is very good, and in a practical design, one could further optimize the antenna experimentally.

A branchline hybrid is an alternative method one may use to generate circular polarization from a circular patch. Figure 3.6 shows a TM_{11} mode patch, fed at orthogonal points, to create circular polarization. This is analogous to the use of a

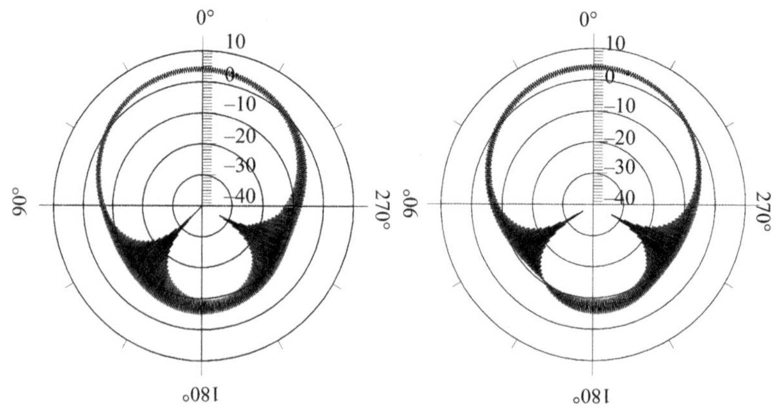

Figure 3.5 Rotating linear plots of an elliptical patch antenna which produces circular polarization, designed using (3.28). On the left, is a cut through the minor axis of the ellipse (x–z), and on the right is a cut through the major axis of the ellipse (y–z). The axial ratio at $\theta = 0$ is 2 dB

branchline hybrid to generate circular polarization with a square patch. The RHCP and LHCP inputs are labeled. In practice, the unused port would be terminated with a load.

3.5.3 The TM_{21} quadrapolar mode

The TM_{21} mode has the next highest frequency of operation (after TM_{11}). This particular mode is useful in creating a monopole radiation pattern which has circular polarization as described by Huang [8]. The electric field pattern for the $n = 2$ mode in Figure 3.2 shows the four electric field reversals which give this mode its name. One can produce circular polarization from this mode by providing two probe feeds to the patch, one is physically located at $\phi = 0°$ and the other at $\phi = 45°$. The feed at $\phi = 0°$ is fed with zero electrical phase. The feed at $\phi = 45°$ is fed with a 90° electrical phase, with an identical amplitude (Figure 3.7). This angular spacing produces two modes driven orthogonal to each other, as is their radiation. The 90° phase difference, with identical amplitude, using orthogonal modes is the usual manner of creating circular polarization. The combination of these feeds produces a resultant

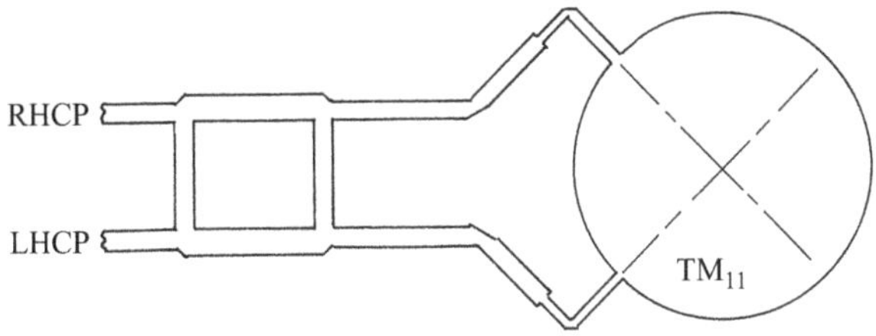

Figure 3.6 Circular polarization using a TM_{11} mode circular patch may be synthesized using a 90° branchline hybrid feeding the patch edges at a spatial angle of 90°

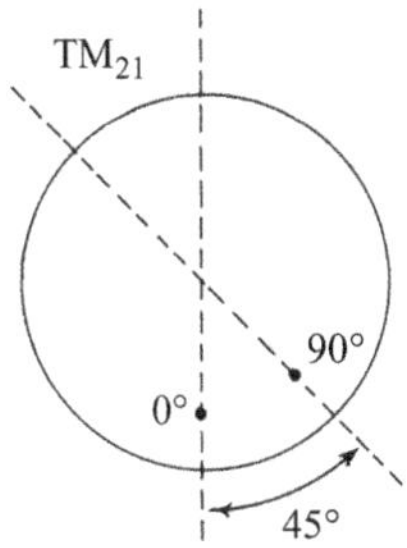

Figure 3.7 A TM_{21} mode circular microstrip antenna driven with two probe feeds (heavy dots) 90° out of phase, with equal amplitudes, spatially separated by 45°. This antenna produces a monopole pattern with circular polarization

quadrapolar electric field as seen in Figure 3.2 which rotates about the center of the patch antenna. This has been verified with FDTD simulation.

One may obtain better circular polarization (axial ratio) by feeding the antenna in four probe locations rather than two. These locations are diametrically across from the two original feedpoints. In the case of an even mode $n = 2, 4, 6, \ldots$ (TM_{21}, TM_{41}, $TM_{61}, \ldots$) the diametric feeds have the same phase as their original counterparts. The feeds have a phase arrangement around the patch counterclockwise of 0°, 90°, 0°, 90°. In the case of an odd mode $n = 1, 3, 5, \ldots$ (TM_{11}, TM_{31}, $TM_{51}, \ldots$), the diametric feeds have a phase arrangement around the patch counterclockwise starting at the top of the patch of 0°, 90°, 180°, 270°. These relationships are related in detail by Huang [8].

As the resonant mode index n increases ($n > 1$), with $m = 1$, the peak directivity of the radiation pattern becomes more and more broadside. The pattern will also move further broadside with increasing relative dielectric permittivity. Huang has reported that the pattern peak may be moved from broadside over a range of 35° to 74° by use of a combination of a chosen higher order mode and substrate relative dielectric permittivity adjustment.

In commercial applications, a complex feed structure with its required feed network may be untenable as a design. It is possible to drive a patch in the TM_{21} mode with a single feed which will produce circular polarization [9]. One may cut a pair of notches in a circular microstrip antenna driven in the TM_{21} mode in accordance with

$$\begin{aligned} \left|\frac{\Delta S}{S}\right| &= \frac{1}{2.5014 \cdot Q} \\ |\Delta S| &= \frac{\pi a^2}{2.5014 \cdot Q} \end{aligned} \tag{3.34}$$

We will use a patch of radius 20.26 mm as a design example. To produce circular polarization, each notch area is $\Delta S/2$, for each of the notches in Figure 3.8(a). The substrate thickness is 1.524 mm, $\epsilon_r = 2.6$, $\tan\delta = 0.0025$, with a resonate frequency of 4.25 GHz computed using FDTD. The feedpoint radius is 16.0 mm. Equation (3.4) predicts 4.278 GHz for the TM_{21} mode. FDTD was used to analyze a circular patch antenna with the previous parameters and produce a return loss plot. The Q was computed to be 22.83 from the return loss plot 3 dB points using (3.30). We find $|\Delta S|$ using (3.34):

$$|\Delta S| = \frac{\pi(20.26\ \text{mm})^2}{2.5014 \cdot 22.83} = 0.022578\ \text{mm}^2$$

which is a square with sides of length $L = 4.75$ mm. Each notch in this example is $L/2$ by L which corresponds to $|\Delta S|/2$ for each notch in Figure 3.8(a). The radiation patterns computed with an FDTD simulation of this design is presented in Figure 3.9. The patterns are synthesized rotating linear plots. Figure 3.8(b) shows an alternative method using tabs and indents which perturb the patch to produce circular polarization from the TM_{21} mode.

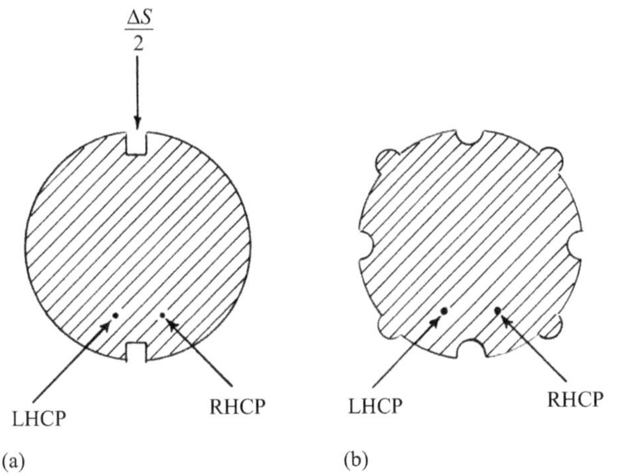

Figure 3.8 *(a) A TM_{21} circular microstrip antenna is modified with a pair of slots using (3.34), with a single probe feed (heavy dots) at $\phi = \pm 22.5°$. This antenna produces a monopole pattern with circular polarization.*
(b) TM_{21} circular microstrip antenna with indents and tabs spaced 45° apart. The feed is at $\phi = \pm 22.5°$, which also produces a monopole pattern with circular polarization

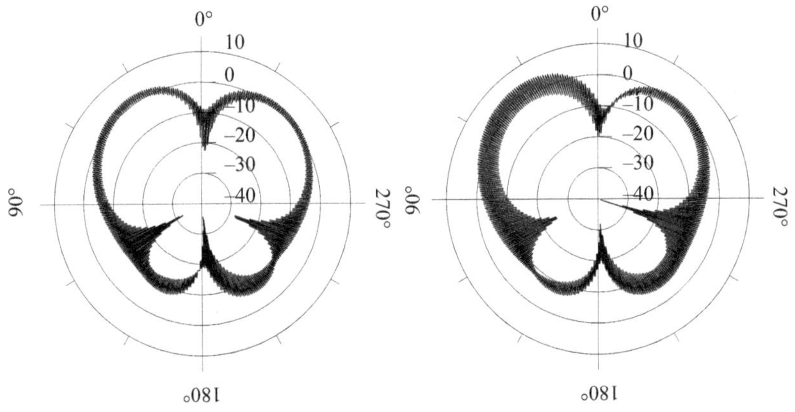

Figure 3.9 *Synthesized rotating linear radiation patterns of TM_{21} circular microstrip antenna modified with a pair of slots using (3.34) with a single probe feed*

3.5.4 The TM_{02} unipolar mode

The next mode in order of increasing frequency is the TM_{02} mode. This mode has the useful characteristic that the electric field around the circular microstrip antenna is uniform. This is seen in Figure 3.2 for the $n = 0$ mode, which we note has no electric field reversal [$\cos(0 \cdot \phi) = 1$ for all ϕ in (3.25)]. The TM_{02} mode has the useful property that it produces a vertically polarized ($\vec{E}_\theta$) monopole type pattern. This can be very useful for replacing a quarter-wave monopole antenna, which can be easily damaged in a hostile mechanical environment, with a conformal version.

We will use a patch of radius 21.21 mm as an example. The substrate thickness is 1.524 mm, $\epsilon_r = 2.6$, $\tan\delta = 0.0025$, on a circular groundplane of radius 33.43 mm, with a resonate frequency of 5.02 GHz, computed using FDTD. The patch is probe fed with a square coaxial transmission line. Equation (3.4) predicts 5.13 GHz for the TM_{02} mode. The feedpoint radius is 7.52 mm. The maximum directivity computed by FDTD is 5.30 dB. The efficiency computed using (3.24) is 87.88% which is a loss of 0.561 dB, for a predicted antenna gain of 4.74 dBi. The computed radiation patterns are presented in Figure 3.10.

The cavity model for the TM_{02} mode circular patch antenna may be used to reveal the radiation pattern change with respect to substrate permittivity. A 4.21 GHz microstrip antenna is analyzed with a 762 μm thick substrate, and the computed patterns are seen in Figure 3.11.

The highest directivity occurs for $\epsilon_r = 1.0$, and as ϵ_r increases, the pattern directivity decreases, and the pattern maximum becomes more toward the horizon (away from broadside).

The values of directivity for given values of ϵ_r are presented in Table 3.14. We note that most of the directivity decrease occurs when even a minimal dielectric loading occurs. Many plastic materials are in the $\epsilon_r = 2.2$–3.0 range. For values of ϵ_r above 3.0, the increase in radiation at the horizon is minimal. The decrease in impedance

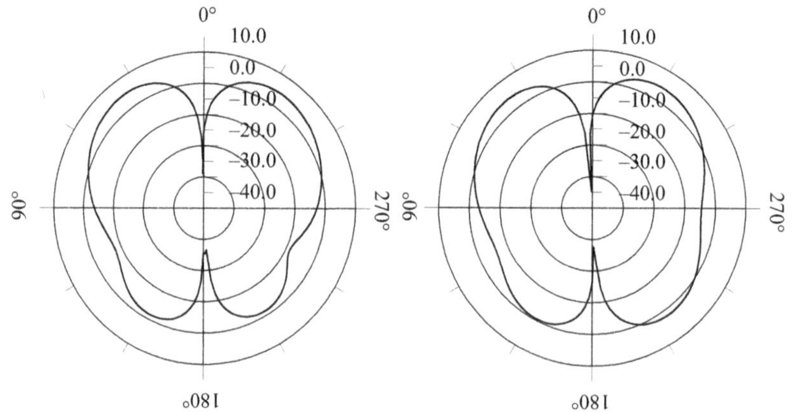

Figure 3.10 TM_{02} circular microstrip antenna pattern as computed by FDTD. The pattern on the left is a cut in a plane perpendicular to the plane which contains the probe feed. On the right is a cut through the plane of the probe. The maximum pattern directivity is 5.30 dB

bandwidth is considerable when compared with the increase in radiation along the horizon as the permittivity increases.

A thermal plot of the total electric field, just above the circular patch element, is presented in Figure 3.12. We see the electric field is uniform around the edge of the element which is consistent with Figure 3.2 for $n = 0$. The small square is the probe feed.

The description of driving point impedance is given in (3.17). The driving point impedance for the TM_{02} mode passes through a short at a radial position where the Bessel function $J_0(kr)$ passes through zero, and then increases to the edge resistance value at $r = a$. Figure 3.13 presents a thermal plot of the total electric field just below the circular patch element. We see a ring of zero electric field corresponding to the short in the driving point impedance predicted by (3.17).

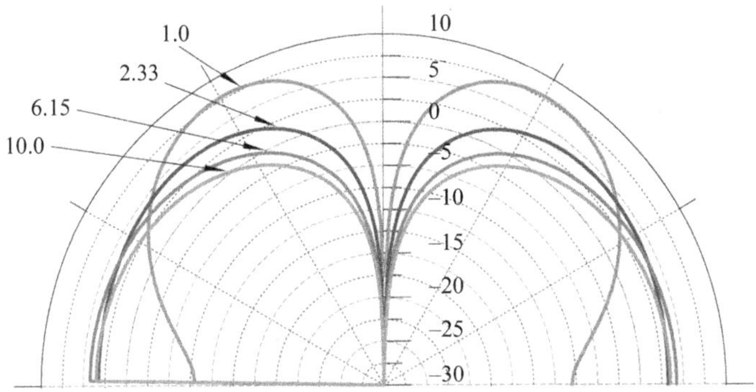

Figure 3.11 TM_{02} circular microstrip antenna radiation pattern with different values of permittivity ϵ_r computed using the cavity model. We can see that as the substrate ϵ_r increases the pattern becomes more broadside

Table 3.14 TM_{02} mode patch directivity versus substrate permittivity

ϵ_r	**Directivity**
1.00	7.31 dB
2.33	3.77 dB
6.15	4.13 dB
10.00	4.37 dB

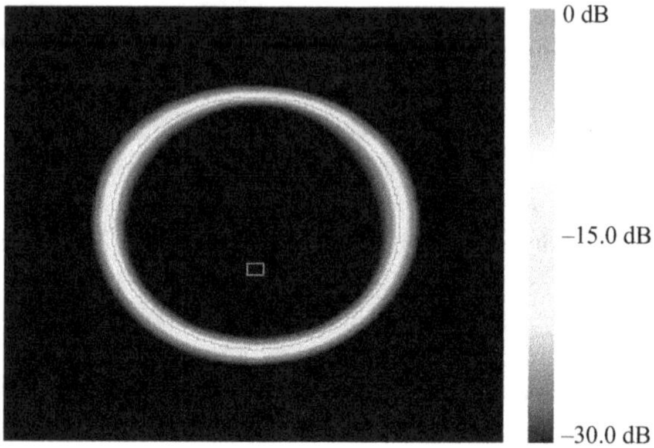

Figure 3.12 *TM_{02} circular microstrip antenna thermal plot of the total electric field just above the element (computed with FDTD). One can see the uniform electric field distribution which is consistent with the $n = 0$ mode of Figure 3.2*

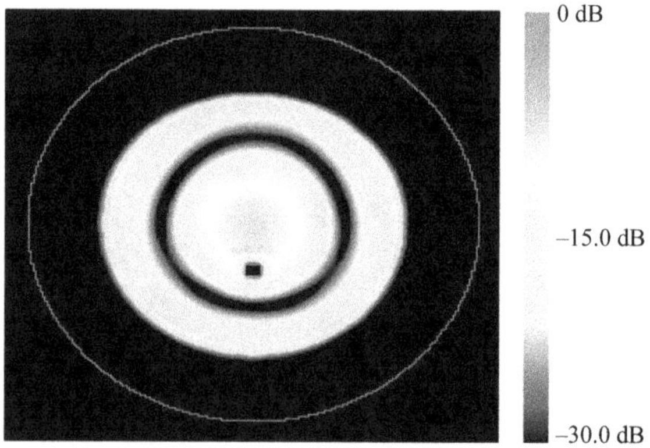

Figure 3.13 *TM_{02} circular microstrip antenna thermal plot of the total electric field distribution just below the element (computed by FDTD). The ring of zero electric field is consistent with the electric field as predicted by Derneryd [2]*

3.6 Circular microstrip antenna cross-polarization

The cross-polarization performance of microstrip antennas is considered to be rather poor. The permittivity and thickness of the substrate used to create a microstrip antenna determines its cross-polarization performance. It has been related previously that the gain and impedance bandwidth of a microstrip antenna can be increased by decreasing the permittivity of the substrate. When a low dielectric constant is used to design a microstrip antenna element, it also increases the radiated cross-polarization level [10].

A higher permittivity substrate will produce better cross-polarization performance, but at the expense of impedance bandwidth. The cross-polarization performance of a linearly polarized patch is dependent on substrate thickness, feed point location, and substrate permittivity.

The origin of the radiated cross-polarization is associated with the generation of higher order modes on the patch [11]. When a circular microstrip patch is driven in the TM_{11} mode with a single driving point, the next highest frequency TM_{21} mode is consistent with the measured cross-polarization patterns [12]. Table 3.1 shows that the modes occur in order of increasing frequency as TM_{11}, TM_{21}, TM_{01}, ... Garcia–Garcia states that when an antenna is driven in the fundamental mode, TM_{11}, this mode is mostly perturbed by the TM_{21} mode. When a patch is designed to be driven in the TM_{21} mode, the radiation purity is disturbed by the dominant TM_{11} and next higher TM_{01} mode.

Figure 3.14 presents sketches of the current of (A) a TM_{21} mode circular patch antenna and (B) a TM_{11} mode circular patch antenna. When a patch is driven in the fundamental TM_{11} mode, and the majority of the cross-polarization radiation is due to TM_{21}, we note that in the x–z plane (H-plane) the co-polarized radiation $\vec{E}_y$ dominates the pattern. The cross-polarized pattern in the H-plane $\vec{E}_z$ has two lobes approximately -15 dB below the co-polarized pattern maximum. The cross-polarization pattern is consistent with the pattern shape expected from the TM_{21} mode. In the E-plane, the radiated field of the driven TM_{11} and the TM_{21} mode are in parallel, which means whatever cross-polarization exists is of uncertain origin. It could be from an imperfect generation of the TM_{21} mode, other modes, or due to a different mechanism.

An illustrative example was analyzed with the FDTD method. The substrate is vacuum ($\epsilon_r = 1$) with a thickness of $h = 1.524$ mm (0.060 inch). The patch has

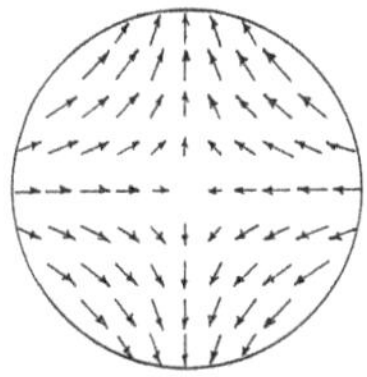

(a) TM_{21} current

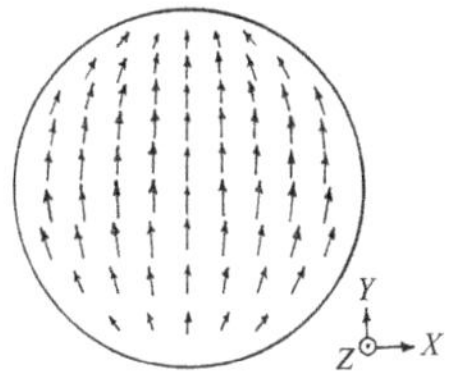

(b) TM_{11} current

Figure 3.14 (a) Sketch of the theoretical current distribution of TM_{21} mode of a circular patch antenna and (b) sketch of the theoretical current distribution of TM_{11} mode of circular patch antenna

a radius of $a = 14.71$ mm. The probe feed is 5.5 mm below the center of the patch which has a resonant frequency of 5.35 GHz. The FDTD analysis results are presented in Figure 3.15(a) and (b). We note the H-plane pattern has the expected TM_{21} mode pattern shape. The E-plane pattern has a small amount of cross-polarized radiation which has a peak magnitude which is approximately −30 dB below the co-polarized maximum. The E-plane cross-polarized pattern has a shape consistent with the TM_{11}

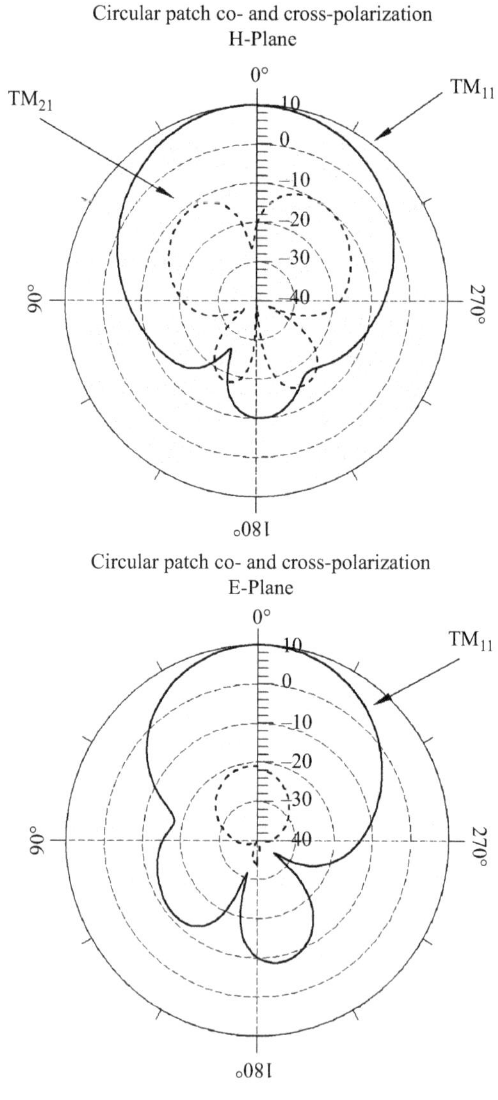

Figure 3.15 TM_{11} circular patch co- and cross-polarization of the (a) H-plane and (b) E-plane

mode. The geometry of a circular patch does not enforce a single direction for the TM_{11} mode as a square patch does for the TM_{10} mode. It is very possible the computed cross-polarization is from the generation of a TM_{11} mode with very small amplitude.

3.7 Annular microstrip antenna

When a concentric circle of conductor is removed from the interior of a circular microstrip antenna, it forms an annulus. The ring-shaped microstrip conductor which is formed has its geometry defined in Figure 3.16. We assume that the thickness of the substrate is small compared with a wavelength $H \ll \lambda_0$ which implies no variation of the electric field in the z direction. The cavity model with magnetic walls at the edges of the annulus may be used to obtain solutions for the fields beneath the annulus. The fields are assumed to be TM_{nm} in cylindrical coordinates [13,14]:

$$E_z = E_0 \left[J_n(kr) \acute{Y}_n(ka) - \acute{J}_n(ka) Y_n(kr) \right] \cos n\phi \tag{3.35}$$

$$H_r = \frac{j\omega\epsilon}{k^2 r} \frac{\partial E_z}{\partial \phi} \tag{3.36}$$

$$H_\phi = \frac{-j\omega\epsilon}{k^2} \frac{\partial E_z}{\partial r} \tag{3.37}$$

The Bessel functions (J_n and Y_n) are of the first and second kind of order n. The prime signs signify the first derivative of the Bessel functions. The wave number k is in the dielectric substrate $k = 2\pi\sqrt{\epsilon_r}/\lambda_0$, where ϵ is the substrate permittivity and λ_0 is the wavelength in free space. The mode integer n is associated with the variation

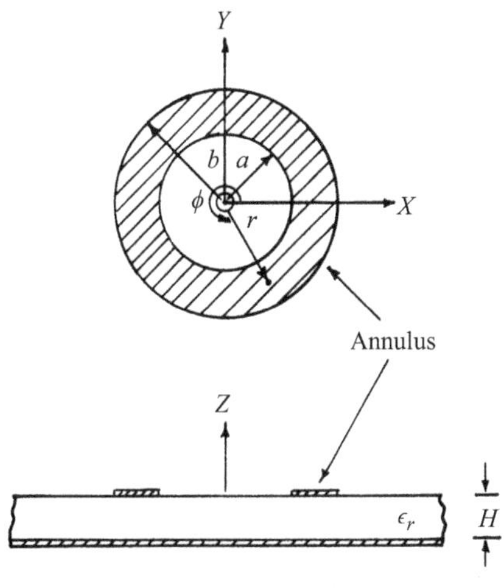

Figure 3.16 Annular microstrip antenna geometry. The outer radius is b, inner radius is a with a probe feed at radius r at angle ϕ

along ϕ, and mode integer m is associated with the variation of fields along the radial direction.

The surface currents on the annulus may be computed using $K_\phi = -H_r$ and $K_r = H_\phi$. The radial component of the surface current will disappear at an edge:

$$K_r(r = b) = H_\phi(r = b) = 0 \tag{3.38}$$

It follows from (3.35), (3.37), and (3.38) that the wave number must obey:

$$\left[\acute{J}_n(kb)\acute{Y}_n(ka) - \acute{J}_n(ka)\acute{Y}_n(kb)\right] = 0 \tag{3.39}$$

Equation (3.39) may be solved to obtain the modes associated with a given inner radius a and outer radius b. $\acute{J}_n$ and $\acute{Y}_n$ are the derivatives of the Bessel functions with respect to kr. Approximate values for k may be obtained with [15]:

$$k_{n1} \approx \frac{2n}{a + b} \tag{3.40}$$

where $\frac{(b-a)}{(b+a)} \leq 0.35$ and $n \leq 5$.

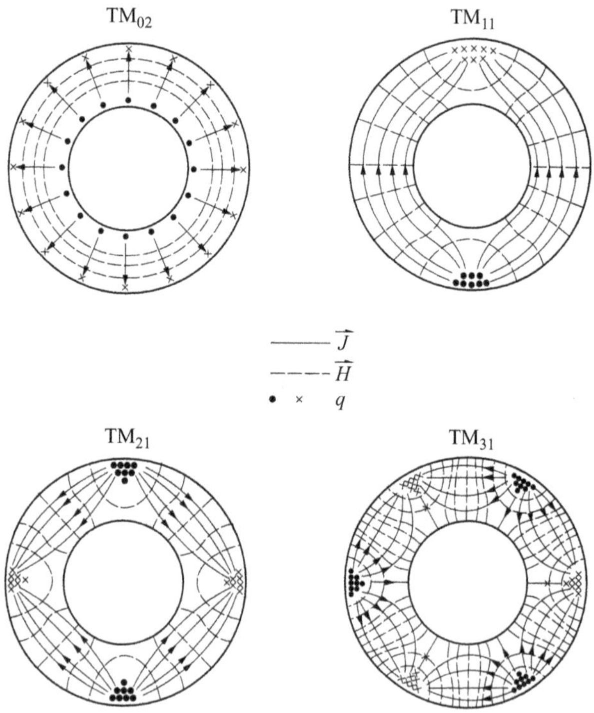

Figure 3.17 Sketch of the theoretical current distribution of TM$_{02}$, TM$_{11}$, TM$_{21}$, and TM$_{31}$ mode of a annular patch antenna (from [13], copyright 1973, IEEE, reprinted with permission)

The resonant frequency of the annular microstrip antenna may be computed with [16]:

$$f_{r_{n1}} = \frac{ck_{n1}}{2\pi\sqrt{\epsilon_e}} \tag{3.41}$$

The relative effective dielectric permittivity ϵ_e is for a microstrip transmission line of conductor width $W = b - a$ (Appendix C). The resonant frequencies predicted by (3.41) is within 3% of experiment.

The first few modes of an annular microstrip antenna are presented in Figure 3.17. We note that they are very similar to the modes of a circular microstrip resonator as shown in Figure 3.2. The patterns produced by the modes are also very similar to those produced by a circular microstrip antenna. Slot insets may be used to create circular polarization in the same manner as was used in Section 3.5.2 [17]. Alternatively, a square tab on the outer edge of the annulus can also produce circular polarization [18, 19]. Annular rings may also be stacked to produce a large impedance bandwidth [20].

3.8 Shorted annular microstrip antenna

The shorted annular microstrip antenna, or SAP, has a concentric shorting plane, which allows a hole to exist in its center (Figure 3.18). This has been used in some aerospace applications to allow cabling or other connections between the back and

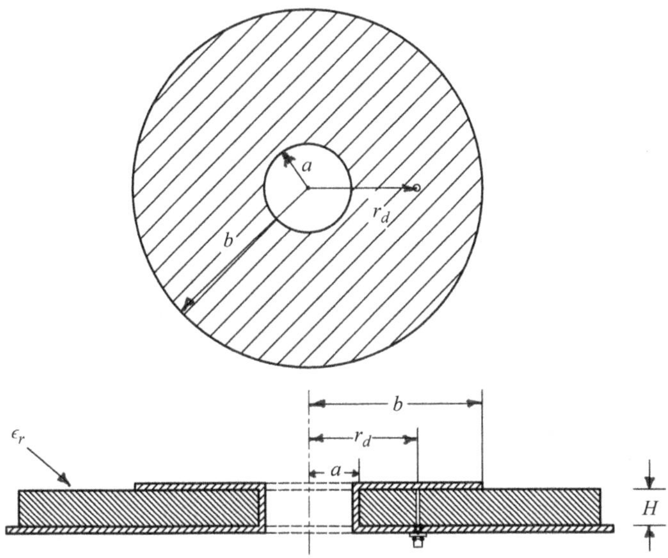

Figure 3.18 Geometry of shorted annular patch (SAP). A shorting wall exists from the groundplane to the annular patch at a. The annulus is from a to b and fed at driving point radius r_d. The thickness of the substrate is H

front side of the antenna. The earliest use of this feature was to design a dual-band set of stacked patches with a shorted annular ring on the bottom layer, and a circular patch antenna on the top. Each antenna was fed with a separate cable. One cable was connected to the shorted annular ring patch on the bottom. A second cable was fed through the center of the shorted annular ring patch to feed the upper circular patch antenna [21].

A cavity model for the annular patch has been derived [22], but accessible general design equations for dimensions of the SAP are not available. Moernaut and Vandenbosh have developed design equations for TM_{11} and TM_{21} modes which produce broadside radiation [23]. These modes have significantly more gain than is realized using an equivalent circular patch design. When the relative permittivity is below 8.3846, the TM_{11} mode is employed, above this value the TM_{21} takes over as a design that is readily implemented:

$$b_{eff} = \frac{1.8412}{k_0} \tag{3.42}$$

where k_0 is the free space wave number $k_0 = \frac{2\pi}{\lambda_0}$.

The fringing field edge extension is given as

$$\Delta b = \frac{H}{\pi \epsilon_r}\left[\ln\left(\frac{\pi b}{2H}\right) + 1.7726\right] \tag{3.43}$$

The physical radius, b, of the interior cutout is given by

$$b = b_{eff} - \Delta b \tag{3.44}$$

The design equations for the outer radius a are

$$a = b_{eff} - \left(\frac{0.237}{\sqrt{\epsilon_r}} + 0.055e^{-\epsilon_r^2}\right)\lambda_0 \;(\epsilon_r < 8.3486) \tag{3.45}$$

$$a = b_{eff} - \left(\frac{0.75}{\sqrt{\epsilon_r}} + 113.633e^{-\epsilon_r^2} + 0.0012\right)\lambda_0 \;(\epsilon_r > 8.3486) \tag{3.46}$$

These equations provide a first-order approximation for the initial design values for the annulus of a SAP antenna near the desired resonant frequency, which can be refined to produce an acceptable design. Determining the location of a 50 Ω driving point requires trial and error. The lowest order mode of the SAP antenna has been shown to have a higher gain than does an equivalent circular patch antenna.

As a design example, we will use a dielectric material with $\epsilon_r = 2.33$ and a 0.0012 $\tan\delta$ with $H = 1524$ μm. The design frequency is chosen as 2.45 GHz. First b_{eff} is computed using (3.42) as 35.882 mm. This value is used in (3.43) to compute $\Delta b = 1.12075$ mm. Using (3.44), we compute the physical inner radius as $b = 34.7612$ mm. We then compute the outer antenna radius $a = 16.8996$ mm with (3.45).

When these design values are modeled with HFSS, the best match is found at $r_d = 21$ mm. The resonant frequency is 2.465 GHz, which is very close to the design frequency of 2.45 GHz. The gain is 9.31 dBi with a broadside pattern as expected.

The shorted annular microstrip antenna can also be driven in a monopole type mode like the TM_{02} mode of a circular patch antenna.

References

[1] Kernweis, N.P., and McIlvenna, J.F., "Liquid Crystal Diagnostic Techniques an Antenna Design Aid," *Microwave Journal*, Vol. 20, October 1977, pp. 47–58.

[2] Derneryd, A.G., "Analysis of the Microstrip Disk Antenna Element," *IEEE Transactions on Antennas and Propagation*, September 1979, Vol. AP-27, No. 5, pp. 660–664.

[3] Burden, R.L., Faires, J.D., and Reynolds, A.C., *Numerical Analysis*, Prindle, Webber and Schmidt, 1978, pp. 31–38.

[4] Jackson, D.R., Williams, J.T., Bhattacharyya, A.K., Smith, R.L., Buchheit, S.J., and Long S.A., "Microstrip Patch Designs that Do Not Excite Surface Waves," *IEEE Transactions on Antennas and Propagation*, August 1993, Vol. 41, No. 8, pp. 1026–1036.

[5] Lo, Y.T., and Richards, W.F., "Perturbation Approach to Design of Circularly Polarized Microstrip Antennas," *Electronics Letters*, May 1981, pp. 383–385.

[6] *Reference Data for Radio Engineers*, 6th ed., Howard W. Sams & CO., Fifth Printing, 1982, pp. 9–7.

[7] Marino, R.A., and Hearst, W., "Computation and Measurement of the Polarization Ellipse," *Microwave Journal*, November 1999, pp. 132–140.

[8] Huang, J., "Circularly Polarized Conical Patterns from Circular Microstrip Antennas," *IEEE Transactions on Antennas and Propagation*, September 1984, Vol. AP-32, No. 9, pp. 991–994.

[9] Du, B., and Yung, E., "A Single-Feed TM_{21}-Mode Circular Patch Antenna with Circular Polarization," *Microwave Optics Technology Letters*, 2002, Vol. 33, pp. 154–156.

[10] Hanson, R.C., "Cross Polarization of Microstrip Patch Antennas," *IEEE Transactions on Antennas and Propagation*, June 1987, Vol. AP-35, No. 6, pp. 731–732.

[11] Lee, K.F., Luk, K.M., and Tam, P.Y., "Crosspolarization Characteristics of Circular Patch Antennas," *Electronics Letters*, March 1992, Vol. 28, No. 6, pp. 587–589.

[12] Garcia-Garcia, Q., "Radiated Cross-Polar Levels and Mutual Coupling in Patch Radiators," *International Journal of RF and Microwave Computer Aided Design*, 2000, Vol. 10, pp. 342–352.

[13] Wu, Y.S., and Rosenbaum, F.J., "Mode Chart for Microstrip Ring Resonators," *IEEE Transactions on Microwave Theory and Techniques*, 1973, Vol. MTT-21, pp. 487–489.

[14] Bahl, I.J., Stuchly S.S., and Stuchly M.A., "A New Microstrip Radiator for Medical Applications," *IEEE Transactions on Microwave Theory and Techniques*, 1980, Vol. MTT-28, pp. 1464–1468.

[15] Garg, R., Bhartia, P., Bahl, I., and Ittipiboon, A., *Microstrip Antenna Design Handbook*, Artech House, 2001, pp. 368–371.

[16] Bahl, I., and Bhartia, P., *Microstrip Antennas*, Artech House, 1980 2nd printing, 1982, p. 114.

[17] Licul, S., Petros, A., and Zafar, I., "Reviewing SDARS Antenna Requirements," *Microwaves & RF*, September 2003.

[18] Bhattacharya, A.K., and Shafai, L., "Annular Ring as a Circularly Polarized Antenna," *IEEE APS Symposium Digest*, 1987, pp. 1020–1023.

[19] Bhattacharya, A.K., and Shafai, L., "A Wider Band Microstrip Antenna for Circular Polarization," *IEEE Transactions on Antennas and Propagation*, February 1988, Vol. 36, No. 2, pp. 157–163.

[20] Kokotoff, D.M., Aberle, J.T., and Waterhouse, R.B., "Rigorous Analysis of Probe-Fed Printed Annular Ring Antennas," *IEEE Transactions on Antennas and Propagation*, February 1999, Vol. 47, No. 2, pp. 384–388.

[21] Goto, N., and Kaneta, K., "Ring Patch Antennas for Dual Frequency Use," *1987 IEEE/AP-S Symposium Digest*, 1987, pp. 944–947.

[22] Massa, G.D., and Mazzarella, G. "Shorted Annular Patch Antenna," *Microwave and Optical Technology Letters*, March 1995, Vol. 8, No. 4, pp. 222–226.

[23] Moernaut, G.J.K., and Vandenbosch, G.A.E., "Simple Pen and Paper Design of Shorted Annular Ring Antenna," *Electronics Letters*, December 2003, Vol. 39, No. 25, pp. 1784–1785.

Chapter 4
Broadband microstrip antennas

4.1 Broadband microstrip antennas

Microstrip antennas are inherently narrow band. The typical bandwidth of a microstrip antenna is around 4%–7%. A considerable number of experimental approaches have been undertaken to develop microstrip antennas which have a broader impedance bandwidth than a single microstrip element achieves without external matching.

The methods employed to increase impedance bandwidth are essentially variations of three approaches: (a) Increasing the antenna volume. This is accomplished by geometry changes which increase the volume under the patch (e.g. increasing the thickness h), decreasing the substrate dielectric constant, or adding additional coupled resonators. (b) The implementation of a matching network. (c) Perturbing the antenna geometry to create or relocate resonances using shorts and slots in the antenna. Kumar and Ray have compiled a considerable number of microstrip antenna design variations which utilize these approaches [1], as has Wong [2].

One approach is the use of a groundplane slot, which is excited by a microstrip line below the groundplane, which in turn couples to a microstrip patch which is above the groundplane. This configuration can be adjusted to produce an elegant matching network, low dielectric constant substrate for the patch, and maximized substrate thickness, which provides a broadband input impedance match, and a large impedance bandwidth. One implementation of this type of antenna is called SSFIP for *Strip–Slot–Foam–Inverted Patch* by Zürcher and Gardiol in the mid-1990s. The design of these antennas is experimental in nature. The researchers state: "Since the various parts of the antenna interact, determining the best design is a long and tedious process, even when carried out systematically." The parameters which may be used to guide this type of design are found in the literature [3–5]. Some possible alternatives to the SSFIP matching-network/thick patch design are explored in this chapter.

4.2 Microstrip antenna broadbanding

The broadbanding of a microstrip antenna is often accomplished by increasing the thickness of a microstrip antenna. This broadbanding reaches a limit when the series inductance produced by higher order modes produces an unacceptable mismatch in the driving point impedance. One can also use a matching network to increase the impedance bandwidth of a microstrip antenna.

The normalized bandwidth of a microstrip antenna can be written as

$$BW = \frac{f_H - f_L}{f_R} \tag{4.1}$$

where f_H is the upper frequency where the impedance match is S:1 VSWR and f_L is the lower frequency where the impedance match is also S:1 VSWR. The VSWR is less than S:1 over $(f_H - f_L)$. f_R is the resonant frequency of the patch. Generally, S is set equal to two ($S = 2$) for most practical applications.

At the resonant frequency of a patch, the driving point impedance is pure real. We will designate this resistance as R_0. When the patch is connected to a transmission line of characteristic impedance Z_0, the impedance bandwidth is predicted using:

$$BW = \frac{1}{Q}\sqrt{\frac{(TS-1)(S-T)}{S}} \tag{4.2}$$

where Q is the total Q of the patch antenna, S is the VSWR S:1 value, and $T = R_0/Z_0$.

When a microstrip antenna is fed with a transmission line where $R_0 = Z_0$, the bandwidth equation reduces to the bandwidth equation for a linear patch antenna [Equation (2.72)]:

$$BW_{Linear} = \frac{S-1}{Q_T\sqrt{S}} \quad (S{:}1\ \text{VSWR})$$

To maximize the impedance bandwidth for a given (S:1) VSWR mismatch between the resonant resistance R_0 and the feeding transmission line characteristic impedance Z_0, we must satisfy this relationship:

$$T_{\text{opt}} = \frac{1}{2}\left(S + \frac{1}{S}\right) \tag{4.3}$$

For example, in the case of a 50 Ω coaxial transmission line probe feeding a rectangular microstrip patch antenna, we can compute the driving point impedance we must choose for a maximized 2:1 VSWR bandwidth. To obtain this value, we compute the optimum T value:

$$T_{\text{opt}} = \frac{1}{2}\left(2 + \frac{1}{2}\right) = 1.25 = \frac{R_0}{Z_0} \tag{4.4}$$

which implies that the resonant driving point resistance for optimum bandwidth is $R_0 = 50\ \Omega \cdot 1.25$, or $R_0 = 62.5\ \Omega$. The driving point location should be chosen where a 62.5 Ω presents itself. This value is near the approximate optimum value of 65 Ω reported by Milligan [6].

When we use (4.4), we trade a perfect match at a single frequency, for wider bandwidth at the expense of a larger mismatch. The equation describes an elementary form of a broadband matching network. The bandwidth obtained when T_{opt} is used may be related to the (S:1) bandwidth by

$$BW_{T_{opt}} = \frac{1}{2Q}\frac{\sqrt{S^4 - 1}}{S} \tag{4.5}$$

For the case of 2:1 VSWR, we can take the ratio of (4.5) to (2.72) to obtain a bandwidth increase factor. The 2:1 VSWR bandwidth is computed to be 1.38 times larger than the bandwidth of a perfectly matched microstrip antenna. Experiment, FDTD, and cavity model data all indicate that in practice the *best* increase in bandwidth is about 1.1 times the original bandwidth of the matched element. This bandwidth increase produced using a simple impedance mismatch is often so small it is not of practical use.

If we allow the use of a perfect matching network, with an unrestricted extent, the maximum impedance bandwidth obtainable is

$$BW_{max} = \frac{1}{Q}\frac{\pi}{\ln[(S+1)/(S-1)]} \tag{4.6}$$

This relationship allows us to compare the maximum bandwidth attainable using a broadband matching network, with the normally obtained bandwidth of (2.72), by taking the ratio of (2.72) to (4.6) which produces a bandwidth increase factor F:

$$F = \frac{\pi\sqrt{S}}{(S-1)\ln[(S+1)/(S-1)]} \tag{4.7}$$

For the case of 2:1 VSWR, the bandwidth increase factor F is 4.044, or approximately four times the bandwidth of a single element without matching.

Hansen [7] has presented a bandwidth increase factor based on the classic impedance matching limitations formulated by Fano and Bode *ca.* 1950:

$$F = \frac{2\pi S}{(S^2-1)\ln\left(\frac{S+1}{S-1}\right)} \tag{4.8}$$

When $S = 2$ is substituted into (4.8), the bandwidth increase factor becomes 3.813 which is slightly lower than the previous formulation. These formulations provide fundamental limits against which we can compare broadband microstrip antenna designs.

4.2.1 Microstrip antenna matching with capacitive slot

Increasing the thickness of a microstrip antenna increases its impedance bandwidth. As the thickness increases, higher order modes produce an equivalent series inductance, which mismatches a rectangular microstrip antenna. The straightforward solution to this problem is to introduce series capacitance to cancel the inductive reactance which appears at the driving point. A cost-effective method is to provide capacitance by modifying the patch geometry. Researchers have used a rectangular or circular slot surrounding the feed probe (Figure 4.1(a) to provide the required series matching capacitance [8,9]). The values of the slot dimensions for the circular or rectangular slots which bound the probe feed are determined experimentally.

Another approach used to provide series capacitance is to place a narrow slot directly in front of the feed point and adjust its length until a match is obtained (Figure 4.1(b)).

As an example of matching a thick patch with a slot, we will use a rectangular microstrip antenna which has a substrate thickness of 6.096 mm, a relative dielectric permittivity of $\epsilon_r = 2.6$, with a 0.0025 tan δ. The patch dimensions are $a = 38.0$ mm $\times$ $b = 34.72$ mm. The 50 Ω driving point location is 6.0 mm from the patch center along the centerline of the width. FDTD computes a resonance at 2.31 GHz (maximum real impedance) with a driving point impedance of $46 + j35.35$ Ω. $h/\lambda_0 = 0.047$ which is electrically thick for a patch antenna. A narrow slot $t = 1$ mm across and $W = 14$ mm in width matches the antenna with a driving point impedance of $49.71 + j5.79$ Ω at 2.31 GHz. The impedance with and without slot matching is plotted on a Smith chart in Figure 4.2. After matching, the antenna has a respectable 7% impedance bandwidth.

4.2.2 *Microstrip antenna broadband matching with bandpass filter*

It must be noted up front, that generally, the design of a broadband impedance matching network is a very difficult network synthesis problem. The prototype element values used in this section for creating an impedance matching network using lumped elements are taken from previous work [10]. The synthesis presented works better for lower frequency microstrip antennas where lumped elements may be incorporated with relative ease. It will be demonstrated that the realization of lumped element models using microwave transmission line structures is quite challenging, but not insurmountable.

A rectangular microstrip antenna fed to excite only one dominant mode (TM_{10} or TM_{01}) has a single resonance which may be modeled as a parallel RLC circuit. These values are designated R_a, L_a, and C_a in Figure 4.3. When the patch is probe fed and becomes electrically thick, a series inductor must also be taken into account, which is designated L_s. In some *rare* cases, feed geometries can produce a series capacitance C_s rather than L_s, but generally it is L_s that exists for a typical patch. The resonant angular frequency, ω_0, is the frequency at which the maximum value of the real part of the driving point impedance occurs. The maximum value of the real part at resonance

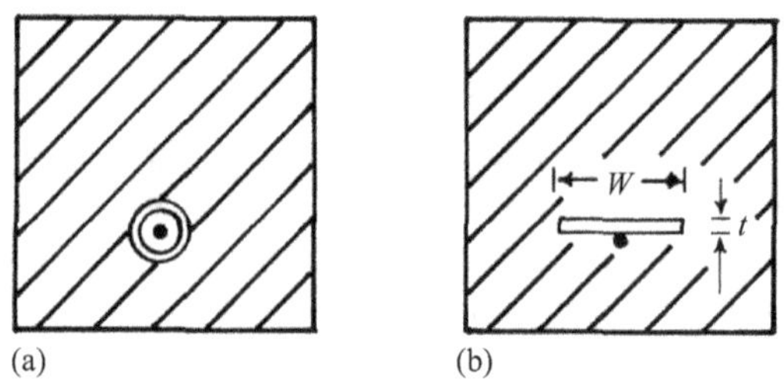

Figure 4.1 (a) Series matching capacitance provided using a circular slot. (b) Series matching capacitance created using a rectangular slot near the driving point

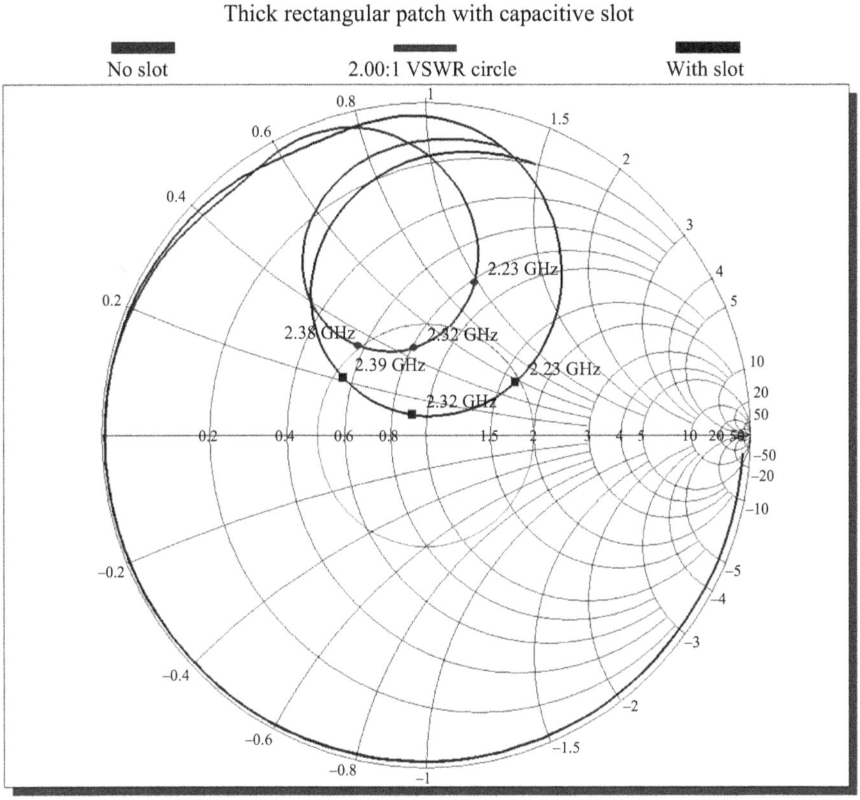

Figure 4.2 *Driving point impedance without (circles) and with (squares) capacitive slot. The series capacitance provided by the rectangular slot cancels the inductive reactance of thick microstrip antennas*

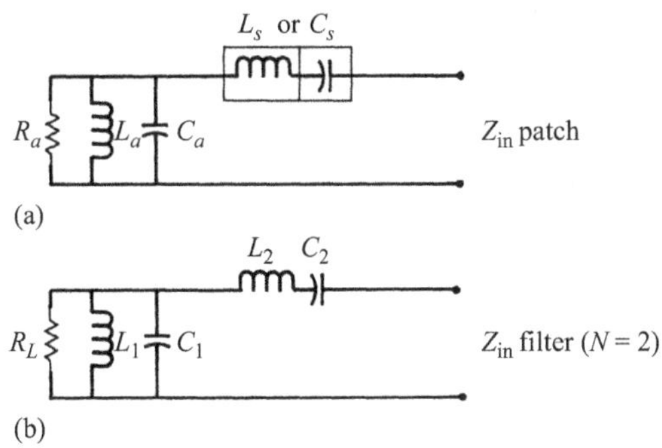

Figure 4.3 *(a) A microstrip antenna may be modeled as a parallel RLC circuit with a series inductance or capacitance. (b) A bandpass filter has a similar model, and may be used to impedance match a microstrip antenna over a larger frequency range than that covered by a single element*

can be obtained directly from a measured impedance plot or a full-wave analysis method. At resonance, the relationship between the resonant angular frequency ω_0, and the patch model values L_a and C_a is

$$\omega_0^2 = \frac{1}{L_a C_a} \tag{4.9}$$

When the patch is resonant, the inductive and capacitive reactance of L_a and C_a cancel each other, and the maximum value of resistance occurs. If the patch is probe fed and thick, the impedance at resonance will have a series inductive reactance term L_s:

$$Z_{\text{in}} = R_a + j\omega_0 L_s \tag{4.10}$$

In order to obtain the values of C_a and L_a from measured or computed data, one must subtract the series inductive reactance from the impedance. Choose two points either side of ω_0 in frequency:

$$\omega_1 = \omega_0 - \Delta\omega_1 \tag{4.11}$$

$$\omega_2 = \omega_0 + \Delta\omega_2 \tag{4.12}$$

With the subtraction of the series inductance, the reactance now changes sign either side of ω_0. The admittance at each frequency may be expressed as

$$Y_1 = \frac{1}{R_a} + j\omega_1 C_a + \frac{1}{j\omega_1 L_a} = G_1 + jB_1 \tag{4.13}$$

$$Y_2 = \frac{1}{R_a} + j\omega_2 C_a + \frac{1}{j\omega_2 L_a} = G_2 + jB_2 \tag{4.14}$$

The susceptance at each frequency is

$$B_1 = \omega_1 C_a - \frac{1}{\omega_1 L_a} \tag{4.15}$$

$$B_2 = \omega_2 C_a - \frac{1}{\omega_2 L_a} \tag{4.16}$$

Solving the equations for C, we obtain

$$C_a = \frac{\omega_1 B_1 - \omega_2 B_2}{\omega_1^2 - \omega_2^2} \tag{4.17}$$

Solving the equations for L:

$$L_a = \frac{1}{\omega_1^2 C_a - \omega_1 B_1} \tag{4.18}$$

We have now computed R_a, L_a, C_a, and L_s (or C_s in some rare cases).

The similarity of the model to a bandpass filter allows one to use filter synthesis techniques to evaluate optimal component values for broadband matching. This method was first articulated by Paschen [11]. Optimum values may be found in the literature [12]. With a requirement of 1.8:1 VSWR (0.35 dB insertion loss), the g_i values for an $N = 2$ bandpass filter network are

$$g_1 = 1.50$$
$$g_2 = 0.455$$
$$g_3 = 1.85$$

The component values of the bandpass filter are given by

$$C_1 = \frac{g_1}{R_L(\omega_U - \omega_L)} \tag{4.19}$$

where ω_U is the upper radian frequency band limit and ω_L is the lower radian frequency band limit.

The filter bandwidth is $F_{\text{bw}} = (\omega_U - \omega_L)$:

$$L_1 = \frac{1}{\omega_0^2 C_1} \tag{4.20}$$

$$L_2 = \frac{g_2 R_L}{(\omega_U - \omega_L)} \tag{4.21}$$

$$C_2 = \frac{1}{\omega_0^2 L_2} \tag{4.22}$$

$$R_{\text{in}} = R_L/g_3 \tag{4.23}$$

We desire a 50 Ω input resistance, which requires the load resistance be $R_L = g_3 \cdot 50\,\Omega = 92.5\,\Omega$ for this case. We now have all the equations required to compute a matching example.

We need to provide $R_a = 92.5\,\Omega$, which is then equated with R_L of the filter circuit. We will do this by using a patch fed in the center of a radiating edge with a microstrip transmission line. The patch width is adjusted to provide a 92.5 Ω edge resistance at resonance. We can compute the value of L_s from measured or simulated data using the inductive reactance value X_L:

$$L_s = \frac{X_L}{\omega_0} \tag{4.24}$$

When the inductive reactance is subtracted from two frequencies either side of resonance, we use (4.17) and (4.18) to compute C_a and L_a. The value of C_a is equated with C_1 which allows one to determine the expected filter bandwidth F_{bw} using (4.19):

$$F_{\text{bw}} = \frac{g_1}{R_a C_a} \tag{4.25}$$

This computation can be used to decide if the bandwidth is acceptable for a given design requirement. If the bandwidth is within the design requirement, we next compute L_2 using (4.21). This value needs to be larger than the series inductance of the microstrip antenna L_s to realize the design. The series inductance consists of two components, L_m and L_p, which are the series inductance attributable to higher order modes, and feed probe self-inductance, respectively. While the majority of the series inductance L_s is due to the excitation of higher order modes, a coaxial probe feed also contributes its self-inductance L_p to the total series inductance. In some cases, the diameter of this probe may be used to adjust the series inductance to produce a realizable design [13]. Next, the value of C_2 is computed from (4.22).

4.2.3 *Example microstrip antenna lumped-element broadband match*

A microstrip antenna was designed, analyzed, and refined using FDTD to create an element fed at a radiating edge with a 50 Ω microstrip transmission line, which has a 94.16 Ω impedance at resonance. This resistance is close enough to realize an impedance matching design as detailed previously. The antenna dimensions are $a = 100.0$ mm, $b = 37.16$ mm, the substrate thickness is $h = 1.524$ mm, $\epsilon_r = 2.6$, $\tan\delta = 0.0025$ with a 130 mm × 75 mm groundplane.

The FDTD data has a maximum resistance value at 2.3317 GHz, and an input impedance of $94.61 + j7.54$ Ω. Figure 4.6 shows the impedance plot for the antenna to be matched. At resonance, we can compute an equivalent series inductance as $L_s = 0.5147$ nH using (4.24).

The effectiveness of this matching method is dependent upon how well the load can be modeled as a parallel RLC circuit. Equations (4.17) and (4.18) provide values of $C_a = 26.67$ pF and $L_a = 175.195$ pH. When plotted with the FDTD data on a Smith chart, the RLC circuit computed can be improved upon. Computer optimization using a random search computes a very good fit to the FDTD data. This is illustrated

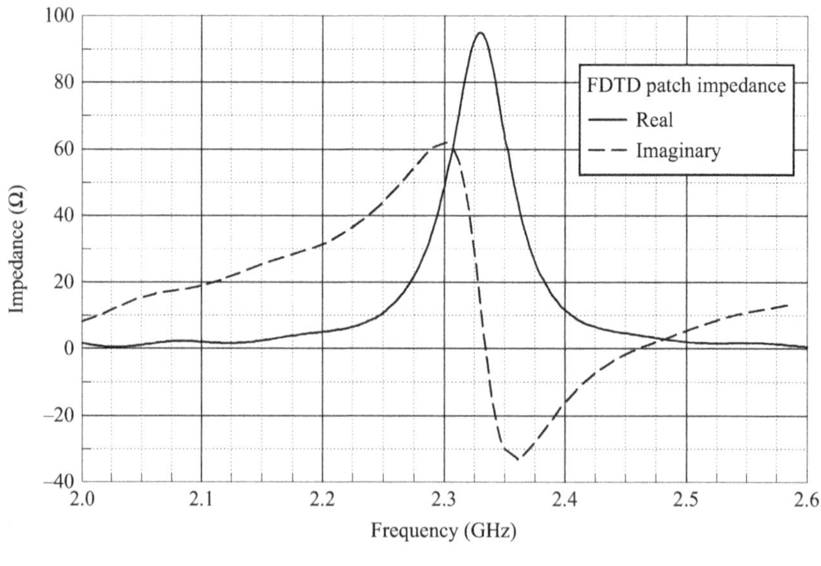

Figure 4.4 Rectangular impedance plot of rectangular microstrip antenna used to illustrate matching network design

in Figure 4.4. The computer optimized values are $R_a = 95.55\ \Omega$, $C_a = 28.64$ pF, $L_a = 163.04$ pH and the series inductance is $L_s = 1.017$ nH. These values clearly model the FDTD data better than the curve fit values. These values are used with (4.21) and (4.22) to compute $L_2 = 77.39$ nH and $C_2 = 0.0602$ pF. The value of C_2 is clearly very difficult to realistically implement with a lumped element. We can still use these values to illustrate the theoretical match produced by this analysis compared with a single frequency match using a quarter-wave transformer.

The expected bandwidth from (4.25) is 88.1 MHz. In Figure 4.5, the bandwidth using a quarter-wave transformer is 41 MHz, the computed bandwidth using the synthesized impedance matching network implemented with discrete elements produces a bandwidth of 92 MHz. The bandwidth increase factor F is a respectable 2.24 in this implementation. The value of C_2 is unrealizable in this example; however, we will continue with the implementation of this method for illustration. One must also keep in mind, this method is sensitive to the tolerance of the components used.

4.2.4 Lumped elements to T-line conversion

At microwave frequencies, it is often desirable to implement a matching network using microstrip transmission line sections in place of lumped elements. One may use admittance and impedance inverters to realize the design of Example 4.2.3 with transmission line.

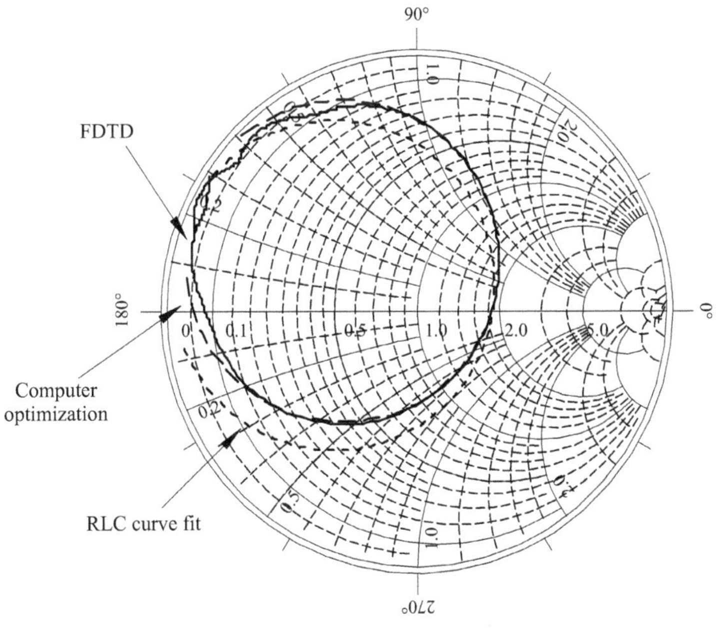

Figure 4.5 Rectangular microstrip antenna impedance from FDTD analysis with a curve fit RLC circuit and a computer optimized fit

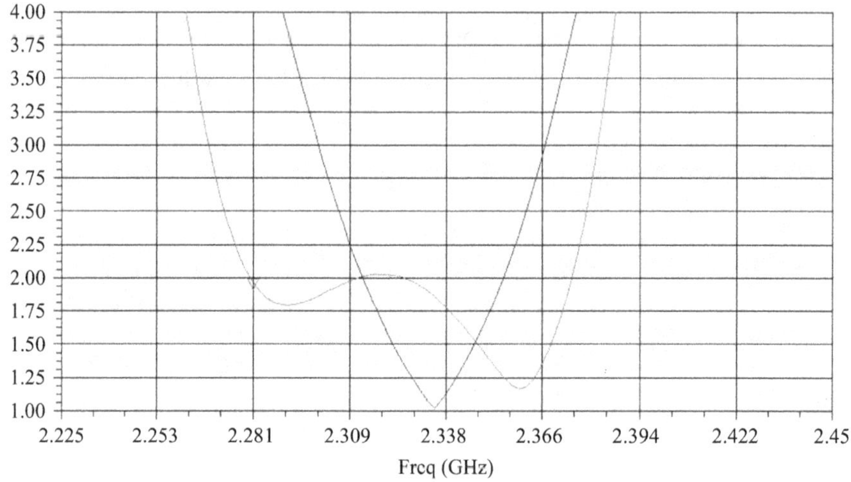

Figure 4.6 Bandwidth of original element matched using a quarter wave transformer and matched using lumped element matching (marked with triangle)

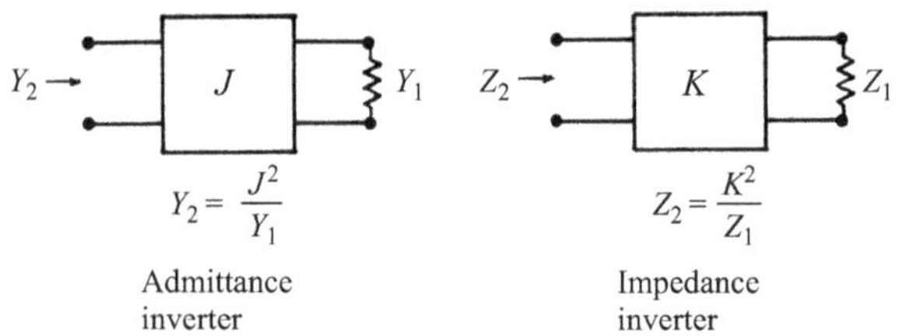

Figure 4.7 Admittance inverter and impedance inverter

An *admittance inverter* is an ideal quarter-wave transmission line section of characteristic admittance J, and an *impedance inverter* is an ideal quarter-wave transmission line section of characteristic impedance K (Figure 4.7)

$$Y_2 = \frac{J^2}{Y_1} \quad \text{(admittance inverter)} \tag{4.26}$$

$$Z_2 = \frac{K^2}{Z_1} \quad \text{(impedance inverter)} \tag{4.27}$$

A series admittance may be represented with a shunt admittance sandwiched between two J inverters. This is illustrated in Figure 4.8. A shunt impedance may be represented with a series impedance sandwiched between two K inverters as shown in Figure 4.9. This property of J and K inverters allows us to take series inductances and capacitances and convert them into shunt capacitances and inductances between a pair of quarter-wave transformers.

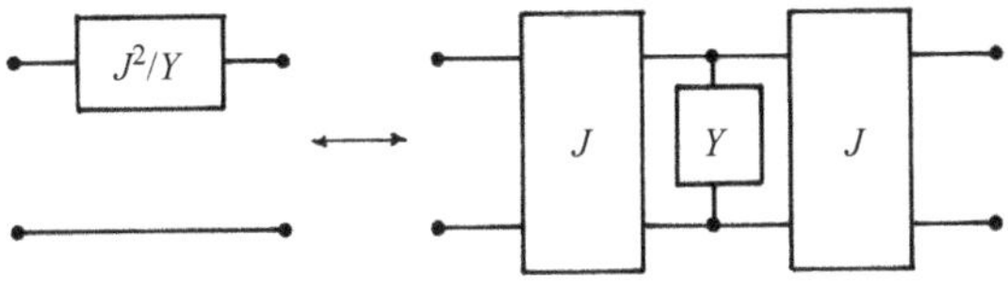

Figure 4.8 Series admittance and its equivalent circuit as a shunt admittance sandwiched between a pair of admittance inverters

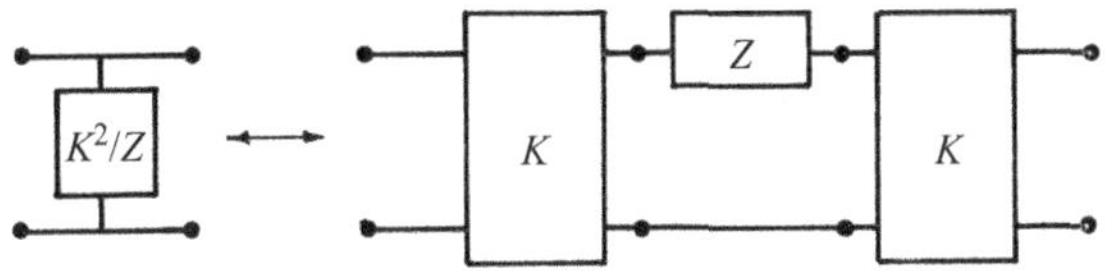

Figure 4.9 Shunt impedance and its equivalent circuit as a series impedance sandwiched between a pair of impedance transformers

Once the series elements have been converted into shunt elements, the shunt capacitance and inductance can be approximated with transmission line sections. Lengths of transmission line provide reactance which is an approximation to a capacitor or inductor, but over a narrower bandwidth than the original lumped elements. The bandwidth of the matching network is decreased because of this difference between transmission line sections and lumped elements. Often these transmission line sections are realized using a microstrip transmission line. One must further compensate for physical microstrip transmission line discontinuities which produce deviations from idealized transmission line theory (Appendix C). With all these factors which can introduce error included, often considerable experimental optimizing must be undertaken to realize a design, which decreases the utility of this matching method considerably. This technique is most useful at low frequencies where lumped elements may be used to implement the matching network directly as in Example 4.2.3 previously.

Figure 4.10(a) shows a microstrip patch antenna which is ideally designed to have a resonant edge resistance of 92.5 Ω. The example we have been using has an input impedance of 94.16 Ω, at a reference plane 1 mm from a radiating edge, along a 50 Ω microstrip transmission line as computed using FDTD ($\Delta = 1$ mm). The patch width is $W = 100.0$ mm, the patch length $L = 37.16$ mm, substrate thickness $h = 1.524$ mm, $\epsilon_r = 2.6$, $W_T = 4.12$ mm (50 Ω). The physical values for the transmission line matching network computed with FDTD refinement are

$$L_1 = 26.56 \text{ mm} \quad L_2 = 11.44 \text{ mm} \quad L_3 = 53.08 \text{ mm} \quad L_4 = 52.84 \text{ mm}$$

These values were obtained by using the transmission line topology developed in this section (synthesized using J and K inverters with the lumped element solution) with the use of computer optimization, to develop an improved transmission line

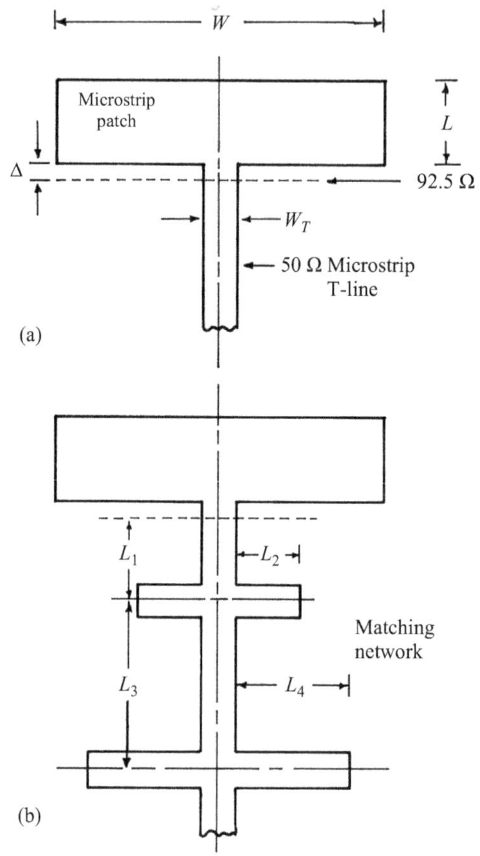

Figure 4.10 *(a) Microstrip patch antenna which has a 92.5 Ω reference plane a* very short *electrical distance from the patch edge impedance.* *(b) Matching network topology synthesized using J inverters*

realization, which requires less experimental intervention than a direct application of J inverters to realize a matching network. A small lengthening of the microstrip transmission line stubs compared with prediction allowed for the development a design which when analyzed with FDTD has less than 2.25:1 VSWR over a 100 MHz bandwidth. A Smith chart plot of the analysis results is found in Figure 4.11.

4.2.5 Real frequency technique broadband matching

Another powerful matching method is the real frequency technique. Carlin realized and demonstrated that traditional broadband matching methods involving Chebyshev functions are optimum for purely resistive loads, but are *not* optimum for complex loads [14]. Carlin and Yarman introduced an alternate version known as the simplified

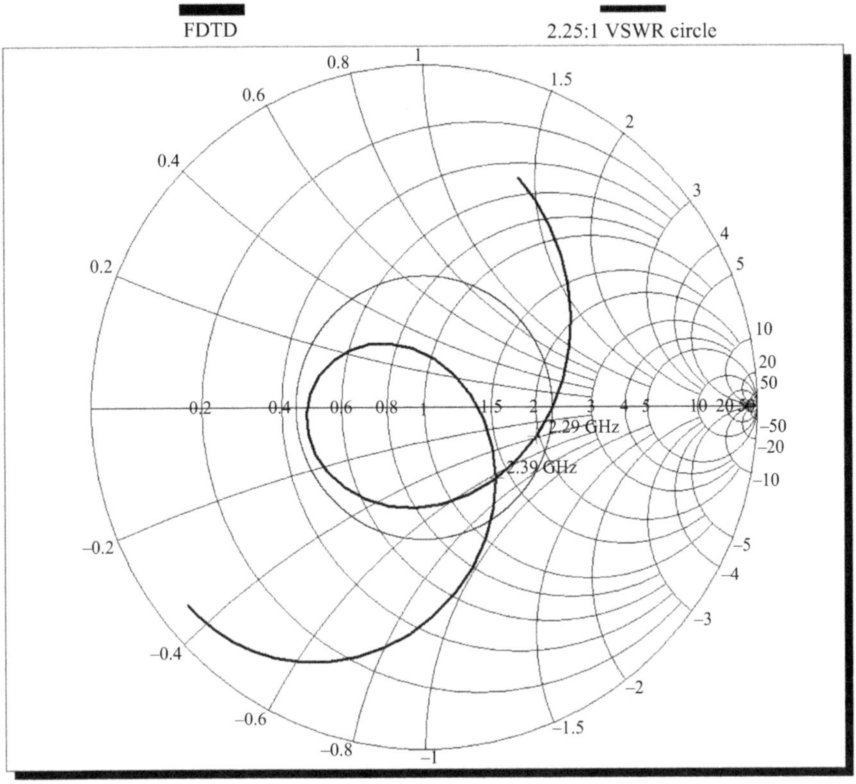

Figure 4.11 Microstrip transmission line matching network design FDTD results. The design is below 2.25:1 VSWR over 100 MHz

real frequency technique which they demonstrated to be a very powerful method for the matching of microwave circuits [15]. The Simplified Real Frequency Technique (SRFT) is much more flexible than using bandpass filter theory as a matching method. With filter theory, one must choose the value of the resistance in the RLC circuit based on achieving a 50 Ω input. The SRFT has the flexibility that it can match any complex load whether it is a theoretical or measured load impedance. Implementation of the SRFT is very involved [16]. It produces lumped element networks which are very effective, but conversion to a transmission line realization remains very difficult. The SRFT has been used to match microstrip antennas by Hongming *et al.* [17].

4.3 Patch shape for optimized bandwidth

We noted in Section 2.5.4 that rectangular microstrip antennas which are linearly polarized have an impedance bandwidth which is half of the impedance bandwidth of a circularly polarized rectangular microstrip antenna. The superposition of two detuned TM_{10} and TM_{01} modes stretches out the bandwidth when compared with a single TM_{10} or TM_{01} mode. In general, a microstrip patch antenna may be of any shape: oval, rectangular, star, cross, circle with slot, pentagon, etc. Consideration of this property of microstrip antennas leads one to a basic question about microstrip antennas which has not been answered by theory is

What microstrip patch shape provides maximum impedance bandwidth?

Subsets of this question are

1. What microstrip patch shape provides maximum impedance bandwidth with maximum linear polarization bandwidth allowing the linear polarization to vary in direction over the bandwidth?
2. What microstrip patch shape provides maximum impedance bandwidth with maximum linear polarization bandwidth without rotation?
3. What microstrip patch shape provides maximum impedance bandwidth with maximum axial ratio bandwidth for a circularly polarized antenna?

One restriction on shape is to have a single continuous patch without apertures (holes). One can relax this restriction and apply the same questions previously stated.

4.3.1 Patch shape bandwidth optimization using genetic algorithm

Delabie *et al.* proposed sectioning a plane in which a microstrip patch antenna is to be created into a set of small square sub-patches [18]. Each sub-patch is metalized, if represented by a one, and no metal is present if represented by a zero. A set of chromosome representations of candidate patches are selected, mated, and mutated using an appropriate electromagnetic analysis technique.

The use of genetic algorithms to develop shapes which have increased impedance bandwidths compared with a square microstrip patch has been investigated by Choo *et al.* [19]. Each antenna is described by a grid of connected squares with no internal voids. The patch antennas they developed use a 1.6 mm thick FR-4 substrate. Their groundplane size is 72 mm $\times$ 72 mm. Two cases were examined, one using a 16 $\times$ 16 grid of squares the second is a 32 $\times$ 32 grid of squares. The latter case is reproduced in Figure 4.12.

This patch is reported to have a fourfold increase in bandwidth improvement when compared with a square microstrip antenna. This increase in bandwidth is very close to the fundamental limit of increased bandwidth factor ($F = 3.81$) when a matching network is implemented (Section 4.2).

The center frequency of the design bandwidth is 2.0 GHz. Simulation by the method of moments predicted 8.04% bandwidth and 8.10% was measured as seen in Figure 4.13.

FDTD analysis reveals that this antenna combines two separate resonances and creates circular polarization (LHCP) at 2.0 GHz. It is interesting to note the genetic algorithm selected circular polarization. It was related in Chapter 2, Section 2.7, that the impedance bandwidth of a patch with circular polarization is two times that of a

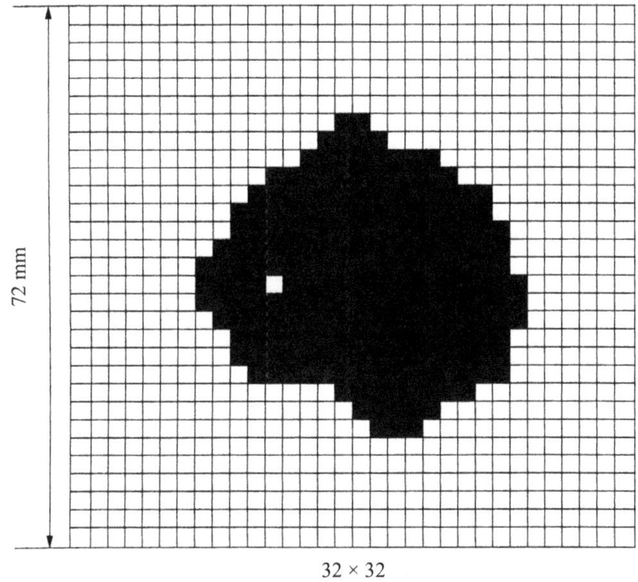

Figure 4.12 Patch shape produced by a genetic algorithm using a 32 × 32 grid of squares on a 72 mm × 72 mm groundplane. The substrate is 1.6 mm thick FR-4 (from [19], copyright 2000, IEE, reprinted with permission)

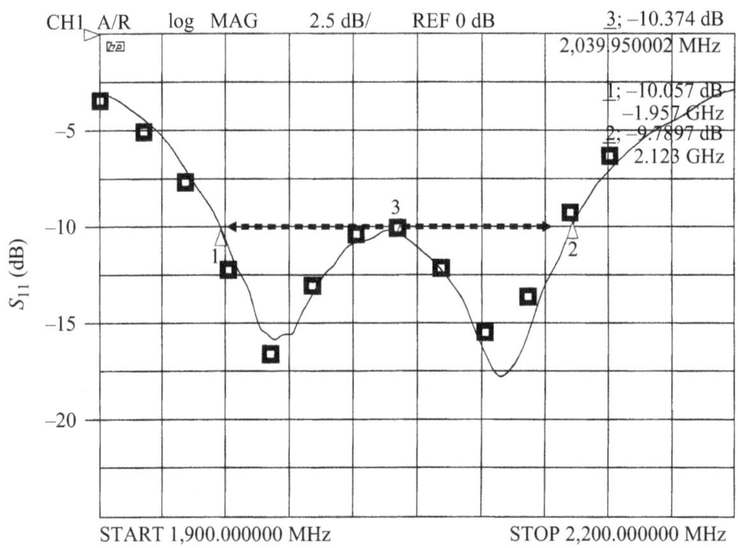

Figure 4.13 S_{11} versus frequency of patch in Figure 4.12 (from [19], copyright 2000, IEE, reprinted with permission)

linear antenna. The patch shaping and area apparently contribute to produce a further doubling of the impedance bandwidth.

4.4 Broadband monopole pattern patch-ring

The use of a circular patch antenna driven in the TM_{02} mode to produce a linearly polarized monopole pattern was explored in Section 3.5.4. Unfortunately, the use of a

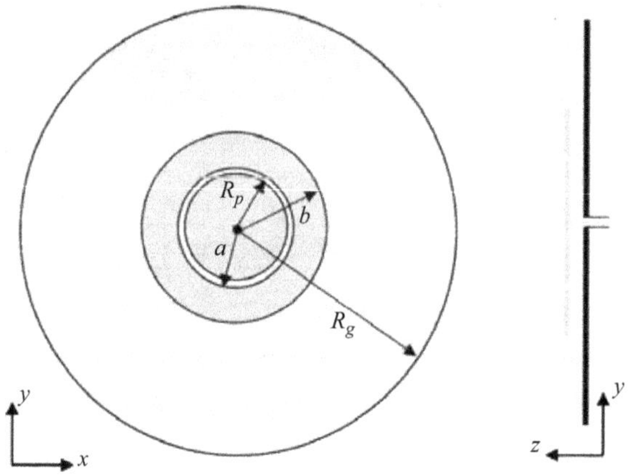

Figure 4.14 Geometry of center-fed circular patch-ring antenna (from [20], copyright 2009, IEEE, reprinted with permission)

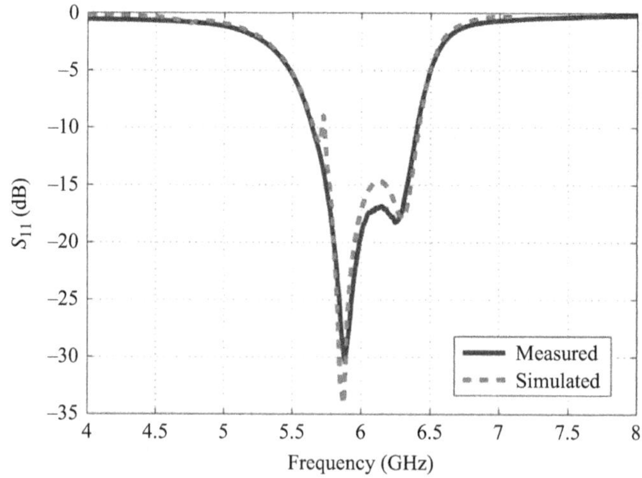

Figure 4.15 Negative return loss of measured and simulated antenna (from [20], copyright 2009, IEEE, reprinted with permission)

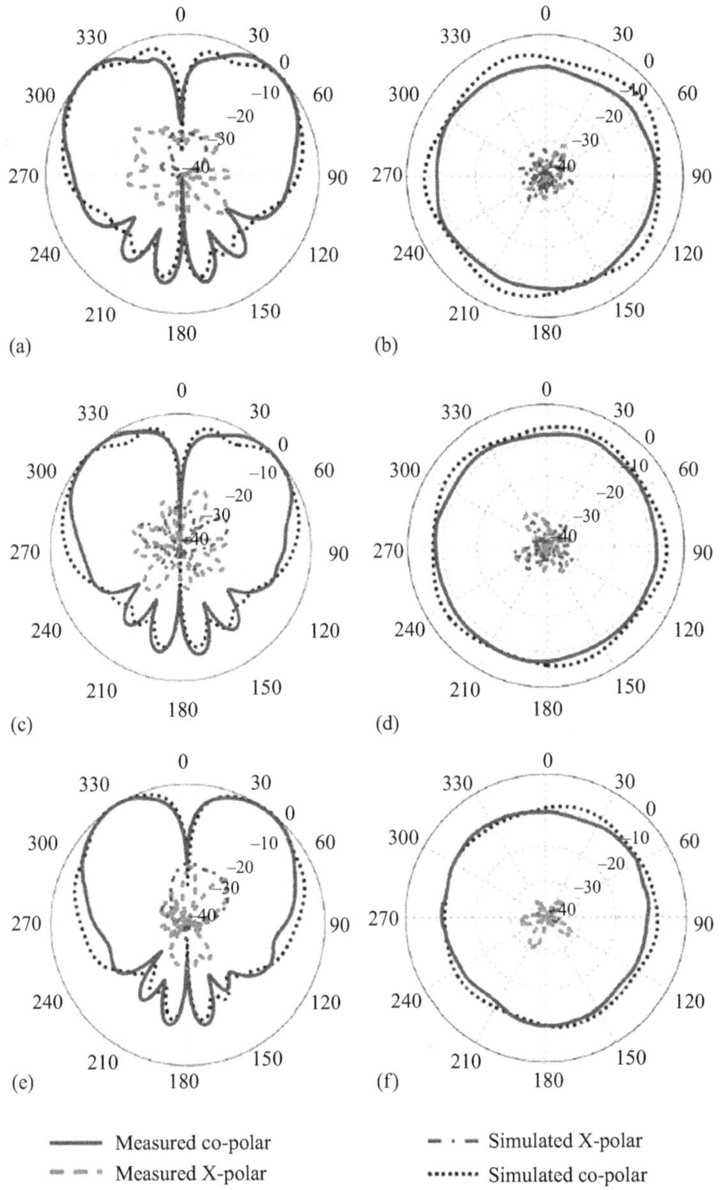

Figure 4.16 Radiation patterns of measured and simulated antenna. E-plane radiation patterns at: (a) 5.7 GHz, (c) 5.8 GHz, and (e) 6.3 GHz. H-plane radiation patterns at: (b) 5.7 GHz, (d) 5.8 GHz, and (f) 6.3 GHz (from [20], copyright 2009, IEEE, reprinted with permission)

higher order mode usually decreases the impedance bandwidth of a circular microstrip antenna significantly.

A center-fed circular patch with an outer ring has been developed, which has a monopole pattern with 12.8% impedance and pattern bandwidth with 5.7 dBi gain [20]. The geometry of this antenna is presented in Figure 4.14. The center patch radius is denoted as R_p. The surrounding annulus has an interior radius designated as a, and the outer radius is b. The groundplane radius is R_g.

The design procedure for this antenna is to design a TM_{02} mode circular patch (Figure 3.2 $n = 0$). This patch was analyzed with the cavity model to resonate at 5.773 GHz. The outer ring also possesses a TM_{02} mode (Figure 3.17). A cavity model analysis produces a resonant frequency of 6.56 GHz without the center patch present. The combination of these two resonators produces a dual resonance in the S_{11} dB plot as seen in Figure 4.15. The design parameters for this prototype antenna are $R_p = 18$ mm, the substrate has a thickness of 1.524 mm, with a relative dielectric permittivity of $\epsilon_r = 2.94$ (RT/Duroid 6002). The coaxial feed probe is placed at the center of the patch and has a probe radius of 0.34 mm. The annulus has an interior radius of $a = 19$ mm and an outer radius of $b = 31$ mm concentric to the interior patch. The radius of the groundplane, R_g, is 75 mm.

In short, the center patch radius R_p affects the first resonant frequency. A larger radius produces a lower value of resonant frequency, a smaller radius, a higher resonant frequency. The interior and exterior radius values a and b influence the second resonant frequency. Increasing a increases the resonant frequency, whereas increasing b will decrease the frequency. The slot between the circular patch and the annulus changes the coupling between the two.

The E-plane and H-plane radiation patterns of the antenna computed and measured at three different frequencies (5.7, 5.8, and 6.3 GHz) are shown in Figure 4.16. The main beam maximum is at approximately 45°.

References

[1] Kumar, G., and Ray, K.P., *Broadband Microstrip Antennas*, Artech House, 2003.

[2] Wong, K.-L., *Compact and Broadband Microstrip Antennas*, John Wiley & Sons, 2002.

[3] Zürcher, J.-F., and Gardiol, F.E., *Broadband Patch Antennas*, Artech House, 1995.

[4] Zürcher, J.-F., "The SSFIP: A Global Concept for High-Performance Broadband Planar Antennas," *Electronics Letters*, November 1988, Vol. 24, No. 23, pp. 1433–1435.

[5] Zürcher, B., Zürcher, J.-F., and Gardiol, F.E., "Broadband Microstrip Radiators The SSFIP Concept," *Electromagnetics*, 1989, Vol. 9, No. 4, pp. 385–393.

[6] Milligan, T., *Modern Antenna Design*, McGraw Hill, 1985.

[7] Hansen, R.C., "Correct Impedance-Matching Limitations," *IEEE Antennas and Propagation Magazine* (Antenna Designer's Notebook), Vol. 51, No. 3, pp. 122–124.

[8] Hall, P.S., "Probe Compensation in Thick Microstrip Patch Antennas," *Electronics Letters*, May 1987, Vol. 23, No. 11, pp. 606–607.

[9] Bernard, R., Tchanguiz, R., and Papiernik, A., "Capacitors Provide Input Matching of Microstrip Antennas," *Microwaves & RF*, July 1994, Vol. 33, No. 7, pp. 103–106.

[10] Matthaei, G., Young, L., and Jones, E.M.T., *Microwave Filters, Impedance-Matching Networks, and Coupling Structures*, McGraw Hill, 1964, pp. 120–130, 681–686.

[11] Paschen, D.A., "Practical Examples of Integral Broadband Matching of Microstrip Elements," *Proceedings of the 1986 Antenna Applications Symposium*, Monticello, IL, September 17–19, 1986, pp. 199–217.

[12] Matthaei, G.L., Young, L., and Jones, G.M.T., *Microwave Filters, Impedance-Matching Networks, and Coupling Structures*, McGraw Hill, 1964, pp. 123–129.

[13] Schaubert, D.H., Pozar, D.M., and Adrian, A., "Effect of Microstrip Antenna Substrate Thickness and Permittivity: Comparison of Theories with Experiment," *IEEE Transactions on Antennas and Propagation*, June 1989, Vol. 37, No. 6, pp. 677–682.

[14] Carlin, H.J., and Amstutz, P., "On Optimum Broad-Band Matching," *IEEE Transactions of Circuits and Systems*, May 1981, Vol. CAS-28, No. 5, pp. 401–405.

[15] Yarman, B.S., "A Simplified Real Frequency Technique for Broadband Matching a Complex Generator to a Complex Load," *RCA Review*, September 1982, Vol. 43, pp. 529–541.

[16] Gerkis, A.N., "Broadband Impedance Matching Using the 'Real Frequency' Network Synthesis Technique," *Applied Microwave & Wireless*, July/August 1998, pp. 26–36.

[17] Hongming, An, Nauwelaers, Bart, K.J.C., and Van de Capelle, Antoine, R., "Broadband Microstrip Antenna Design with the Simplified Real Frequency Technique," *IEEE Transactions on Antennas and Propagation*, February 1994, Vol. 42, No. 2, pp. 129–136.

[18] Delabie, C., Villegas, M., and Picon, O., "Creation of New Shapes for Resonant Microstrip Structures by means of Genetic Algorithms," *Electronics Letters*, August 1997, Vol. 33, No. 18, pp. 1509–1510.

[19] Choo, H., Hutani, A., Trintinalia, L.C., and Ling, H., "Shape Optimization of Broadband Microstrip Antennas Using Genetic Algorithm," *Electronics Letters*, December 2000, Vol. 36, No. 25, pp. 2057–2058.

[20] Al-Zoubi, A., Yan, F., and Kishk, A., "A Broadband Center-Fed Circular Patch-Ring Antenna with a Monopole Like Radiation Pattern," *IEEE Transactions on Antennas and Propagation*, March 2009, Vol. 57, No. 3, pp. 789–792.

Chapter 5
Dual-band microstrip antennas

Dual-band microstrip antennas are generally of two types. (1) Separate microstrip resonators coupled to a transmission line. (2) Perturbed microstrip resonators, where their original resonant frequencies are shifted by geometrical alteration of a basic resonator [1,2]. Dual-band microstrip antenna designs which allow for independent frequency selection have the most design utility. A good overview of dual-band microstrip antennas is given by Maci and Gentili [3].

5.1 Rectangular microstrip dual-band antenna

If one requires a single-element dual-band microstrip antenna, which has a broadside radiation pattern at each of the dual-band design frequencies with the same polarization, one must drive the TM_{10} and TM_{30} modes. When these restrictions are chosen, the upper frequency must be approximately three times the lower frequency of operation. For a rectangular microstrip antenna, the pattern and polarization restrictions, which have been previously imposed, severely limit the number of applications to which this design can be utilized.

In Section 2.5.1, a rectangular microstrip antenna is used to create circular polarization by allowing the TM_{10} and TM_{01} modes to overlap in frequency. A similar approach may be used to create single resonator (element) dual-band microstrip antenna by separating the modes until they are isolated. If a designer can allow orthogonal linear polarization for each of the dual-band frequencies and desires broadside radiation patterns, one can choose rectangular patch dimensions a and b so the TM_{10} and TM_{01} modes correspond to a desired upper and lower frequency pair (F_U, F_L). This allows one to choose the two frequencies of operation in an independent manner. One can choose the patch dimensions a and b to produce the desired frequency pair using methods from Chapter 2. The feedpoint location (X_P, Y_P) is the place which will optimally match the two modes simultaneously. The transmission line model of a rectangular microstrip antenna allows one to feed a patch antenna anywhere along a plane which is 50 Ω. A 50 Ω driving point impedance plane exists for each of the two modes generated. Where the impedance planes intersect is an optimum point to feed a dual-band patch of this type [4]. The geometry of a dual-band patch antenna of this type is illustrated in Figure 5.1.

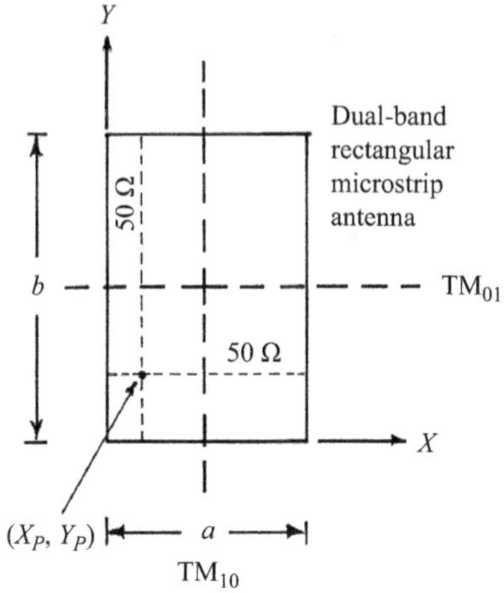

Figure 5.1 Optimization using a genetic algorithm with the cavity model allows for the design of a rectangular dual-band antenna with a single feed point. This is accomplished by matching the TM_{10} and TM_{01} modes with the desired upper and lower design frequencies and simultaneously determining a matched driving point location at (X_P, Y_P)

In order to avoid inaccuracies which can occur using the transmission line model, one can use the cavity model with genetic optimization to design a dual-band patch based on the use of TM_{01} and TM_{10} modes. This approach allows us to study the feasibility of designs using different frequency separations

For ease of analysis, we can define a midpoint frequency F_m which is centered between the dual-band frequencies F_L and F_U:

$$F_m = \frac{F_L + F_U}{2} \tag{5.1}$$

We can define a frequency separation factor:

$$F_s = \left| \frac{F_m + \Delta_F}{F_m - \Delta_F} \right| \tag{5.2}$$

with

$$\Delta_F = \frac{F_U - F_L}{2} = F_m \frac{(F_s - 1)}{(F_s + 1)} \tag{5.3}$$

Multiplying the lower frequency by this factor produces the upper frequency:

$$\begin{aligned} F_U &= F_s \cdot F_L = F_m + \Delta_F \\ F_L &= F_m - \Delta_F \end{aligned} \tag{5.4}$$

As a design example, we choose a midpoint frequency (F_m) of 2.0 GHz and use a genetic algorithm optimization with the cavity model to obtain designs for $F_s = 1.05$,

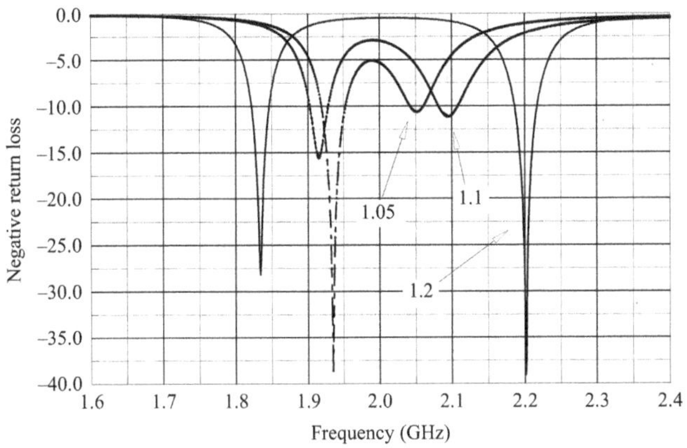

Figure 5.2 Single feed dual-band solutions for a rectangular microstrip antenna using a genetic algorithm optimization with the cavity model

1.1, 1.2 with $\epsilon = 4.1$ and $h = 1.524$ mm. The fitness function used for the genetic optimization is

$$Fit = \frac{|\text{Min}(\Gamma_L, \Gamma_U)|}{|\text{Max}(\Gamma_L, \Gamma_U)|} \cdot [(1 - |\Gamma_L|) + (1 - |\Gamma_U|)] \tag{5.5}$$

where $|\Gamma_L|$ is the magnitude of the driving point reflection coefficient at the lower center frequency and $|\Gamma_U|$ is the magnitude of the driving point reflection coefficient at the upper center frequency.

We note in Figure 5.2 that as the frequency separation becomes larger, the genetic algorithm produces designs that are very close to the desired design frequencies and well matched (>20 dB return loss). An F_s of 1.05 is a design which genetic optimization is unable to match both frequencies. When $F_s > 1.2$, genetic optimization produces well matched dual band designs for the rectangular patch geometry.

5.2 Multiple resonator dual-band antennas

5.2.1 Coupled microstrip dipoles

A second dual-band antenna design option is to place a pair of parallel narrow microstrip elements (microstrip dipoles) in close proximity with a separation of (d) and feed them with a single coupled microstrip line as illustrated in Figure 5.3. The two resonators are in the same plane which makes this a co-planar dual band design. This type of design allows one to more easily control F_L and F_U and maintain a good match. A microstrip line extends under the microstrip dipoles, ending at the center of each microstrip dipole (i.e. at $L_U/2$ and $L_L/2$). The width of each dipole affects the

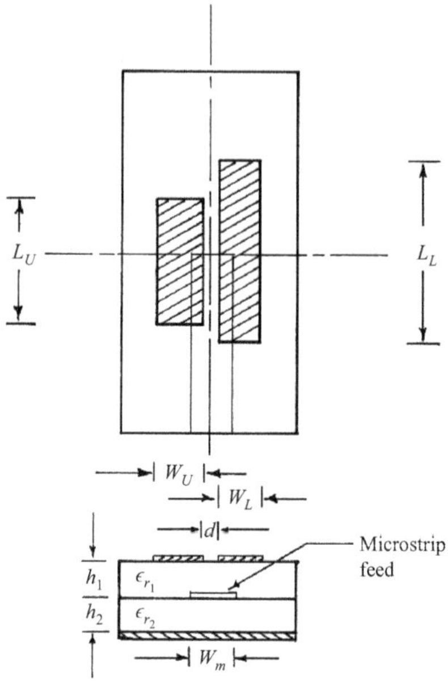

Figure 5.3 Two microstrip dipoles coupled to an imbedded microstrip transmission line

match of each antenna. The lengths L_U and L_L change the upper and lower frequencies. The match is fairly stable as the resonator lengths are altered, when compared to the sensitive nature of a single resonator dual-band antenna. This is illustrated by the negative return loss plot of Figure 5.4 and Table 5.1 which describes the design. The length of each element can be adjusted to produce frequency separations F_s from 1.25 to 2.0 without altering other dimensions. The antenna patterns at each frequency are squinted by the presence of the non-driven element, which is the trade-off one must endure when using this antenna design.

5.2.2 Stacked rectangular microstrip antennas

One can stack microstrip resonators to produce a dual-band antenna. The geometry of this design is illustrated in Figure 5.5. The upper patch is the high-frequency element of the dual-band antenna. The lower patch is larger than the upper patch and acts as a groundplane for the upper patch when it is resonant at F_U. When the lower patch is excited at the lower dual-band frequency, F_L, the upper patch has little affect on the lower patch. The groundplane of the structure acts as the groundplane for the lower patch.

Generally, a single feed probe passes through the bottom patch without connection (a small circle of the lower patch is removed for this purpose), and connects to the upper patch. This excitation geometry is sometimes called a common feed. When

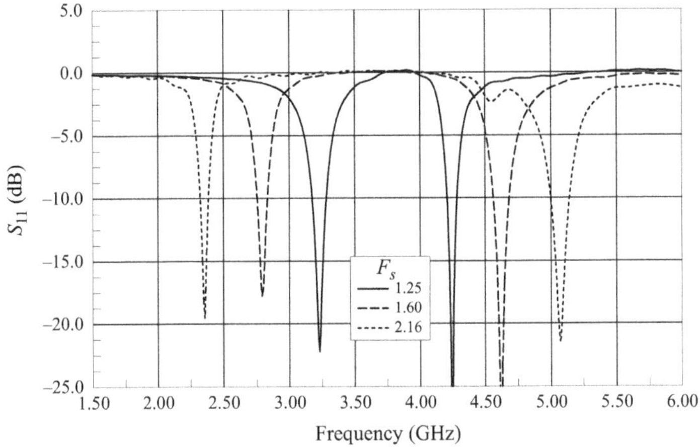

Figure 5.4 Microstrip dipole dual-band designs of Table 5.1 analyzed with finite difference time domain method (FDTD)

Table 5.1 Coupled dual-band microstrip antenna parameters $\epsilon_{r1} = \epsilon_{r2} = 4.1$, tan $\delta = 0.005$, $h_1 = h_2 = 1.524$ mm, d = 2 mm, microstrip line width = 2.98 mm

F_s	L_L (mm)	L_U (mm)	W_L (mm)	W_U (mm)
2.15	31.20	13.52	9.50	6.20
1.60	25.73	15.00	9.50	6.20
1.25	21.78	16.83	9.50	6.20

the upper patch is resonant, the lower patch produces negligible reactance and vice versa. Another approach is parasitic feeding. The feed probe connects to the lower patch, and the upper patch is electromagnetically coupled. Parasitic feeding is often used to broaden the bandwidth of the upper patch, rather than produce a dual-band antenna. When used to broaden antenna bandwidth, the upper patch is larger than the lower patch [5].

When the upper patch is directly fed, the feedpoint location for a set of stacked linear rectangular microstrip antennas is close to the location of each patch separately. A shorting pin can be placed through the center of both patches to the groundplane. This helps to enforce the driven mode and eliminate many ESD problems.

The upper and lower patches may have their aspect ratios adjusted to produce circular polarization at each of the two dual-band frequencies. One can also stack quarter-wave patch elements to create dual-band stacked element configurations with a smaller footprint than required by half-wave elements.

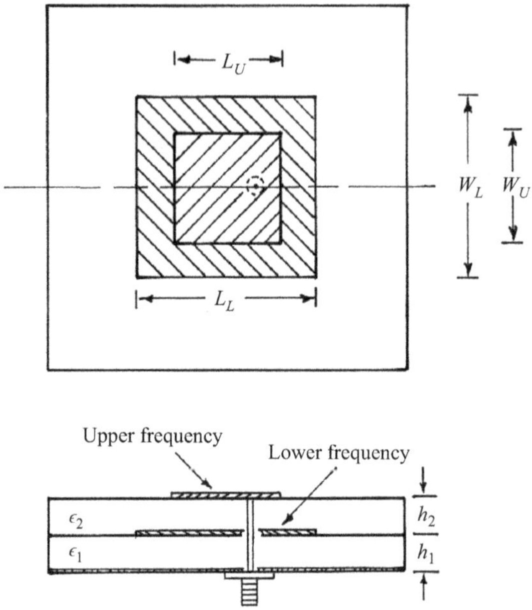

Figure 5.5 Stacked patch antennas

As stated previously, the upper and lower patches may have their aspect ratios adjusted so the upper and lower patches have their TM_{01} and TM_{10} modes driven at distinct frequencies. This allows one to create a pair of stacked patches which will operate at four separate frequencies.

One may create a quad-band stacked patch antenna using the cavity model with genetic algorithms, as detailed in Section 5.1 previously. A dual-band patch design as shown in Figure 5.1 is the starting point. The two lowest frequencies are assigned to the bottom patch antenna, and the two highest frequencies are assigned to the upper rectangular patch antenna. Lengths a_L and b_L are chosen to resonate the TM_{10} and TM_{01} modes of the lower patch at the desired lower set of design frequencies. Lengths a_H and b_H are chosen to resonate the TM_{10} and TM_{01} of the upper patch at the desired upper set of design frequencies.

A pair of 50 Ω impedance planes, as shown in Figure 5.1, will exist for each of the two antennas. The point of intersection of the impedance planes of the upper patch antenna (X_{P_U}, Y_{P_U}) is aligned with the lower patches intersection point (X_{P_L}, Y_{P_L}). The two points are co-located as (X_P, Y_P) in Figure 5.6. Experimental optimizing of the antenna is required to produce a final design.

5.3 Dual-band microstrip antenna design using a diplexer

An alternative to using a stacked patch to create a dual-band microstrip antenna is to use a diplexer with two single frequency elements. A diplexer separates a pair of

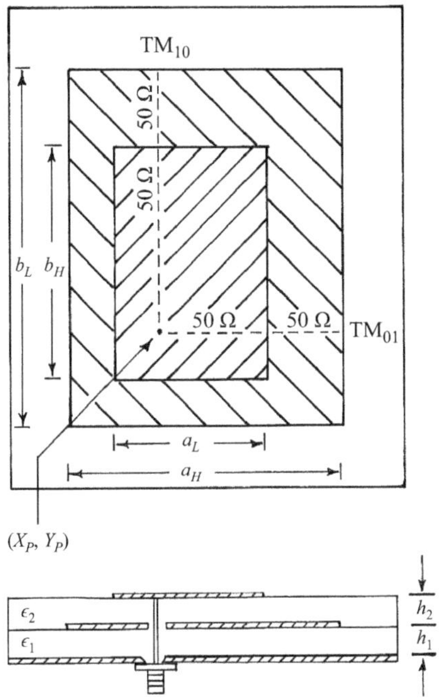

Figure 5.6 Stacked patch antenna fed to drive four separate modes to create a quad-band antenna

frequency bands which are originally combined at the input port and presents each frequency at each of a pair of output ports which are isolated from one another. The classic lumped element version of a diplexer consists of a high-pass filter and low-pass filter with a common input.

The exact and approximate design of diplexers, and multiplexers, is presented in the literature [6,7]. This design process can be rather extensive when developing microwave diplexers. If the frequency ratio between bands is approximately 2:1, one can use a diplexer circuit introduced by de Haaij *et al.* [8] and presented in Figure 5.7.

The three-port microstrip circuit has a 50 Ω input port with a length of transmission line, which forms a T-junction, with a pair of 50 Ω transmission lines. The upper and lower frequency ports are designated F_U and F_L for upper and lower frequency bands, respectively.

When the center frequency F_L of the lower frequency band is presented to the input port, the quarter wavelength open-circuit stub ($\lambda_L/4$) produces a short circuit one-quarter wavelength ($\lambda_L/4$) from the T-junction which becomes an open circuit at the T-junction. λ_L is the guide wavelength of the microstrip transmission line at the center of the lower frequency band. The short-circuited stub at the lower frequency band becomes an open circuit at the microstrip transmission line. This allows the low-frequency signal to pass through to the F_L port unrestricted. This occurs because $\lambda_U/2 = \lambda_L/4$ when a 2:1 frequency ratio is assumed.

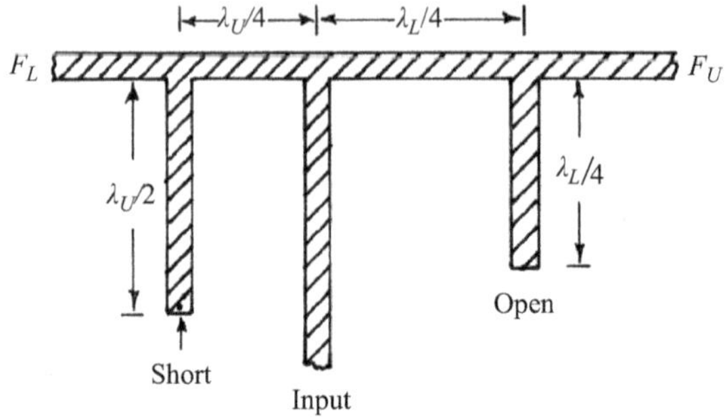

Figure 5.7 Diplexer for combining two bands which have an approximate 2:1 center frequency ratio

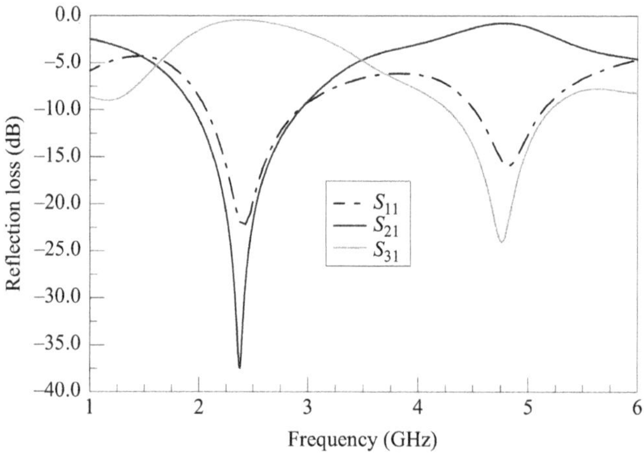

Figure 5.8 Diplexer designed to feed a pair of rectangular patches to produce a dual-band radiating structure for Example 5.3.1

When the center frequency, F_U, of the upper frequency band is presented to the diplexer's input port, the open-circuit stub on the right-hand side is one-half guide wavelength ($\lambda_L/4 = \lambda_U/2$) and presents an open circuit to the transmission line. This allows the signal to pass to the F_U output port. The short-circuited stub becomes a short circuit at the microstrip transmission line at a location one-quarter wavelength from the T-junction which produces an open circuit at the T-junction.

5.3.1 Example dual-band microstrip antenna using a diplexer

In Figure 5.8, a diplexer of the form found in Figure 5.7 is realized using Ansoft HFSS, which passes 2.38 and 4.77 GHz. This diplexer has limited utility in practice, but offers a succinct design for illustrating the use of a diplexer in a dual-band microstrip antenna design. The diplexer is used to feed a pair of rectangular microstrip antennas, which are wide enough to directly match a 50 Ω microstrip transmission line. Two ultrawide rectangular microstrip antennas, with 50 Ω edge resistance, were designed to have resonant frequencies of 2.38 and 4.77 GHz, respectively.

The two patches integrated with the diplexer are shown in Figure 5.9. The physical values of the distance from the low-frequency (F_L) port to the input port (D_1) of the low frequency patch is 41.7 mm. The low-frequency patch has a width of $W_1 = 75.0$ mm and a length of $L_1 = 37.576$ mm. The length from the upper frequency (F_U) port to the input port (D_2) of the high-frequency patch is 35.0 mm. The high-frequency patch has a width of $W_2 = 60$ mm and a length of $L_2 = 17.86$ mm. $\lambda_L = 83.4$ mm and $\lambda_U = 41.7$ mm. The width of the interconnecting 50 Ω transmission lines is 4.17 mm. The substrate parameters are $\epsilon_r = 2.6$ with 0.0019 tan δ and the substrate thickness is 1.524 mm. The length and width of the substrate are $L_G = 125$ mm and $W_G = 200$ mm.

The gain values predicted by Ansoft HFSS are 6.7 dBi for the low-frequency patch and 9.52 dBi for the high-frequency patch. The negative return loss plot of the ultrawide elements separately, and integrated with the example diplexer as predicted by HFSS is shown in Figure 5.10.

More complex diplexers have larger and more flexible passbands, which allow for an easier antenna design implementation. The previous example illustrates a simple diplexer design, which can be used to produce a dual-band planar antenna with a frequency separation of ≈ 2.0, which is useful for illustration. When a more elaborate diplexer design is used with broad bandwidth planar antenna elements, diplexer

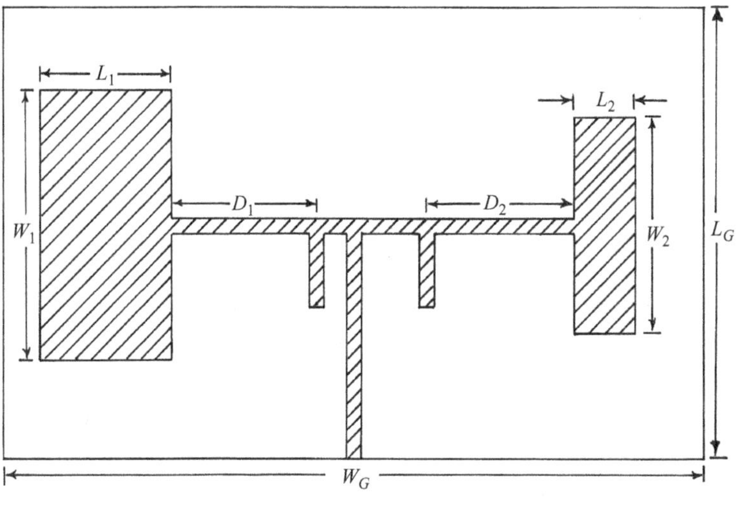

Figure 5.9 Geometry of diplexer with integrated rectangular patch antennas used to produce a dual-band structure

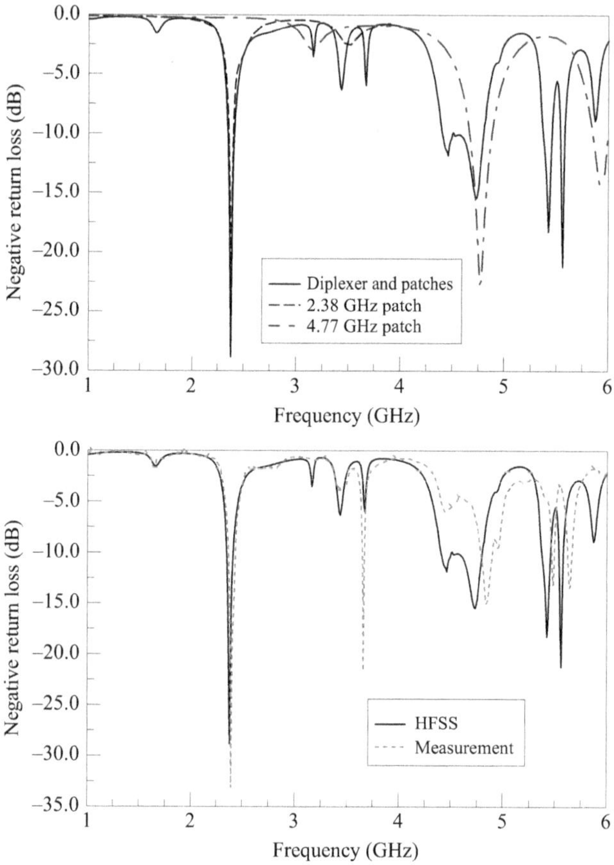

Figure 5.10 (Top) Negative return loss plots of high- and low-frequency patches analyzed separately, and the response of the diplexer and patches predicted by HFSS. (Bottom) Negative return loss plots predicted by HFSS and measured

designs that allow up to 20% bandwidth per band are possible. The planar printed antennas employed are monopole-type elements with the groundplane removed.

5.4 Multiband patch shaping using a genetic algorithm

In Section 4.3.1, the creation of microstrip antennas with a large impedance bandwidth, using patch shaping, guided with a genetic algorithm was discussed.

This method has been used by Choo and Ling to design a microstrip antenna patch shape, using the method of moments, which produces multiband antennas on FR4 [9,10]. An antenna shape was created which operates at 900 MHz (GSM),

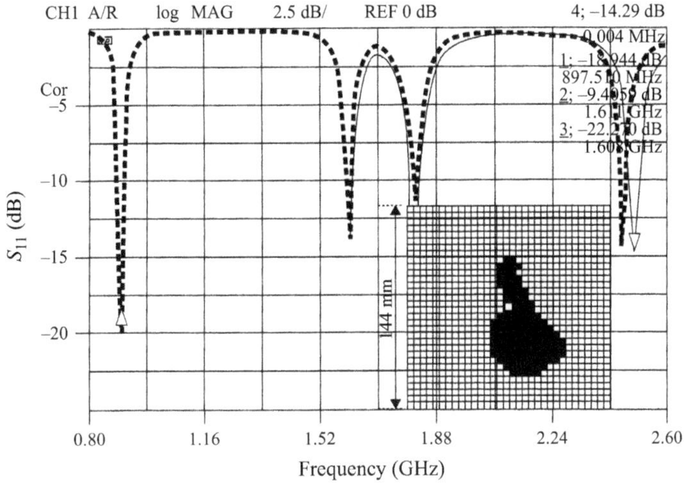

Figure 5.11 Patch shape of a microstrip antenna which has been optimized for quad-band operation. The dashed lines are simulation, and the solid line is measurement (from [9], copyright 2002, IEEE, reprinted with permission)

1.6 GHz (GPS/L1), 1.8 GHz (DCS), and 2.45 GHz (ISM/Bluetooth). This design is reproduced in Figure 5.11. This method offers the possibility of developing low-cost single layer multiband antennas with arbitrary frequencies.

References

[1] Kumar, G., and Ray, K.P., *Broadband Microstrip Antennas*, Artech House, 2003.

[2] Wong, K.-L., *Compact and Broadband Microstrip Antennas*, John Wiley & Sons, 2002.

[3] Maci, S., and Gentili Biffi, G. "Dual-Frequency Patch Antennas," *IEEE Antennas and Propagation Magazine*, December 1997, Vol. 39, No. 6, pp. 13–20.

[4] Chen, J.-S., and Wong, K.-L., "A Single-Layer Dual-Frequency Rectangular Microstrip Patch Antenna Using a Single Probe Feed," *Microwave and Optical Technology Letters*, February 1996, Vol. 11, No. 2, pp. 83–84.

[5] James, J.R., and Hall, P.S., *Handbook of Microstrip Antennas Volume 1*, Chapter 6, Peter Peregrinus Ltd., 1989, pp. 324–325.

[6] White, J.F., *High Frequency Techniques*, John Wiley, 2004, pp. 364–369.

[7] Malherbe, J.A.G., *Microwave Transmission Line Filters*, Artech House, 1979, Chapter 7.

[8] de Haaij, D.M., Joubert, J., and Odendaal, J.W., "Diplexing Feed Network for Wideband Dual-Frequency Stacked Microstrip Patch Antenna," *Microwave and Optical Technology Letters*, January 2003, Vol. 36, No. 2, pp. 100–103.

[9] Choo, H., and Ling, H., "Design of Multiband Microstrip Antennas Using a Genetic Algorithm," *IEEE Microwave and Wireless Components Letters*, September 2002, Vol. 12, No. 9, pp. 345–347.

[10] Choo, H., and Ling, H., "Design of Dual-Band Microstrip Antennas Using the Genetic Algorithm," *Proceedings of the 17th Annual Review of Progress in Applied Computational Electromagnetics* (Session 15), Monterey, CA, May 19–23, 2001, pp. 600–605.

Chapter 6
Microstrip arrays

A single microstrip antenna utilizing an air dielectric substrate ($\epsilon_r \approx 1$) is able to provide a maximum gain of nearly 10 dBi. When larger gain is required, and a microstrip antenna solution is the best choice, a number of microstrip elements may be connected together to form an array of antennas. The array of elements provide a much larger effective aperture and therefore gain compared to a single microstrip element. This chapter will discuss elementary methods used to design microstrip antenna arrays.

6.1 Planar array theory

Classic linear and planar array analysis papers were presented by Elliot in the early 1960s, which are very useful for the analysis of rectangular microstrip antenna arrays [1–3]. In Figure 6.1, a number of rectangular microstrip antennas are located in the x–y plane. The z-axis is directed out from the paper. Each microstrip antenna can be modeled as a pair of radiating slots in a groundplane. Assuming a TM_{01} mode, the antennas are polarized along the y-axis. A patch with its center located at (X_n, Y_n) is effectively modeled as a pair of slots located at $(X_n, Y_n + L/2)$ and $(X_n, Y_n - L/2)$ of width W and thickness H, with identical excitation amplitude for each pair as shown in Figure 6.2.

The array factor for a number of point sources (N_p) in free space is

$$A(\theta, \phi) = \sum_{n=1}^{N_p} I_n e^{jk \sin\theta [X_n \cos\phi + Y_n \sin\phi]} \tag{6.1}$$

To analyze the radiation pattern of a rectangular microstrip array, we sum the array factor over all the slots, taking into account the excitation factor of each slot: $I_n = E_n \cdot w \cdot h$, where E_n is the electric field across each slot of patch n, w is the patch width, and h is the substrate thickness. θ is measured from the z-axis. The electric field radiated is proportional to the array factor multiplied by an element (antenna) factor $R(\theta, \phi)$. The electric field decreases by $1/r$:

$$\vec{E}(\theta, \phi) = A(\theta, \phi) \cdot R(\theta, \phi) \tag{6.2}$$

The θ and ϕ components of the electric field are

$$\vec{E}_\theta(\theta, \phi) = A(\theta, \phi) \sin\phi \cdot R(\theta, \phi) \tag{6.3}$$

$$\vec{E}_\phi(\theta, \phi) = A(\theta, \phi) \cos\theta \cos\phi \cdot R(\theta, \phi) \tag{6.4}$$

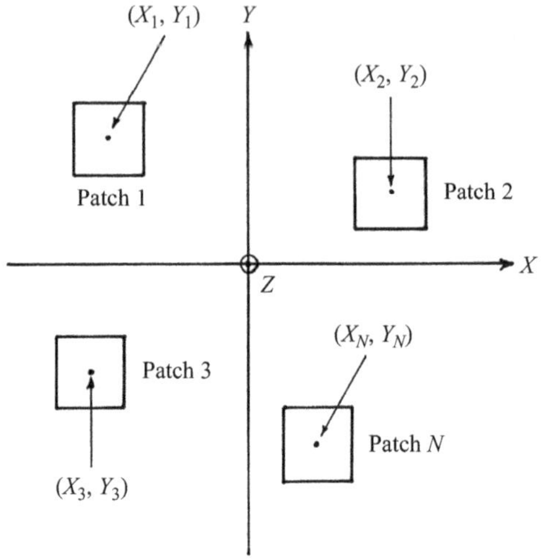

Figure 6.1 Geometry of a set of rectangular microstrip antennas 1, 2, 3, . . . , N in the x–y plane. The center of each patch is used for reference to locate a pair of equivalent slots

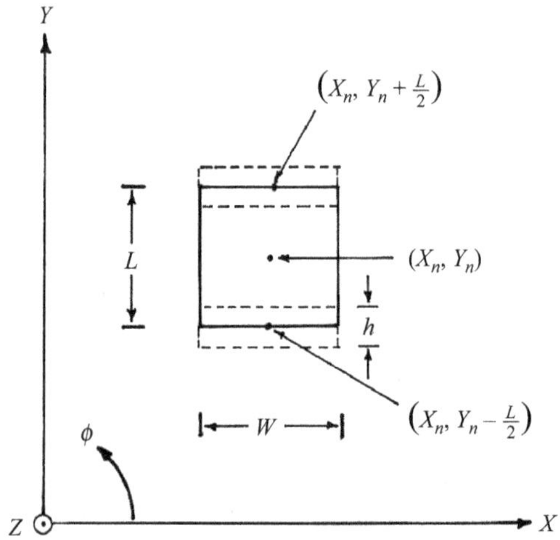

Figure 6.2 The radiation from each patch of Figure 6.1 is modeled as a pair of slots

The element factor for a rectangular aperture is [4]:

$$R(\theta, \phi) = \frac{\sin[(kw/2)u]}{(kw/2)u} \frac{\sin[(kh/2)v]}{(kh/2)v} \tag{6.5}$$

where

$$u = \sin\theta\cos\phi \tag{6.6}$$

$$v = \sin\theta\sin\phi \tag{6.7}$$

The power at any point in space is

$$P(\theta,\phi,r) = \frac{|E_\theta|^2}{r^2} + \frac{|E_\phi|^2}{r^2} \tag{6.8}$$

We can multiply (6.8) by r^2, which gives us the power per unit solid angle or the radiation intensity U:

$$U(\theta,\phi) = |E_\theta|^2 + |E_\phi|^2 \tag{6.9}$$

The expression for radiation intensity can be used to compute the array directivity using numerical techniques [5]:

$$D_0 = \frac{4\pi U_{max}}{P_{rad}} \tag{6.10}$$

$$P_{rad} = \left(\frac{\pi}{N}\right)\left(\frac{2\pi}{M}\right)\sum_{j=1}^{M}\left[\sum_{i=1}^{N} U(\theta_i,\phi_j)\sin\theta_i\right] \tag{6.11}$$

where

$$\theta_i = i\left(\frac{\pi}{N}\right) \tag{6.12}$$

$$\phi_j = j\left(\frac{2\pi}{M}\right) \tag{6.13}$$

The radiation sphere is divided into N and M sections. Microstrip antennas only radiate into the upper half space, $\theta < \pi/2$, so the radiation intensity is zero for $\theta > \pi/2$.

These equations produce reliable estimates of the directivity of a planar microstrip array.

6.2 Rectangular microstrip antenna array modeled with slots

The gain of an antenna is directly proportional to its effective aperture. As the gain of an antenna increases, so does its effective aperture. When microstrip antennas are arranged in an x–y plane, one can assume each of the separate antennas radiates only into the region above the x–y plane. If we have a single microstrip antenna with a gain of say 6.0 dBi, and then connect a second identical antenna which is spaced at one wavelength center to center ($\approx$ 0.5 wavelength edge to edge, $\epsilon_r = 1$), we will increase the effective aperture by approximately a factor of 2. This aperture doubling translates into approximately 3 dB increase in the maximum gain of the two elements. The gain of the two antennas combined is approximately 9.0 dBi. If we continue this line of reasoning, we have a rule of thumb to predict the approximate gain of a planar array of microstrip antennas. If a single element has 8.0 dBi of gain, then a pair of these elements can have approximately 11.0 dBi maximum gain. We need to double the aperture again to increase the gain by 3 dB, so we add two more elements for a total of 4. This 4 × 4 array should have up to 14.0 dBi gain. To obtain 3 dB more, we double from 4 elements to 8 elements to obtain 17.0 dBi. One can quickly estimate the maximum theoretical gain (or directivity) of a uniformly fed microstrip array by noting the approximate gain of a single element and adding 3 dB for each doubling of the number of elements until one reaches the total number of elements. This procedure is illustrated in Figure 6.3.

As an example, we will use the equations of Section 6.1 to compute the directivity of the array configurations depicted in Figure 6.3. The frequency chosen is 5.3 GHz, with a patch width and length (L and W) of 15.70 mm. The center-to-center distance between patches is 29.66 mm. The substrate thickness (h) is 1.524 mm.

In Table 6.1, we have computed and estimated directivity of the arrays of Figure 6.3. The direction of the E-plane of the antenna array is from top to bottom of the page. The H-plane is from left to right as illustrated in Figure 6.2. The difference between the directivity computations for a uniformly illuminated planar array and the estimates based on aperture are less than 1 dB. The computed patterns of Figure 6.4 offer some insight into how microstrip antennas array. Pattern (a) is a single element. The E-plane pattern is hemispherical and the H-plane diminishes to zero as θ approaches 90°. In (b), when two elements are placed side by side along the antennas H-planes, we note that they array along the H-plane and narrow the beamwidth. The E-plane pattern remains unchanged. When two more patch antennas

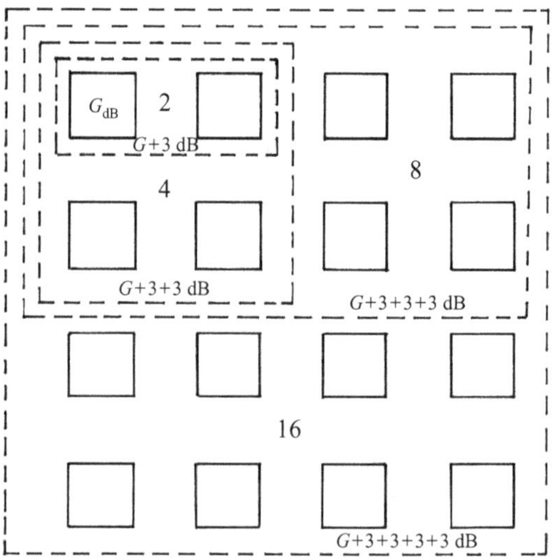

Figure 6.3 Relationship between array aperture and directivity. Each time the aperture is doubled (for a properly spaced array), the directivity is increased by approximately 3 dB

Table 6.1 Computed directivity versus estimated directivity

Elements	Directivity (dB)	Estimate (dB)
1	6.25	6.25
2	8.32	9.25
4	11.81	12.25
8	14.67	15.25
16	17.64	18.25
32	20.57	21.25

are added below, along the E-plane, for a total of four, we see the E-plane now array, leaving the H-plane unchanged in (c). Repeating the same selections in (d), (e), and finally (f), we get an understanding how element and array factors interact to narrow the beamwidth as the effective aperture increases, which increases the antenna gain.

6.3 Aperture excitation distribution

Figure 6.4 shows that a number of sidelobes appear in an array of 32 elements with equal amounts of current (or equivalent voltage) driven in each element. This type of distribution is often referred to as *uniform*. When the array excitation distribution of a planar microstrip array is uniform and electrically large, the gain of the main beam of the array is maximized, and the first sidelobe is 13.2 dB below the gain of the main beam. The value of gain for the first sidelobe below the gain of the main beam of an array, is known as the first sidelobe level (SLL). The maximum sidelobe value compared with the main beam is the SLL of the array.

In many applications, a -13.2 dB SLL is not acceptable. The SLL of an array can be controlled by tapering the excitation amplitudes of the array elements from a maximum value in the center to lower levels as one reaches the outside (Figure 6.5).

When the array excitation values are chosen to correspond with binomial coefficients, all sidelobes can be completely suppressed in theory. The beamwidth of the main beam of the array will widen, which decreases the gain of the array. Between the maximum gain of a uniform array with -13.2 dB sidelobes and the minimized gain of a binomial array with $-\infty$ dB sidelobes is an optimum choice known as a Dolph–Chebyshev distribution [6]. The realization of a Dolph–Chebyshev distribution can prove to be very challenging. Often in practice non-optimum aperture distributions relinquish only a small amount of gain and are more readily realizable than an optimum distribution.

A normalized linear taper on a pedestal is one which has a maximum value of one at the center of the array and tapers to a value of C at the maximum extent of the array. Because the distribution tapers to a non-zero value at its edge, it is said to be on a pedestal. A cosine on a pedestal distribution superimposes a cosine curve which

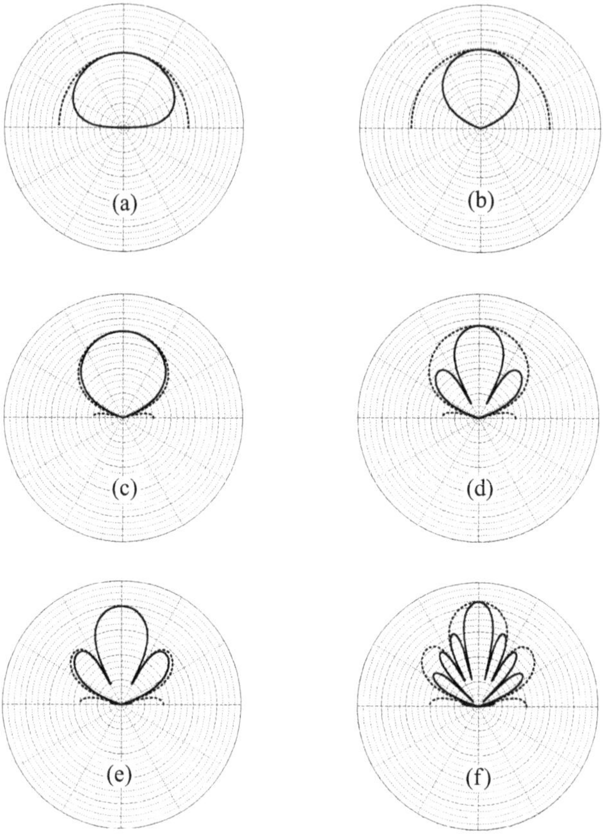

Figure 6.4 The computed patterns for the geometry of Figure 6.3. The E-plane patterns are represented as dashed lines and the H-plane patterns are solid: (a) Single antenna element; (b) two antenna elements; (c) four antenna elements; (d) eight antenna elements; (e) sixteen antenna elements; and (f) thirty-two antenna elements

has a non-zero value of C at the edges of an array. The expressions that describe these distributions are summarized below:

- Uniform

$$I_n = 1 \tag{6.14}$$

- Linear taper on a pedestal

$$I_n = C + (1 - C)\left[1 - \frac{|x_n|}{L_A}\right] \tag{6.15}$$

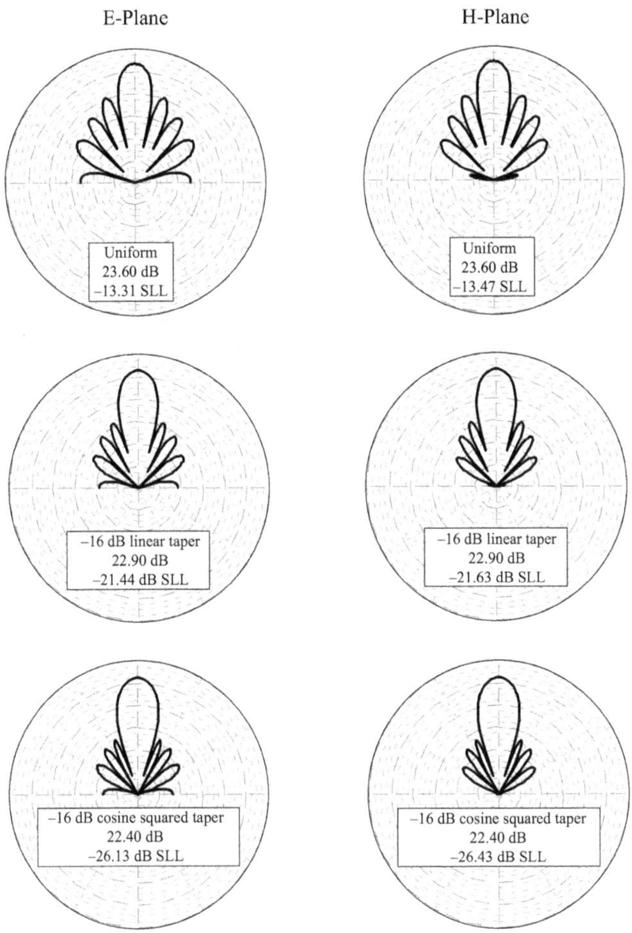

Figure 6.5 The computed E- and H-plane patterns for a 64-element antenna array (8 × 8) with uniform, −16 dB linear taper on a pedestal and −16 dB cosine squared taper on a pedestal

- Cosine taper on a pedestal

$$I_n = C + (1 - C)\left[1 - \frac{|x_n|}{L_A}\right] \tag{6.16}$$

- Cosine squared taper on a pedestal

$$I_n = C + (1 - C)\cos^2\left[\frac{\pi}{2}\frac{x_n}{L_A}\right] \tag{6.17}$$

- Quadratic taper on a pedestal

$$I_n = C + (1 - C)\left[1 - \left(\frac{x_n}{L_A}\right)^2\right] \tag{6.18}$$

x_n is the location of the nth element along the x-axis as shown in Figure 6.6 for a linear taper. The length of the array is $2L_A$. I_n is the excitation value for each element n. For an array with an odd number of elements, the element at the center of the array is located at x_0 and has a normalized value equal to one ($I_0 = 1$). When an array has an even number of elements, the I_0 element is removed, and only the even elements remain. Figure 6.6 illustrates a linear taper on a pedestal of value C.

The value of C is may be expressed as a decibel level with respect to the unity excitation value in the center:

$$C = 10^{(C_{\text{dB}}/20.0)} \tag{6.19}$$

For a -16 dB taper, we compute $C = 0.1585$ which may be used in (6.15)–(6.18). Figure 6.5 contains patterns with a uniform, linear, and cosine taper, computed with (6.15) and (6.16) and using the values of the 5.3 GHz patches in the previous section. We can see that for a 64-element, (8 × 8) array, the uniform distribution provides a maximum main beam gain of 23.60 dBi (assuming 100% efficiency). We can reduce the SLL to -21.7 dB with a -16 dB linear taper. The gain of the main beam decreases by 0.70 dB which is the trade off one makes for a lower SLL. A -16 dB cosine squared on a pedestal reduces the main beam gain by -1.20 dB compared with a uniform distribution, but the SLL is now approximately -26 dB.

Figure 6.7 has an illustration of how the aperture functions are sampled to determine the excitation values I_n. The origin of the coordinate system is on the center of patch 1. A vector to the center-most patch is formed. For an even array, 2 × 2, 4 × 4,

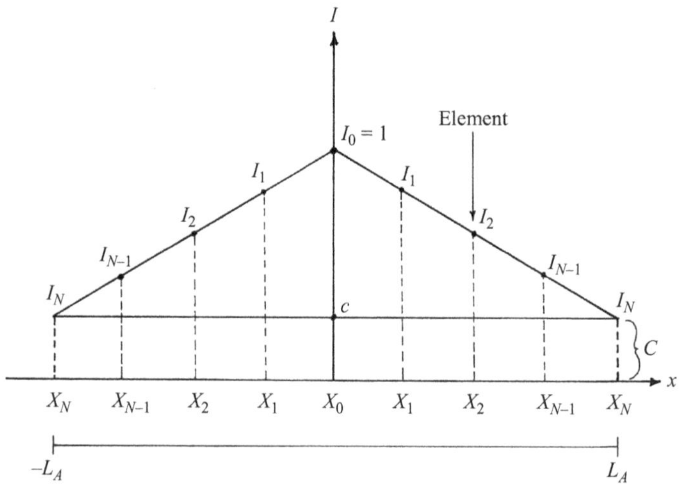

Figure 6.6 Linear taper on a pedestal

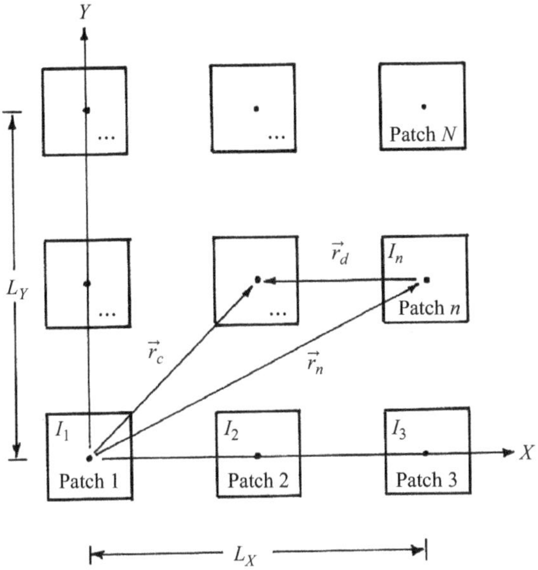

Figure 6.7 The distribution values I_n for a linear taper on a pedestal and cosine on a pedestal may be obtained by using (6.15) and (6.16). The distance from the center of the array $\vec{r}_c$ to the center of patch n ($\vec{r}_n$) determines the scalar value $d_n = |\vec{r}_c - \vec{r}_n|$ the value of $L_A = |\vec{r}_c|$

etc., $\vec{r}_c$ is a vector to the center of the array. In the geometry of Figure 6.7, (6.20) describes the location of $\vec{r}_c$. The vector from each patch to the array center patch $\vec{r}_d$ (6.21) is used to provide the distance $d_n(=x_n)$ (6.22). The value L_A is defined with (6.23). The values of d_n and L_A are placed into (6.15) for a linear taper on a pedestal and (6.16) for a cosine taper on a pedestal excitation for each n patch element:

$$\vec{r}_c = \frac{L_x}{2}\hat{i} + \frac{L_y}{2}\hat{j} \tag{6.20}$$

$$\vec{r}_d = \vec{r}_c - \vec{r}_n \tag{6.21}$$

$$d_n = |\vec{r}_d| \tag{6.22}$$

$$L_A = |\vec{r}_c| \tag{6.23}$$

The distributions of (6.15)–(6.18) are useful for most array implementations. When choosing a distribution, one first obtains designs that meet the directivity and sidelobe requirements. The pedestal value C determines the SLL of the array in each case. Generally, it is easiest to realize a distribution with the minimum taper which meets the pattern requirements. This choice also minimizes the beamwidth, which keeps the pattern directivity maximized. The quadratic and cosine squared distributions have the most gentle variation of the non-uniform distributions presented. The linear distribution has an aperture taper which decreases at the highest rate of the given distributions.

6.4 Microstrip array feeding methods

6.4.1 Corporate-fed microstrip array

The array feed methods which are most popular are corporate feeding and series feeding. Series feeding has a number of difficulties involved in its implementation. The beam direction is sensitive to frequency, and producing a desired amplitude taper is proven to be difficult. The design details of series-fed arrays will be addressed in the next section [7].

To illustrate an elementary corporate feed network, we will feed a linear array of four patch antennas as illustrated in Figure 6.8. Each of the identical square patch antennas has an element input resistance R_{in} at resonance $[R_{in} = R_e/2 = 1/(2G_e)]$. This element resistance at each patch may be matched to connecting transmission line impedances Z_1, Z_2, Z_3, Z_4, which will be used to provide a desired power split. This is accomplished with a number of quarter-wave transformers $Z_q^1, Z_q^2, Z_q^3, Z_q^4$.

To simplify this design, we will feed the linear array with a 50 Ω microstrip transmission line into a pair of 100 Ω lines. This will split the power in an equal manner to Z_a and Z_b quarter-wave transformers. These transformers are used to match between the 100 Ω feedline and the pair of transmission lines used to divide the incident power between each pair of patches (i.e. 1 and 2, 3 and 4). In the case of patch 1 and patch 2, we wish to provide them with current I_1 and I_2. We obtain the desired currents by realizing the voltage at the junction of Z_1 and Z_2 is common. The power propagated

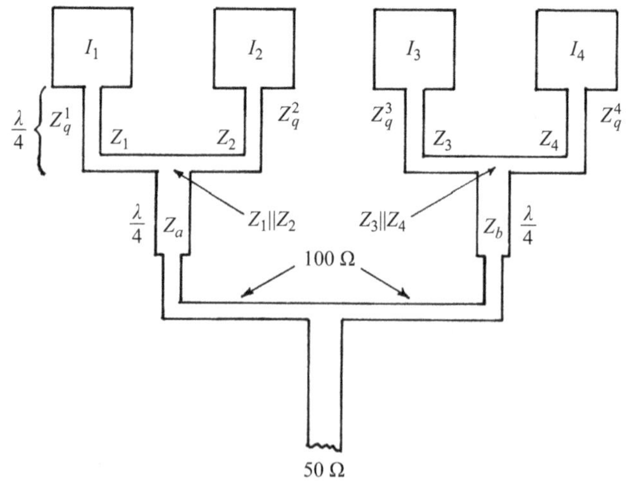

Figure 6.8 Four patch linear array fed with a microstrip corporate feed network

in each transmission line can be equated to the power in each patch to produce the desired current I_1 and I_2:

$$\frac{I_1^2 R_e}{I_2^2 R_e} = \frac{V_0^2/Z_1}{V_0^2/Z_2} \tag{6.24}$$

The ratio of I_1 to I_2 is controlled by the ratio of Z_2 to Z_1:

$$\frac{I_1^2}{I_2^2} = \frac{Z_2}{Z_1} \tag{6.25}$$

Once we have chosen the desired ratio of current, we can choose the ratio of the transmission line impedances.

As an example, let us choose an operating frequency of 5.25 GHz. A square patch with $a = b = 15.7$ mm on 1.524 mm thick substrate, with $\epsilon_r = 2.6$, has an element edge resistance of $R_{in} \approx 271.21\ \Omega$.

When etching microstrip circuits, most printed circuit board shops do not like line widths smaller than 152 μm (0.006 inch). This provides us with a maximum line impedance which may be used. For this substrate, this value is 180 Ω. If $I_1 = I_4 = 0.50$ and $I_2 = I_3 = 0.75$:

$$\frac{Z_2}{Z_1} = \frac{(0.50)^2}{(0.75)^2} = \frac{1}{2.25}$$

We know the element resistance of the patch antenna at resonance (R_{in}) and may choose the quarter-wave transformer for patch 1 (Z_q^1) to be the maximum impedance of 180 Ω, we can compute the value of Z_1:

$$\frac{(Z_q^1)^2}{R_{in}} = Z_1 = 119.5\ \Omega$$

This allows us to compute $Z_2 = 53.11\ \Omega$ with (6.25), which then may be used to compute Z_q^2:

$$Z_q^2 = \sqrt{Z_2 \cdot R_{in}} = \sqrt{53.11 \cdot 271.21} = 120.02\ \Omega$$

The impedance at the power split is $Z_1||Z_2 = 36.77\ \Omega$. The quarter-wave transformer Z_a is found as $\sqrt{36.77 \cdot 100.0} = 60.64\ \Omega$.

In this case, we have chosen a symmetric array distribution so the values of the impedances for both sides have been evaluated. The loss in a corporate-fed array will increase as the substrate height increases and with decreasing dielectric permittivity. The loss also increases as the feedline impedances decrease [8]. As the dimensions of an array increase, the length of the corporate feed network extends further and further. The microstrip line losses increase and in turn decrease the realized gain of the array and also add to the antenna noise figure. One can reach a point of diminishing return. As the number of array elements (and thus the effective aperture) is increased, the losses from the feed network become larger and larger. The increase in gain produced by a larger aperture can be balanced by the losses in the feedlines or overtaken.

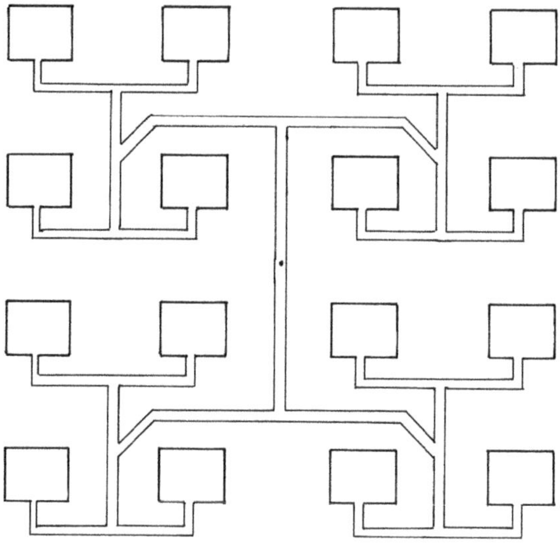

Figure 6.9 Four-by-four patch planar array fed with a microstrip corporate feed network. Heavy dot in the center is the feedpoint

This method of creating a feed network can be used to design the feed network of a planar array. Figure 6.9 has a four-by-four (16 element) planar array fed with a corporate network. This antenna can be sectioned into four 2×2 subarrays. The distribution can be computed for a subarray and the subarrays may be arrayed to create a planar array.

6.4.2 Series-fed microstrip array

A set of microstrip patches may be fed in series with microstrip transmission line connecting their radiating edges [9,10]. This is illustrated in Figure 6.10(a). The patches are separated by a microstrip transmission line, which has a $\lambda/2$ guide wavelength. In theory, the characteristic impedance of the transmission line is not critical, because each of the loads created by the edges of a microstrip radiator is all one-half wavelength in spacing. This is shown in the center illustration (b) of Figure 6.10. At resonance, the pair of slots of each rectangular microstrip element may be combined to form an equivalent admittance. A load that represents each patch is separated by half-wavelength sections of a microstrip transmission line. This is the right-hand (c) illustration of Figure 6.10. This separation allows one to compute the input resistance of the series-fed array at resonance as

$$R_{in} = \left(2\sum_{n=1}^{N} G_n\right)^{-1} \tag{6.26}$$

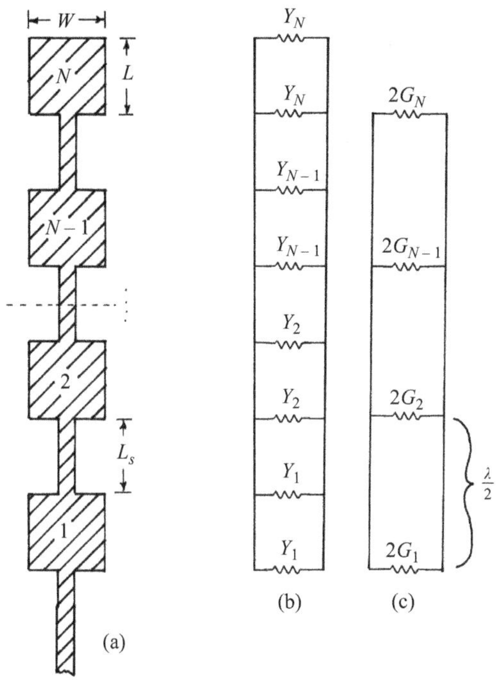

Figure 6.10 (a) Series-fed microstrip array. (b) Transmission line representation of series microstrip array. (c) Microstrip elements represented as lumped resistive loads between half-wave sections of transmission line

where G_n is the edge conductance at each edge of each patch. For example, if all the patches in a series-fed array have the same slot conductance, G_e, at resonance, we may express the input resistance as

$$R_{in} = \frac{1}{2NG_e} \tag{6.27}$$

This expression allows one to choose a value of G_e that allows for the design of a series-fed array with uniform distribution, which has a desired input resistance.

As an example, we choose a series-fed array with four elements ($N = 4$). When a 50 Ω input resistance is chosen, the edge resistance of each slot ($R_e = 1/G_e$) is 400 Ω. We can then use (2.4) to approximate the width of a patch that produces this edge resistance. One can use full-wave analysis methods to refine this patch width to produce an element which has an edge resistance closer to the desired value of 400 Ω ($R_{in} = 200$ Ω) than these approximate expressions can provide. When the antennas are connected, there will be a slight shift in frequency from the loading of the radiating edges.

For the example above, we use a dielectric substrate with $\epsilon_r = 2.6$ and a substrate thickness of $H = 1.524$ mm. The width (W) of each patch is 19.4 mm, with a resonant length of 17.0 mm. The patches are interconnected with 100 Ω transmission line. This is done to minimize the influence of the interconnects on the design.

The design of a single patch was accomplished using finite difference time domain (FDTD) analysis. A single patch antenna was developed, which was of an approximate size to produce a 200 Ω element resistance ($R_e = 400\ \Omega$, $R_{in} = R_e/2 = 1/2G_e$). This was connected through a 100 Ω quarter-wave transformer to a 50 Ω feedline. When this patch is well matched to the transformer and feedline, it should have a 200 Ω element resistance R_{in}.

A four patch series array was designed using the patch design obtained using the FDTD analysis. The array is matched at 5.09 GHz with a 1.35% (2:1) voltage standing wave ratio (VSWR) bandwidth. The substrate thickness is 1.524 mm (0.060 inch) with $\epsilon_r = 2.6$ and a 0.0025 tan δ. The length of each patch is $L = 17.0$ mm with a width $W = 19.4$ mm. The four patches are connected with a 0.8 mm wide ($Z_0 = 100\ \Omega$) microstrip transmission line of length 19.08 mm. The bottom-most patch is fed with one of these line sections which are $\lambda/2$ and driven with a 4.12 mm ($Z_0 = 50\ \Omega$) feedline. The groundplane is 44 mm × 128 mm.

The E-plane and H-plane pattern computed using FDTD is presented in Figure 6.11. The maximum directivity is 12.74 dB. One can see the elements array along the E-plane and maintain their individual element patterns in the H-plane.

With proper spacing, the main beam of a series array is broadside to the array at resonance. The main beam will squint from broadside with changing frequency. The bandwidth of this type of array is rather narrow and in general only about 1%–2% [11]. As additional patches are added, the impedance bandwidth of a series array narrows.

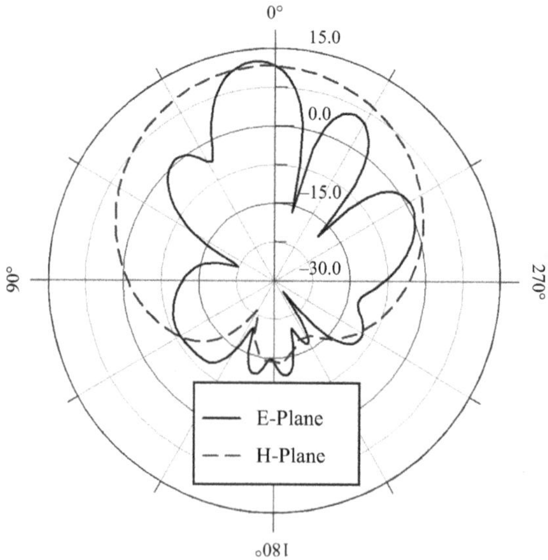

Figure 6.11 Four patch series fed planar array with uniform element excitation

The previous design example has patches of identical width. When this is the case, all the elements have the same excitation amplitude.

If we wish to produce an amplitude taper to decrease the array SLL, we may change the widths of each patch to accept a designated amount of power. The power accepted by the nth element of a series array is

$$P_n = 2g_n V^2 \tag{6.28}$$

The value g_n is the conductance of each antenna n normalized to the desired driving point resistance, which in this case is 50 Ω:

$$g_n = 2G_n \cdot Z_{drv} = \frac{2G_n}{G_{drv}} \tag{6.29}$$

V is the voltage across each element. The power radiated by an element is proportional to the square of the electric field. The electric field is proportional to the excitation coefficient I_n. We can produce an amplitude distribution along a series-fed array by choosing the patch conductivities to be proportional to the desired amplitude excitation coefficients:

$$g_n = KI_n^2 \tag{6.30}$$

where K is a constant of proportionality. The elements are all spaced between half-wavelength microstrip transmission line sections, so the input conductance (normalized) of the array is the sum of the element conductances:

$$g_{in} = \sum_{n=1}^{N} g_n \tag{6.31}$$

where $g_n = 2.0 \cdot 50.0\ \Omega \cdot G_{e_n}$ and N is the number of elements in the series array. For normalized conductances, the condition for input match is

$$g_{in} = \sum_{n=1}^{N} g_n = 1 \tag{6.32}$$

we can substitute (6.30) into (6.32) and obtain K as

$$K = \left(\sum_{n=1}^{N} I_n^2\right)^{-1} \tag{6.33}$$

The value of K allows one to compute the element conductances g_n using the desired amplitude values I_n. The element conductances allow one to compute patch widths required to present these desired conductances and in turn the designated amplitude distribution values.

A simple illustration of how to introduce an aperture taper into a series array is to begin with the uniform array example examined previously. The uniform array has four elements, each element contributes the same conductance (i.e. $G_n = constant$). A realistic taper for the physical realization of a four-element series array is to widen the two center elements by a factor of 1.5 and reduce the outer elements to one-half of the original width. When this is done, the sum of the conductances remains constant,

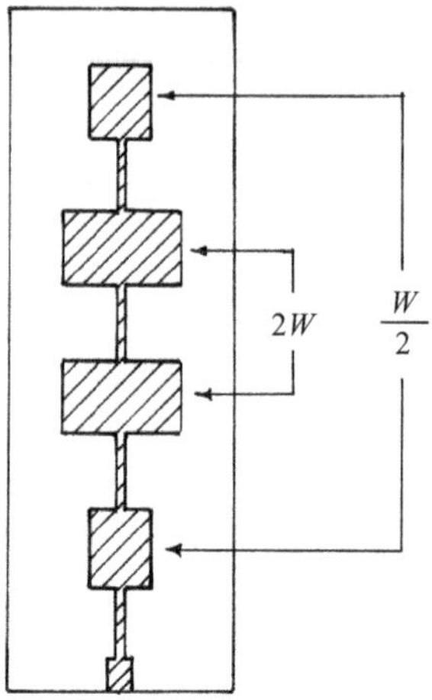

Figure 6.12 Four patch series-fed planar array of Figure 6.10 with the center two elements widened to 2W and the outer-most elements reduced to a width of W/2. This modification creates a −4.04 dB amplitude taper. The radiation patterns of this array are presented in Figure 6.13

and the series array remains matched as it was in the uniform conductance example. FDTD analysis results demonstrate the input match, and the resonant frequency of 5.09 GHz remains constant when the array is modified in this way. The modified array is seen in Figure 6.12. This tapering of the patch widths translates to a −4.04 dB linear taper. The array directivity is 12.86 dB.

The pattern computed by FDTD is shown in Figure 6.13. We can see the sidelobe on the left side at approximately 45° has all but vanished. The sidelobe on the right, at approximately 30°, has been reduced considerably.

If one chooses to widen the inner two patches further and narrow the outer two patches so a 50 Ω driving point impedance is maintained, the narrow patches will be more affected by the feedlines as their dimension decreases.

A second option for a series-fed array is to feed the antennas along non-radiating edges. One can choose the position of each input and output location on each patch to provide a desired amplitude taper. The design of this type of array is very involved mathematically and details may be found in the literature [12–14].

6.4.3 Series/parallel standing wave feed

A third method of feeding a microstrip array is to combine series and parallel transmission lines [15]. We will examine three versions of this topology. The first is a standing wave feed design. The geometry and transmission line equivalent is given in Figure 6.14. The series transmission line has a characteristic impedance of Z_s. Parallel sections of transmission line are placed one guide wavelength ($1\lambda_g$) apart along the series transmission line. The length of each of these parallel sections of transmission

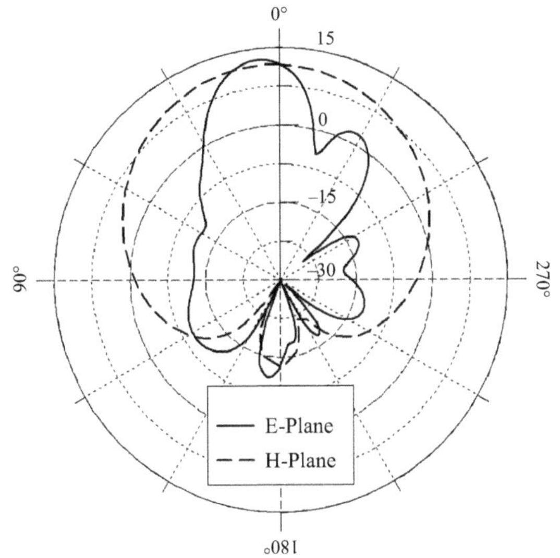

Figure 6.13 Four patch series fed planar array modified to have a −4.04 dB linear taper

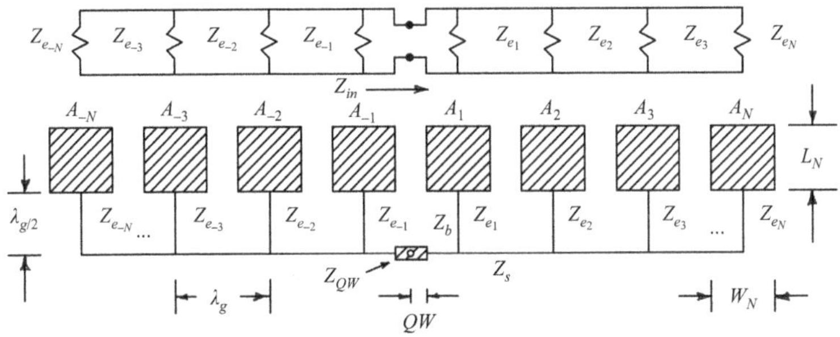

Figure 6.14 Series/parallel standing wave feed of a microstrip antenna array. The array consists of 2N elements. A single series microstrip transmission line of characteristic impedance Z_s has shunt microstrip lines spaced λ_g apart that feed each individual patch antenna. The line lengths are $\lambda_g/2$, and so the edge impedance of each patch at resonance, Ze, appears across the series feedline. A quarter-wave transformer is used on each side to produce a 100 Ω impedance, which add in parallel to produce a 50 Ω driving point impedance

line is one-half a guide wavelength ($\lambda_g/2$) long. Each section of parallel line terminates at the edge of a microstrip antenna. Each antenna on the right is designated $A_1, A_2, A_3, \ldots, A_N$. For each antenna on the right, there is a mirror image element $A_{-1}, A_{-2}, A_{-3}, \ldots, A_{-N}$. The total number of elements in this linear array is $2N$.

Because the length of the parallel transmission line feeding each patch is one-half guide wavelength, the edge impedance, Z_e, appears across the series transmission line.

For the general case, each antenna element can have its own edge impedance, designated $Z_{e_1}, Z_{e_2}, Z_{e_3}, \ldots, Z_{e_N}$. These edge impedances are mirrored on the left side and designated as $Z_{e_{-1}}, Z_{e_{-2}}, Z_{e_{-3}}, \ldots, Z_{e_{-N}}$. Each antenna has a length $L_1, L_2, L_3, \ldots, L_N$ on the right side and $L_{-1}, L_{-2}, L_{-3}, \ldots, L_{-N}$ on the left. The widths of each element are $W_1, W_2, W_3, \ldots, W_N$ on the right and $W_{-1}, W_{-2}, W_{-3}, \ldots, W_{-N}$ on the left.

A quarter-wave transformer is introduced at the center of the array to change the input impedance Z_{in} from the left and right sections to 100 Ω at the middle to produce a 50 Ω driving point impedance.

We first examine the case where all the antenna widths and lengths are identical. When this is the case, all the edge resistances R_e (i.e. $Z_e = R_e$) at resonance are all equal. The impedance Z_{in} into the right hand section is

$$Z_{in} = Z_e/N = R_e/N \tag{6.34}$$

The characteristic impedance of a quarter-wave transformer that will produce 100 Ω at the right-hand side of the driving point is

$$Z_{QW} = \sqrt{Z_e/N \cdot 100\ \Omega} \tag{6.35}$$

The equations are valid for the left-hand side of the array as it is symmetrical.

As an example, we will choose six identical elements at 5 GHz ($N = 3$). We will use a dielectric substrate with a thickness of 1.524 mm, $\epsilon_r = 2.2$, with $\tan\delta = 0.0009$. At resonance, the edge impedance becomes pure real and $Z_e = R_e = 184.48\ \Omega$. As all the elements are identical, we can use (6.34) and obtain an approximate impedance before the quarter-wave transformer of 61.49 Ω. A 78.42 Ω quarter-wave transformer will produce a 100 Ω input on the left side section of the array. The left side is identical to create a 50 Ω driving point impedance.

These values are good for reference, but a full-wave software package provides accurate values for a more exact design. In this case, the patch edge resistance (R_e) at resonance was about 295 Ω, and when computed at the end of a half-wavelength branch line, increased to about 331 Ω. The input impedance on the right-hand side was serendipitously 25 Ω, which for a 100 Ω transformation is a quarter-wave transformer of 50 Ω, producing a 50 Ω driving point impedance at the center.

The example design values for the equal patch dimension array are $W = L =$ 18.843 mm (square patch). The length of the 100 Ω connecting branch (tap) lines is $\lambda_g/2 = 22.640$ mm with a trace width of 1.353 mm. The branch line separation along the 50 Ω series feed is $\lambda_g = 43.684$ mm with a trace width of 4.637 mm.

These types of arrays have narrow bandwidth. HFSS analysis predicts a 1.76% bandwidth (88.2 MHz) around the 5 GHz design frequency. The predicted gain of the equal patch array is 14.38 dBi, with an efficiency of 91.90%.

As was done in Section 6.4.2, we can alter the uniform amplitude distribution by narrowing the patch widths. Altering patch widths changes the resonant length, and this must be compensated for on each patch. The load presented at each branch along the series feed will be different, and the impedance presented to the center quarter-wave transformer will be the sum of the edge conductances G_e ($1/G_e = R_e$).

We will use the same design configuration as in the previous example, with $N = 3$. The innermost array element will be square, the outermost will be half that of

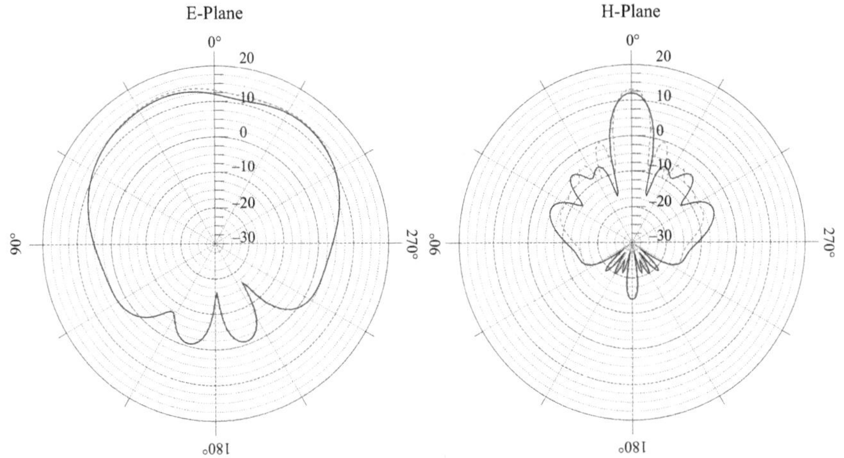

Figure 6.15 Comparison of radiation pattern for uniform patch widths (dashes) and tapered patch widths (solid) examples

the square patch. The center patch will be the geometric mean or 0.707 of the square patch width. The resistance presented at the bottom of each branch feed is 335 Ω, 547 Ω, and 880 Ω. A 40 Ω quarter-wave transformer was used to produce a 100 Ω impedance on the right-hand side. The array was mirrored on the left-hand side.

The design details are patch A_1 has $W_1 = L_1 = 18.771$ mm, A_2 has $W_2 = 13.277$ mm and $L_2 = 19.012$ mm, and A_3 has $W_3 = 9.366$ mm with $L_3 = 19.328$ mm. The half-wave branch lengths are 22.726 mm, and the 50 Ω microstrip guide wavelength is $\lambda_g = 43.850$ mm. The width of the 50 Ω line is 4.637 mm, and the 100 Ω line is 1.353 mm. The quarter-wave transformer has a width of 6.359 mm and a length of 11.112 mm.

The analysis predicts a 1.27% bandwidth (63.7 MHz) with a resonant frequency of 5 GHz. The predicted gain of the tapered patch array is 13.73 dBi, with an efficiency of 89.35%.

The radiation patterns of the two example array designs are presented in Figure 6.15. We see a 5 dB improvement in the SLL for the patch array with tapered patch widths versus uniform patch widths. A series/parallel-fed array of this type has the disadvantage of very narrow bandwidth.

6.4.4 Series/parallel matched tapped feed array

An alternative to the series-parallel standing wave feed of Section 6.4.3 is to use back-to-back quarter-wave transformers for matching and aperture taper. A six-element tapped S-P array is shown in Figure 6.16.

The center of the array is fed with a 50 Ω SMA probe feed. A pair of 100 Ω microstrip lines from the right and left sides of the array add in parallel at the driving point to provide a 50 Ω match at the center. Each antenna/matching unit, except for the final antenna on each end, has the geometry, as shown in Figure 6.17.

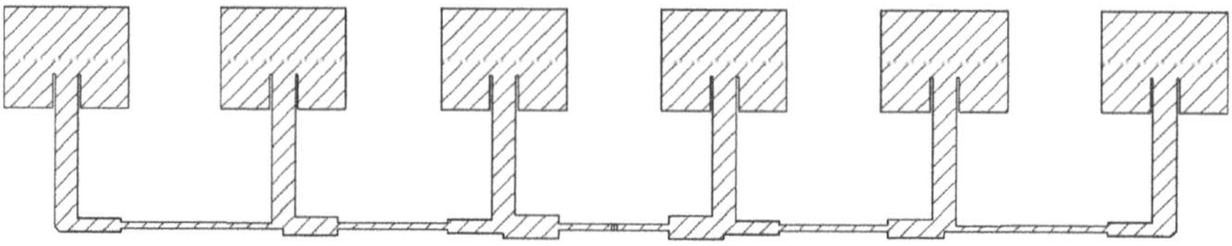

Figure 6.16 A series/parallel matched tapped feed array uses a 100 Ω microstrip transmission line between antenna/matching units. Each unit determines the amount of current provided to each antenna along the array

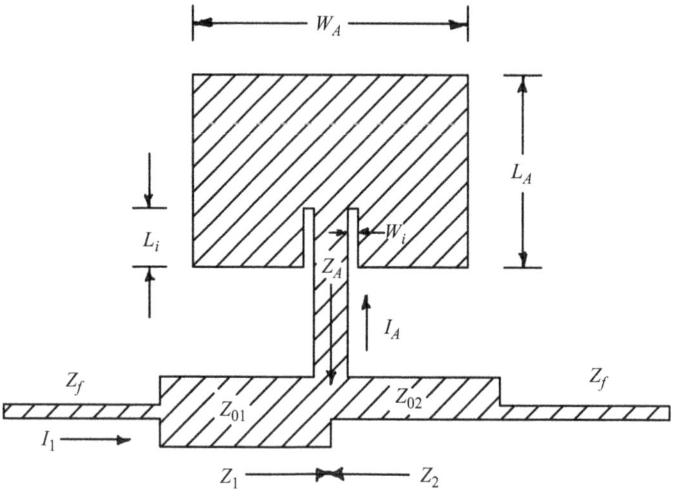

Figure 6.17 Series/parallel antenna matching unit

The ratio of the current arriving from the left, I_1, to the amount of current diverted to the antenna element I_A is given by

$$\frac{I_1}{I_A} = \frac{Z_A}{Z_1} = N \tag{6.36}$$

where Z_A is the driving point impedance of the inset patch antenna at the junction of the back-to-back quarter-wave transformers. $1/N$ is the split ratio.

When the antenna driving point impedance is given, and the number of elements needed on each side is known, we can determine the value of Z_1 using:

$$Z_1 = \frac{Z_A}{N} \tag{6.37}$$

The value of Z_2 may then be determined using:

$$Z_2 = \frac{Z_1}{\left(1 - \frac{1}{N}\right)} \tag{6.38}$$

The characteristic impedances of the quarter-wave transformers, Z_{01} and Z_{02}, are given by

$$Z_{01} = \sqrt{Z_f \cdot Z_1} \tag{6.39}$$

and

$$Z_{02} = \sqrt{Z_2 \cdot Z_f} \tag{6.40}$$

The final antenna on the right and left ends is matched with a single quarter-wave transformer.

As an example, we will design a six-element array with the same inter-element spacing as the array of Section 6.4.3. We will choose a 100 Ω microstrip line for interconnecting between units. Each inset patch will have a driving point impedance of 50 Ω and be connected with 50 Ω microstrip transmission line to the matching unit. There will be two symmetric matching units on the left and right sides of the array, or $N = 2$. The first antenna will have 0.5 current (1/2) $N = 2$, the second antenna 0.707 current (1/1.4142) $N = 1.4142$, and the last element will receive 0.25 (1/4) $N = 4$. The first unit will have $Z_1 = 25$ Ω, $Z_2 = 50$ Ω, with $Z_{01} = 50$ Ω and $Z_{02} = 70.71$ Ω. The second unit will have $Z_1 = 35.3$ Ω, $Z_2 = 120.07$ Ω, with $Z_{01} = 59.41$ Ω and $Z_{02} = 109.58$ Ω. The final quarter-wave transformer at the end transforms 100 Ω to 50 Ω and is therefore 70.71 Ω. The dielectric material has $\epsilon_r = 2.2$ with $\tan\delta = 0.0009$ and a thickness of 1.524 mm. The width of each inset patch is 25 mm (W_A) with a length of 19.5129 mm (L_A). The inset distance (L_i) is 6.5 mm from the radiating edge with a 0.5 mm gap (W_i) on each side the total length of the 50 Ω branchline is 28.92 mm. The spacing between elements is 43.684 mm.

An HFSS analysis predicts a 2:1 VSWR bandwidth of 157.5 MHz or 3.15%. This is considerably better than the 1.27% impedance bandwidth of the standing wave array. The gain of the array is 14.4 dBi, with an efficiency of 93.40%.

The E- and H-plane radiation pattern of the tapped array is shown in Figure 6.18. The sidelobes are around −19 dB below the main beam.

Each antenna in this design may possess a different driving point resistance at resonance, which offers an extra design parameter. The patches could also be rotated and fed along a non-radiating edge, which would also rotate the polarization by 90°.

6.4.5 Feedline radiation and loss

The feedline radiation and loss from microstrip antenna arrays, which are corporate fed, was studied by Levine *et al.* [16]. Their general conclusions are

1. The transmission line losses are inversely proportional to the characteristic impedance of the microstrip transmission line $\left(\frac{1}{Z_0}\right)$.
2. The transmission line radiation loss and surface wave excitation are weakly dependent on the characteristic impedance (Z_0) of the microstrip transmission line. This implies that the transmission line losses, which are inversely proportional to Z_0, dominate. This implies that as large of values of characteristic impedance should be chosen for the feedlines as is practical.

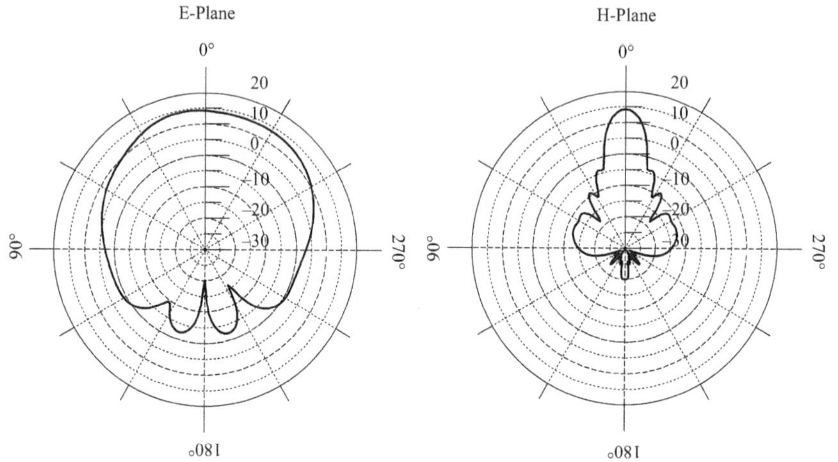

Figure 6.18 E- and H-plane pattern of the example tapped array

3. Microstrip transmission lines with fixed dimensions (length and width) have radiation losses, which increase as $(H\sqrt{\epsilon_r}/\lambda_0)^2$. Surface wave losses increase as $(H\sqrt{\epsilon_r}/\lambda_0)^3$. These are approximate relationships which are good for substrates where $H\sqrt{\epsilon_r}/\lambda_0 \approx 0.1$ and $\epsilon_r \approx 2.2$. In order to decrease radiation losses from the feedlines requires minimizing the value of $H\sqrt{\epsilon_r}/\lambda_0$.
4. Microstrip transmission lines with fixed thickness and characteristic impedance have radiation losses which depend on their length. When the length is from $0 < L < \lambda_0$ the loss grows in proportion to $(L/\lambda_0)^2$. When the line length is greater than three free space wavelengths $L > 3\lambda_0$, they are not sensitive to length. The surface wave loss as a function of line length becomes oscillatory.
5. The radiation losses from end-fed microstrip transmission lines are larger than those from center-fed microstrip lines. The reason for this is that center-fed lines have currents which are oppositely directed and tend to have radiation which cancel each other in the far-field. The surface wave losses are nearly identical for end-fed and center-fed lines.
6. With $\epsilon_r \approx 2.2$ and $H\sqrt{\epsilon_r}/\lambda_0 \approx 0.08$ and a Z_0 of 200 Ω, typical losses are approximately 3% for center-fed microstrip transmission lines, and 5% for end-fed microstrip transmission lines. As stated previously, these losses are inversely proportional to the characteristic impedance, directly proportional to the square of the substrate thickness, and insensitive to length.

The contributions from ohmic and dielectric losses are as described in Appendix C. The authors also indicate that

> It is interesting to notice that, once the substrate is chosen, the dissipation losses can be reduced by choosing feedlines with low impedances, but the radiation and surface wave losses would become higher. As a result, the total loss is not sensitive to the chosen impedances within the range of 100–200 Ω.

Table 6.2 Microstrip corporate array losses versus dish antenna (50% efficiency) all values in dB. 10 GHz microstrip modular arrays with $\epsilon_r = 2.2$, substrate thickness = 1.6 mm with $0.8\lambda_0$ spacing [16]

Number of elements	16	64	256	1024	4096
Directivity	20.9	27.0	33.0	39.2	45.1
Radiation loss	0.8	1.0	1.3	1.9	2.6
Surface wave loss	0.3	0.3	0.2	0.2	0.1
Dielectric loss	0.1	0.3	0.5	1.0	2.1
Ohmic loss	0.1	0.3	0.6	1.2	2.4
Total loss	1.3	1.9	2.6	4.3	7.2
Calculated gain	19.5	25.0	30.0	34.5	37.5
Gain of reflector	18.0	24.0	30.0	36.0	42.0

Making an antenna array which is as symmetric as possible is generally a good practice for reducing unintentional radiation from feedlines. The authors also note that losses from the feedlines of a corporate-fed array increase substantially as the array size increases.

They also compare the performance of microstrip arrays with corresponding dish antennas for 16, 64, 256, 1024, and 4096 element square arrays (4×4, 8×8, 16×16, 32×32, and 64×64). The results of this analysis are found in Table 6.2. The 16, 64, and 256 element microstrip arrays are similar in performance to comparable dish antennas. The array with 1024 elements is 1.5 dB worse than a dish, but by 4096 elements the difference increases to 4.5 dB giving the dish a considerable advantage.

6.4.6 Microstrip transmission line radiation

The amount of radiation from microstrip transmission lines has been analyzed, and equations that estimate the amount were derived in 1979 [17]. The attenuation constant, α_r, for the radiation loss of a microstrip transmission line in nepers/meter is

$$\alpha_r = 60\left(\frac{2\pi H}{\lambda_0}\right)^2 F(\epsilon_e) \tag{6.41}$$

For the case of a terminated microstrip transmission line, $F(\epsilon_e)$ is

$$F(\epsilon_e) = 1 - \frac{\epsilon_e - 1}{2\sqrt{\epsilon_e}} \log_{10}\left(\frac{\sqrt{\epsilon_e}+1}{\sqrt{\epsilon_e}-1}\right) \tag{6.42}$$

For the case of an open-circuited microstrip line,

$$F(\epsilon_e) = \frac{\epsilon_e + 1}{\epsilon_e} - \frac{(\epsilon_e - 1)^2}{2\epsilon_e^{3/2}} \log_{10}\left(\frac{\sqrt{\epsilon_e}+1}{\sqrt{\epsilon_e}-1}\right) \tag{6.43}$$

where ϵ_e is the effective relative permittivity of the microstrip transmission line, H is the substrate thickness, and λ_0 is the free space wavelength.

Tables 6.3, 6.4, and 6.5 provide an illustrative touchstone. The computed radiation loss in dB/100 mm is given for a 50 Ω line at 2.5, 5, 10, 28, and 60 GHz. This is

Table 6.3 50 Ω microstrip transmission line estimated radiation loss [17], $\epsilon_r = 1.1$ (17 μm thick copper)

50 microstrip transmission line radiation loss (foam)						
H	**2.5 GHz**	**5 GHz**	**10 GHz**	**28 GHz**	**60 GHz**	**Loss**
126 μm	0.00	0.01	0.03	0.27	1.22	dB/100 mm
254 μm	0.01	0.03	0.14	1.10	4.94	dB/100 mm
508 μm	0.03	0.14	0.55	4.30	19.74	dB/100 mm
762 μm	0.08	0.31	1.24	9.67	44.16	dB/100 mm
1524 μm	0.31	1.24	4.94	38.47	176.64	dB/100 mm

Table 6.4 50 Ω microstrip transmission line estimated radiation loss [17], $\epsilon_r = 2.6$ (17 μm thick copper)

50 microstrip transmission line radiation loss ($\epsilon_r = 2.6$)						
H	**2.5 GHz**	**5 GHz**	**10 GHz**	**28 GHz**	**60 GHz**	**Loss**
126 μm	0.00	0.01	0.03	0.18	0.94	dB/100 mm
254 μm	0.01	0.03	0.11	0.83	3.75	dB/100 mm
508 μm	0.03	0.11	0.42	3.27	14.79	dB/100 mm
762 μm	0.06	0.24	0.95	7.29	33.08	dB/100 mm
1524 μm	0.24	0.95	3.74	28.84	131.81	dB/100 mm

Table 6.5 50 Ω microstrip transmission line estimated radiation loss [17], $\epsilon_r = 10.2$ (17 μm thick copper)

50 microstrip transmission line radiation loss ($\epsilon_r = 10.2$)						
H	**2.5 GHz**	**5 GHz**	**10 GHz**	**28 GHz**	**60 GHz**	**Loss**
126 μm	0.00	0.01	0.02	0.17	0.80	dB/100 mm
254 μm	0.01	0.02	0.09	0.71	3.21	dB/100 mm
508 μm	0.02	0.09	0.36	2.80	12.72	dB/100 mm
762 μm	0.05	0.20	0.81	6.26	28.54	dB/100 mm
1524 μm	0.20	0.81	3.21	24.87	113.94	dB/100 mm

done for typical substrate thicknesses. Table 6.3 has the radiation from a microstrip with a typical foam substrate. At 2.5 GHz, the losses are small. When the frequency is increased to 5 GHz, the thickest substrate has losses that are large enough for concern. At 10 GHz, the two thinnest thicknesses would be the best choice for minimizing radiation from microstrip transmission lines. At 28 GHz, only the thinnest is a candidate. By the time 60 GHz is reached, a substrate which is thinner than the thinnest example thickness would be sought.

Table 6.4 shows that when a substrate's relative permittivity is increased to 2.6, the radiation loss decreases. The amount is not large, and while it would decrease the amount of radiation loss, it would probably not be significant enough to justify moving from a foam substrate to a solid dielectric substrate.

Table 6.5 increases the substrate relative permittivity to 10.2. The decrease in transmission line radiation is again probably not large enough to move the useful thickness up by a standard thickness.

Generally, it is of interest to minimize the spurious radiation from the microstrip network that feeds a phased array. It is also generally desired to maximize the efficiency and bandwidth of microstrip radiators. In order to minimize feed network radiation, one should decrease the thickness of the substrate as much as possible, and increase its relative permittivity. Using a thinner substrate decreases the bandwidth and efficiency of microstrip antennas, which is the opposite of what is generally desired. A rule of thumb for the maximum substrate thickness that mitigates microstrip transmission line radiation is about $\lambda_0/25$ or less.

Often this conflicting set of design parameters are mitigated with a two-layer PCB. An upper substrate with a low relative permittivity may be used to realize the microstrip antennas for maximum bandwidth. A thin high dielectric permittivity substrate can be used on the back side (non-element side) to minimize feed network radiation. The amount of non-radiating transmission line loss increases as the substrate is thinned, so an optimum balance of radiated, ohmic/dielectric loss, and surface wave generation exists for feed network thickness. The top layer microstrip antennas and microstrip feed network have a shared groundplane between them in this scenario. Vias can be introduced to provide probe feeding from the feed network side to the patch antenna side. One can alternatively use stripline on the back side to further minimize unwanted feedline radiation.

6.5 Mutual coupling

When multiple microstrip antenna elements are arrayed, the elements will couple to one another. One mechanism by which coupling occurs is surface wave generation. One can use the analysis presented in Section 2.6 to minimize surface wave generation.

Often with element spacings encountered in practice, the amount of coupling between microstrip elements is small enough to be neglected. When inter-element coupling is significant enough to be included, often in practice, measured values of coupling are used in place of analysis. The availability of full-wave analysis tools allow one to compute mutual coupling with relative ease compared with the approximate analysis offered here.

One can analyze the effects of mutual coupling using network methods [18,19]. The voltage and current at the driving point of each element in an antenna array with the coupling of all the other elements included is related by (6.44):

$$\begin{vmatrix} V_1 \\ V_2 \\ V_3 \\ \vdots \\ V_{N-1} \\ V_N \end{vmatrix} = \begin{vmatrix} Z_{11} & Z_{12} & Z_{13} & \dots & Z_{1N} \\ Z_{21} & Z_{22} & Z_{23} & \vdots & Z_{2N} \\ Z_{31} & Z_{32} & Z_{33} & \vdots & Z_{3N} \\ \vdots & \vdots & \vdots & \vdots & \vdots \\ Z_{N,1} & Z_{N,2} & Z_{N,3} & \dots & Z_{N,N} \end{vmatrix} \begin{vmatrix} I_1 \\ I_2 \\ I_3 \\ \vdots \\ I_{N-1} \\ I_N \end{vmatrix} \tag{6.44}$$

Each row of the matrix equation can be written out. The voltage at the driving point of element 1 of an array of N elements becomes

$$V_1 = Z_{11}I_1 + Z_{12}I_2 + Z_{13}I_3 + \ldots + Z_{1N}I_N \tag{6.45}$$

We can divide both sides of (6.45) by I_1 which produces an equation which relates the driving point impedance of element 1 in terms of the ratio of currents in each of the other elements to the current in element 1. This equation is called the *active impedance* of element 1 (Za_1):

$$Za_1 = Z_{11} + Z_{12}\frac{I_2}{I_1} + Z_{13}\frac{I_3}{I_1} + \cdots + Z_{1N}\frac{I_N}{I_1} \tag{6.46}$$

In general, for each element n, with $m = 1, 2, 3, \ldots, \acute{N}$, we write

$$Za_n = \frac{V_n}{I_n} = \sum_{m=1}^{\acute{N}} \frac{I_m}{I_n} Z_{mn} + Z_{nn} \tag{6.47}$$

The $\acute{N}$ in (6.47) means the sum which *excludes* the $m = n$ term.

The currents at each of the elements is unknown initially, but we may use (6.47) to iterate to a convergent solution starting with an initial guess at the driving point currents.

The starting current for the array can be computed by dividing the driving point voltages by each antenna's self-impedance neglecting coupling:

$$|I_n^{s1}| = \left|\frac{E_n}{Z_{nn} + Z_0}\right| \tag{6.48}$$

We can then use (6.47) to calculate a new active impedance. After the active impedance has been calculated, we calculate a new current distribution, keeping the voltage distribution constant. The current at the kth iteration is

$$|I_n^{fk}| = \left|\frac{E_n}{Z_{an} + Z_0}\right| \tag{6.49}$$

The superscript f is the final current for iteration k which drives element n. A new starting current for iteration $k + 1$ is found with

$$|I_n^{s(k+1)}| = \frac{1}{2}\left|I_n^{sk} + I_n^{fk}\right| \tag{6.50}$$

At each iteration an error is evaluated using:

$$\text{Error} = \sum_{i=1}^{N} \left|\text{Re}\left[I_i^{fk}\right] - \text{Re}\left[I_i^{sk}\right]\right|^2 + \left|\text{Im}\left[I_i^{fk}\right] - \text{Im}\left[I_i^{sk}\right]\right|^2 \tag{6.51}$$

The mutual coupling terms may be computed using the cavity model as related in Section 6.5.1. After the currents have been calculated, the techniques of Sections 6.1,

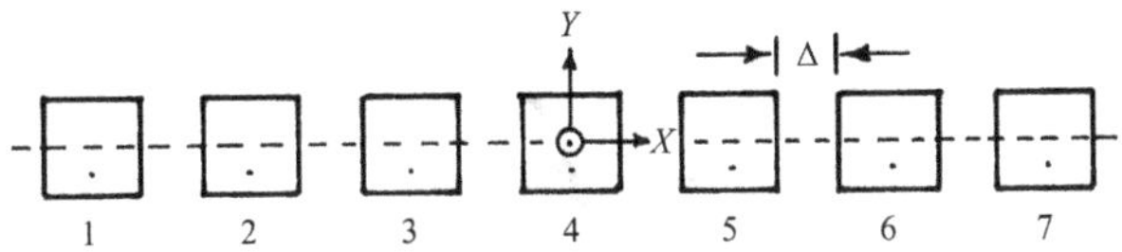

Figure 6.19 *Seven-element rectangular microstrip array with H-plane mutual coupling (electric field in the y direction) (dots show feed points)*

6.2, and 6.3, presented previously, are used to compute the radiation pattern of the array.

Example: We will use a seven-element linear array of rectangular microstrip antennas to illustrate the effects of mutual coupling.

The geometry of the elements is illustrated in Figure 6.19. The polarization is directed along the y-axis. The patches all have identical dimensions. The resonant length of each patch is $a = 50.0$ mm with a width of $b = 60.0$ mm. The substrate thickness is $H = 1.575$ mm with $\epsilon_r = 2.50$ and a 0.0018 $\tan\delta$. The frequency is 1.560 GHz. These values are consistent with those presented by Jedlinka and Carver.

The computation of mutual coupling between rectangular microstrip elements using the cavity model is very sensitive to the value of the wall admittance used in the computation. In many practical cases, one can measure the mutual coupling of a fabricated prototype array, or employ a full-wave analysis method, to determine more accurate mutual coupling values.

We will illustrate the effects of mutual coupling on the radiation pattern of a seven-element array using the cavity model. Section 6.5.1 provides an outline of the computation of mutual coupling between a pair of rectangular microstrip antennas using the cavity model.

In Figure 6.20(Top), the seven-element array is excited with a uniform voltage distribution. When no mutual coupling is present the directivity is maximum. When mutual coupling is included with an edge-to-edge spacing of $\Delta = 0.8a$ (where a is the patch width) the directivity decreases, as does the SLL. This trend continues as the spacing between non-radiating edges is decreased to $0.6a$.

Figure 6.20(Bottom) presents the patterns of the seven-element array with a voltage excitation which has a -6 dB linear taper on a pedestal. The pattern computed with no mutual coupling is shown and when the coupling with an edge-to-edge separation of $0.8a$ is included we see that as before the directivity decreases, but the SLL increases. At $\Delta = 0.6a$, the coupling is such that the current distribution is almost identical to that without mutual coupling. When the spacing is decreased to $0.4a$, the directivity decreases.

6.5.1 Mutual coupling between square MSAs

The mutual coupling between two microstrip antennas may be calculated using the following relationship [20]:

$$Z_{21} = \frac{1}{I_1 I_2} \oint \vec{H}^{patch1} \cdot \vec{M}^{patch2} \, dl_2 \tag{6.52}$$

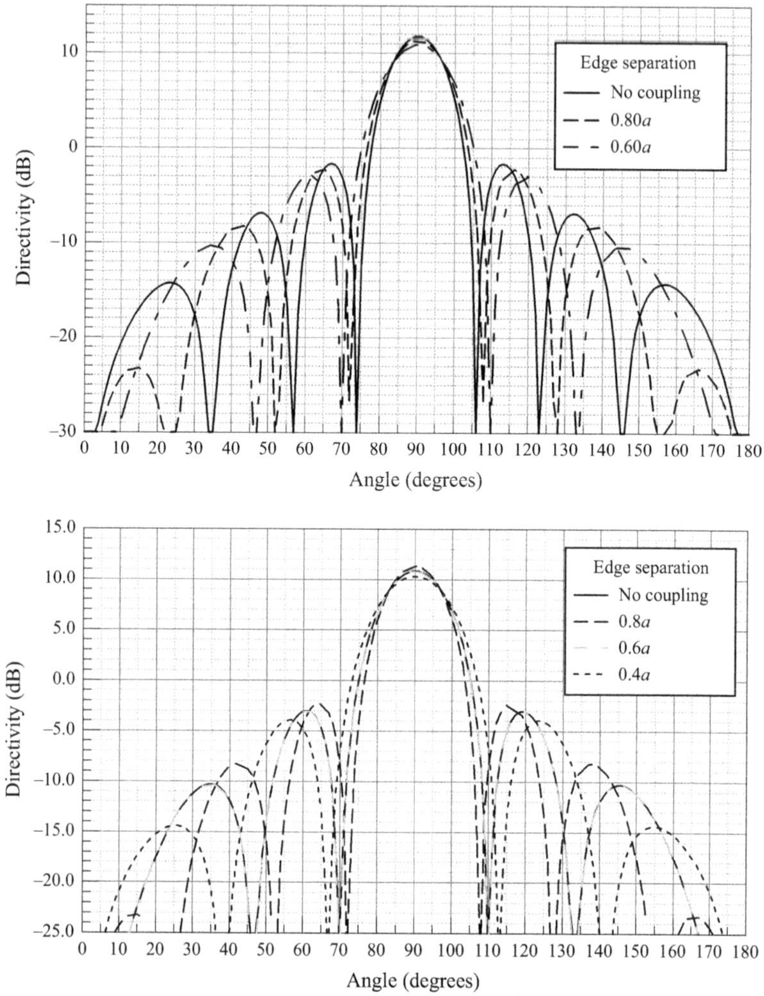

Figure 6.20 (Top) Seven patch rectangular microstrip array with uniform excitation with effects of mutual coupling computed with the cavity model. (Bottom) Seven patch rectangular microstrip array with -6 *dB tapered excitation with effects of mutual coupling computed with the cavity model*

One may use the cavity model with (6.52) to obtain the mutual impedance between probe-fed microstrip antennas. I_1 and I_2 are the currents at the feeds of the patches. $\vec{H}^{patch1}$ is the magnetic field on antenna number 2 produced by antenna number 1. $\vec{M}^{patch2}$ is the linear magnetic current density on antenna number 2 when it has been self-excited. The integration is over the perimeter of antenna number 2.

The dimensions and geometry for this analysis are defined in Figure 6.21.

The patches are located in the x–y plane. The center of patch 1 is the origin of the coordinate system. The center of patch two is at $\vec{r}_0 = X_0\hat{i} + Y_0\hat{j}$. We calculate $\vec{H}^{patch1}$ from the magnetic current at the edge of patch 1. The equivalent magnetic line currents at the edge of the cavity are related to the field at the boundary of the cavity by

$$\vec{M}^{patch2} = 2HE_z(x,y)\hat{z} \times \hat{n} \tag{6.53}$$

The unit vector $\hat{n}$ is an outward normal at the cavity boundary H is the substrate thickness. The interior electric field is calculated using the cavity model. This relationship is given by

$$E_z(x,y) = \frac{2jI_0\omega \cos\frac{\pi y}{b} \cos\frac{\pi y_p}{b}}{\epsilon^* ab\left[\omega^2 - (\omega_r + j\omega_i)^2\right]} \tag{6.54}$$

where $\epsilon^* = \epsilon_r(1 - j\tan\delta)$, ω is the angular frequency (rad/s), a is the width of patch in the x-plane, b is the width of patch in the y-plane, y_p is the y-coordinate of the feed probe position, and I_0 is the driving point (i.e. feedpoint) current, and $(\omega_r + j\omega_i)$ is the complex resonant frequency.

The normal vectors, magnetic current directions and numbering of patch edges is illustrated in Figure 6.21. The resulting magnetic current about the patches is shown in Figure 6.22.

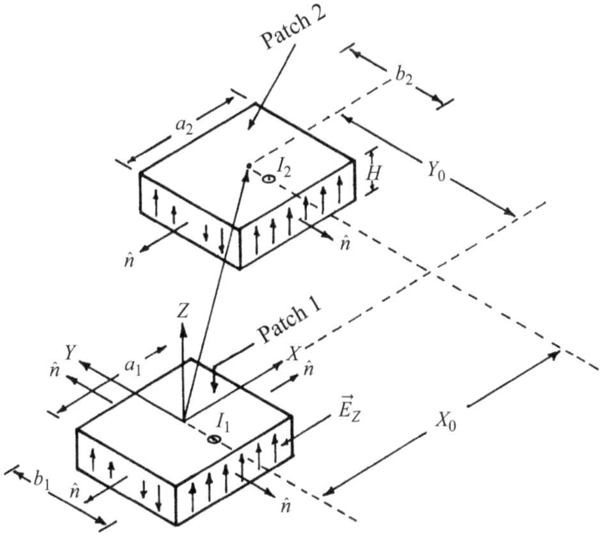

Figure 6.21 Geometry of the cavity model of two rectangular microstrip patch antennas in the x–y plane, with distance between antenna centers, used to compute the mutual coupling using the cavity model

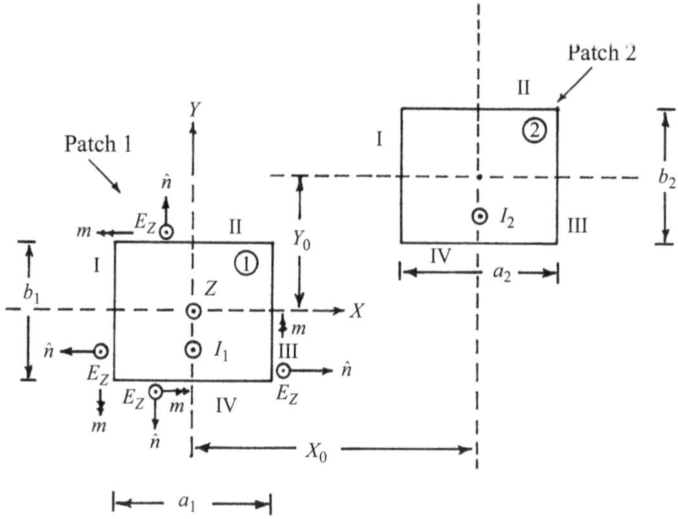

Figure 6.22 Direction of magnetic current on each side of a rectangular microstrip patch antenna defined by (6.53) with the cavity model

The magnetic field radiated by a small $\hat{z}$-directed magnetic current $\vec{M}$ of length Δz is given by [21]:

$$\vec{H}(r,\theta) = j\omega\epsilon\frac{\vec{M}\,\Delta z}{4\pi}\left[1+\frac{1}{jkr}+\frac{1}{(jkr)^2}\right]\frac{e^{-jkr}}{r}\sin\theta\,\hat{\theta} + j\omega\epsilon\frac{\vec{M}\,\Delta z}{2\pi}\left[1+\frac{1}{(jkr)^2}\right]\frac{e^{-jkr}}{r}\cos\theta\,\hat{r} \tag{6.55}$$

For a magnetic current of differential length dz, the differential magnetic field simply becomes

$$d\vec{H}(r,\theta) = j\omega\epsilon\frac{\vec{M}dz}{4\pi}\left[1+\frac{1}{jkr}+\frac{1}{(jkr)^2}\right]\frac{e^{-jkr}}{r}\sin\theta\,\hat{\theta} + j\omega\epsilon\frac{\vec{M}dz}{2\pi}\left[1+\frac{1}{(jkr)^2}\right]\frac{e^{-jkr}}{r}\cos\theta\,\hat{r} \tag{6.56}$$

The magnetic field radiated by patch 1 along some vector $\vec{r}$ may be calculated by integrating the field contributions of the magnetic current from each of its sides:

$$\vec{H}(r,\theta) = j\frac{\omega\epsilon}{4\pi}\int_{patch\ 1\ edges}\vec{M}(z)\left[1+\frac{1}{jkr}+\frac{1}{(jkr)^2}\right]\frac{e^{-jkr}}{r}\sin\theta\,dz\,\hat{\theta} + j\frac{\omega\epsilon}{2\pi}\int_{patch\ 1\ edges}\vec{M}(z)\left[1+\frac{1}{(jkr)^2}\right]\frac{e^{-jkr}}{r}\cos\theta\,dz\,\hat{r} \tag{6.57}$$

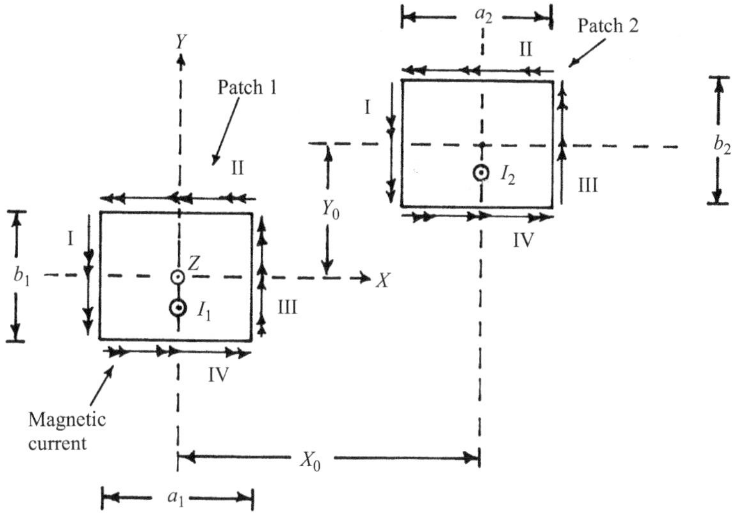

Figure 6.23 Magnetic current around rectangular patch 1 which with (6.53) describes the magnetic field at any point along the perimeter of patch 2. With the magnetic field at patch 2 due to patch 1 and the magnetic current of patch 2 (6.52) is used to compute Z_{21}

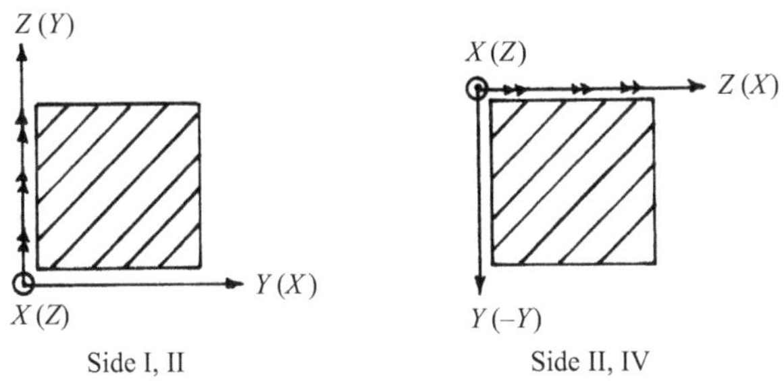

Figure 6.24 Coordinate transformations which facilitate the computation of mutual coupling using (6.52)

The $\hat{z}$-directed current has fields expressed in spherical coordinates (Figures 6.23 and 6.24). In order to facilitate the dot product with the magnetic current around patch 2, rectangular coordinate values are calculated using the expressions below with ($H_\phi = 0$ and $\phi = 90$ when $\vec{r}$ is in the magnetic currents $\hat{z}$–$\hat{y}$ plane):

$$H_y = H_r \sin\theta + H_\theta \cos\theta \tag{6.58}$$

$$H_z = H_r \cos\theta - H_\theta \sin\theta \tag{6.59}$$

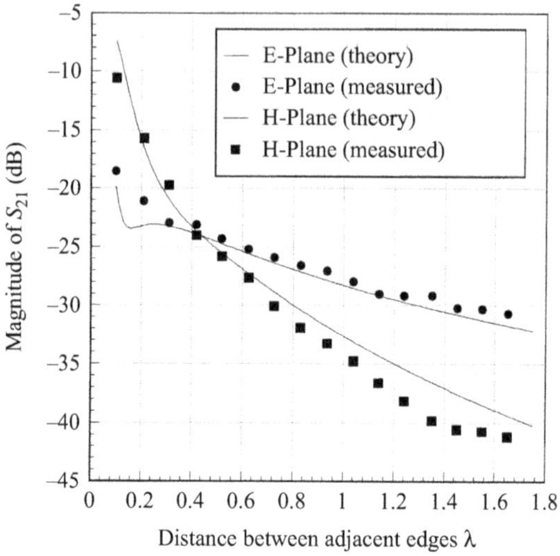

Figure 6.25 Comparison of experimental results of Jedlicka and Carver [22] with those computed with (6.52) using the cavity model. $f = 1.56$ GHz, $\epsilon_r = 2.50$, $a_1 = a_2 = 50.0$ mm, $b_1 = b_2 = 60.0$ mm, $H = 1.57$ mm. The probe feed is located at the center $x = a/2$ of each patch with the probe fed at 8.25 mm from center ($R_{in} = 50.125\ \Omega$)

For side I or III of patch 1 the $\hat{z}$-directed current is placed along the edge in the direction of the $\hat{y}$-axis of the coordinate system for patch 1. In this situation,

$$\vec{H}_x^{\vec{M}} \rightarrow \vec{H}_z$$
$$\vec{H}_y^{\vec{M}} \rightarrow \vec{H}_x$$
$$\vec{H}_z^{\vec{M}} \rightarrow \vec{H}_y$$

The coupling between a pair of rectangular microstrip antennas has been evaluated and compared with the experimental results of Jedlicka and Carver [22]. The theoretical results are plotted with the experimental results of Jedlicka and Carver in Figure 6.25. The correlation between experiment and theory is good overall, and very good from 0.3 to 0.6λ edge separation.

References

[1] Elliot, R.S., "Bandwidth and Directivity of Large Scanning Arrays, First of Two Parts," *Microwave Journal*, December 1963, Vol. 6, No. 12, pp. 53–60.

[2] Elliot, R.S., "Beamwidth and Directivity of Large Scanning Arrays, Last of Two Parts," *Microwave Journal*, January 1964, Vol. 7, No. 1, pp. 74–82.

[3] Hansen, R.C., *Significant Phased Array Papers*, Artech House Reprint Volume, 1973.

[4] Stutzman, W.L., and Thiele, G.A., *Antenna Theory and Design*, John Wiley & Sons, 1981, pp. 385–391.

[5] Balanis, C.A., *Antenna Theory Analysis and Design*, Harper & Row, 1982, pp. 37–42.

[6] Dolph, C.L., "A Current Distribution for Broadside Arrays Which Optimizes the Relationship between Beamwidth and Sidelobe Level," *Proceedings IRE*, June 1946, Vol. 34, No. 6, pp. 335–348.

[7] Sainati, R.A., *CAD of Microstrip Antennas for Wireless Applications*, Artech House, 1996, pp. 191–199.

[8] Hall, P.S., and Hall, C.M., "Coplanar Corporate Feed Effects in Microstrip Patch Array Design," *IEE Proceedings*, June 1988, Vol. 135, Pt. H, No. 3, pp. 180–186.

[9] Sainati, R.A., *CAD of Microstrip Antennas for Wireless Applications*, Artech House, 1996, pp. 210–220.

[10] Collin, R.E., *Antennas and Radiowave Propagation*, McGraw-Hill, 1985, pp. 266–268.

[11] Derneryd, A.G., "Linearly Polarized Microstrip Antennas," *IEEE Transactions on Antennas and Propagation*, November 1976, Vol. 24, pp. 846–851.

[12] Derneryd, A.G., "A Two Port Rectangular Microstrip Antenna Element," *Scientific Report No. 90*, July 1987, Electromagnetics Laboratory, University of Colorado, Boulder, CO.

[13] Gupta, K.C., and Benalla, A., "Transmission-Line Model For Two-Port Rectangular Microstrip Patches With Ports At The Nonradiating Edges," *Electronics Letters*, August 1987, Vol. 23, No. 17, pp. 882–884.

[14] Gupta, K.C., and Benalla, A., "Two-Port Transmission Characteristics of Circular Microstrip Patch Antennas," *IEEE Antennas and Propagation International Symposium Digest*, June 1986, pp. 821–824.

[15] Pozar, D.M., and Shaubert, D.H., "Comparison of Three Series Fed Microstrip Array Geometries," *IEEE Antennas & Propagation Symposium*, Ann Arbor, July 1993, pp. 728–731.

[16] Levine, E., Malamund G., Shtrikman, S., and Treves, D., "A Study of Microstrip Array Antennas with the Feed Network," *IEEE Transactions on Antennas & Propagation*, April 1989, Vol. 37, No. 4, pp. 426–434.

[17] Abouzahra, M.D., and Lewin, L., "Radiation from Microstrip Discontinuities," *IEEE Transactions on Microwave Theory and Techniques*, August 1979, Vol. MTT-27, No. 8, pp. 722–723.

[18] Waterhouse, R., *Microstrip Patch Antennas A Designer's Guide*, Kluwer Academic Publishers, 2003, pp. 361–364.

[19] Malherbe, A., and Johannes, G., "Analysis of a Linear Antenna Array Including the Effects of Mutual Coupling," *IEEE Transactions on Education*, February 1989, Vol. 32, No. 1, pp. 29–34.

[20] Huynh, T., Lee, K.F., and Chebolu, S.R., "Mutual Coupling Between Rectangular Microstrip Patch Antennas," *Microwave and Optical Technology Letters*, October 1992, Vol. 5, No. 11, pp. 572–576.

[21] Stutzman, W.L., and Thiele, G., *Antenna Theory and Design*, John Wiley & Sons, 1981, p. 98.

[22] Jedlicka, R.P., and Carver, K.R., "Mutual Coupling Between Microstrip Antennas," *Proceedings of Workshop on Printed Circuit Antenna Technology*, New Mexico State University, Physical Science Laboratory, October 17–19, 1979.

Chapter 7
Printed antennas

Microstrip antennas have a large number of applications despite their limitations. In some cases, pattern or bandwidth requirements can only be met with planar antennas that are not a traditional microstrip configuration. We will generally refer to these as printed or planar antennas. In some instances, microstrip transmission lines may be integrated with an antenna and so it is often still called a microstrip antenna. In this chapter, we will investigate a number of useful printed/microstrip antenna designs.

7.1 Omnidirectional microstrip antenna

An antenna with an omnidirectional pattern is desired for a number of wireless applications [1]. An omnidirectional antenna design which is easily scaled to produce a range of gain values, does not require a balun when fed with a coaxial transmission line, and has a 50 Ω driving point impedance was presented by Bancroft and Bateman [2,3]. Design details for dual-shorted rectangular omnidirectional microstrip antennas (OMAs) are presented in the literature [4]. Aspects of this design were anticipated by Jasik *et al.* [5,6] in the early 1970s and by Hill [7] as a traveling wave antenna in the late 1970s. A similar geometry was related by Ono *et al.* in 1980 [8]. In 2012, Wei *et al.* stacked two OMAs to produce a dual-band design [9].

The geometry of the OMA is presented in Figure 7.1. The antenna consists of a bottom trace which begins with width W_e and length L_e long. The trace narrows to W_m and length L_m and alternates between wide and narrow until the final wide section. Both wide-end sections are shorted in their center to the upper trace. The upper trace begins at the bottom short with a narrow trace which is of width W_m, which alternates between wide and narrow sections, complementing the upper trace. The final upper trace terminates at the upper short. The short at each end connects the upper and lower traces. The driving point is shown in Figure 7.1. The outer shield of a coaxial line is soldered to the wide bottom trace, and the center conductor drives the upper trace.

The OMA may be viewed as a set of $\lambda/2$ microstrip transmission lines which is illustrated in Figure 7.2. The top illustration is of a microstrip transmission line with its currents. Each half-wavelength section of microstrip transmission lines is flipped, so the groundplane is connected to the trace, and the trace to the groundplane of the next section. Each section is a 50 Ω microstrip transmission line, but at each junction,

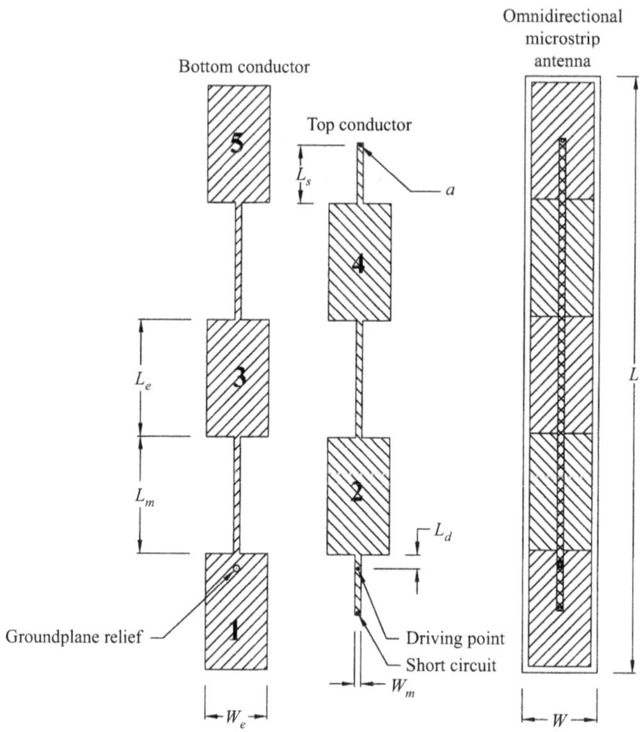

Figure 7.1 Five section rectangular dual-short omnidirectional microstrip antenna

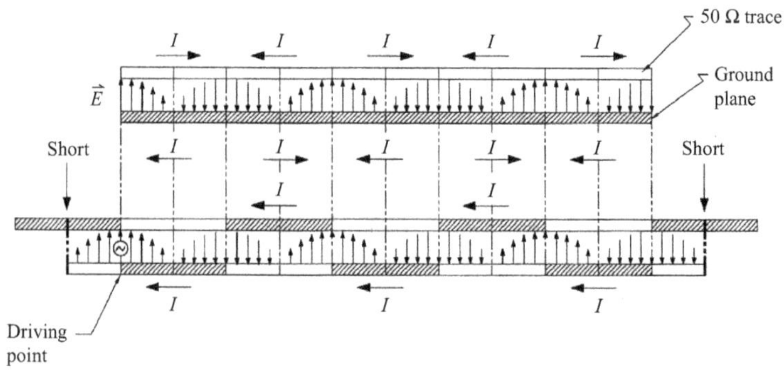

Figure 7.2 Current on a microstrip transmission line (top). Current on flipped sections of microstrip transmission line which make up a seven section omnidirectional microstrip antenna (bottom)

the reversal of the groundplane and trace produces a mismatch of the field mode desired by each section. This set of discontinuities encourages radiation. The electric field is maximum at each of the junctions, and the surface current is maximum in the center of each wide section (along the groundplane edges).

The shorting pin at the bottom of the antenna adds a negative 180° phase shift to a downward travelling wave produced by the voltage source ($L_d = 0$), which is 90° behind the driving point phase at the short, and as it travels back to the driving point, it adds another 90° (360 total). This causes the wave reflected from the lower short to arrive in phase with a wave which is traveling upward along the antenna generated at the driving point. The upper short operates in the same manner, so that upward and downward traveling waves are in phase. This creates a resonant structure, where the current on each wide groundplane (and the traces) is all in phase, which produces an omnidirectional antenna pattern.

The shorting pins also minimize the amount of current which appears below the short on the driving point end. The outer shield of the coaxial feed line is generally soldered from the driving point edge on the groundplane side to the short. This short decouples the feed line from the antenna below the short (single conductor) so that only a minute amount of current is driven on the outer conductor of the coaxial cable, and no balun is required. It can also act to mitigate electrostatic discharge (ESD).

The driving point impedance is maximum at the junction of elements 1 and 2 in Figure 7.1. The magnitude of this maximum impedance is inversely proportional to the width (W_e) of the elements. As the width of the elements is decreased, the maximum impedance at the junction increases, when W_e is increased, the maximum impedance decreases. Generally, one can find a 50 Ω driving location between the bottom short (0 Ω) and the maximum driving point resistance at the junction.

The total number of sections making up the total length of the antenna may be altered to provide a desired gain. The gain of a rectangular dual-short OMA versus the number of elements for $W_e = 10$ and 20 mm is shown in Figures 7.3 and 7.4, respectively. The antennas were analyzed using HFSS with $H = 0.762$ mm, $\epsilon_r = 2.6$, $W_m = 2$ mm, $a = 0.5$ mm, operating at 2.45 GHz. One can see that the gain steadily increases as the total number of elements increases.

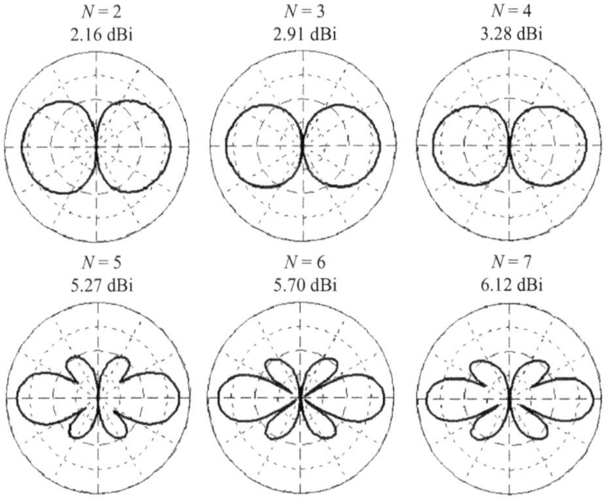

Figure 7.3 Gain of 10 mm wide (W_e) elements

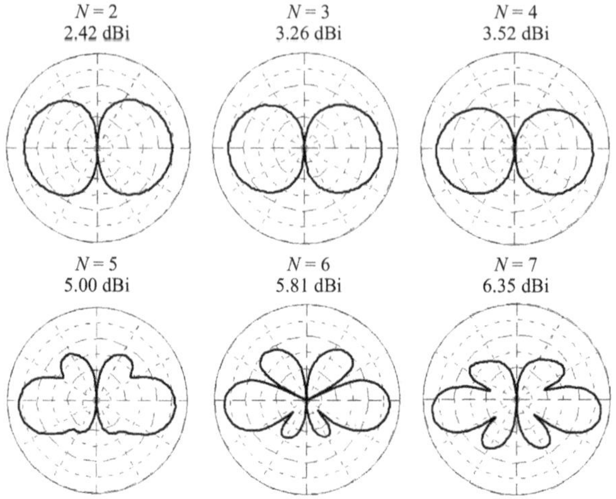

Figure 7.4 Gain of 20 mm wide (W_e) elements

The wider the elements, the higher the antenna efficiency, but at the expense of a pattern shape. When the antenna elements are narrow (10 mm), the antenna patterns are symmetric and omnidirectional. As the width is increased (20 mm), a lower frequency resonance moves upward and produces a superposition of modes. The lower frequency mode has a butterfly type of radiation pattern, which increases the sidelobe level (SLL) of the design. This is illustrated in Figures 7.3 and 7.4.

The impedance bandwidth for the dual-short rectangular OMA driven in a pure omnidirectional mode is about 3%–4% almost independent of the length.

The antenna efficiency is very stable for $W_e = 20$ mm versus number of elements (96.5% for $N = 2$ and 94.7% for $N = 7$ as predicted by HFSS), but drops with length as the number of elements is increased for $W_e = 10$ mm (91.3% for $N = 2$ and 87.6% for $N = 7$).

A seven section OMA was designed to operate at 2.45 GHz on 0.762 mm (0.030 inch) laminate material. The relative dielectric permittivity of the substrate is $\epsilon_r = 2.6$ with a 0.0025 $\tan\delta$. The dimensions of the antenna are $W_m = 2.06$ mm, $W_e = 16.25$ mm, $L_e = L_m = 36.58$ mm. Shorting pins located on either end of the antenna have a 0.5 mm radius (a). The antenna is fed with a probe at the junction of the first narrow line and the next wide section meet (i.e. $L_d = 0$) in Figure 7.1. The dielectric material extends out 2.0 mm from each side and 2.0 mm from each end.

The Finite Difference Time Domain Method (FDTD) was used to compute the expected radiation patterns [10]. A sinusoidal 2.586 GHz source was utilized to compute the radiation patterns of the antenna. They are presented in Figure 7.5 with corresponding measured radiation patterns. The best antenna performance is at the high end of the band. The measured patterns are slightly squinted downward compared with the FDTD analysis. It appears the attached feeding cable slightly affects

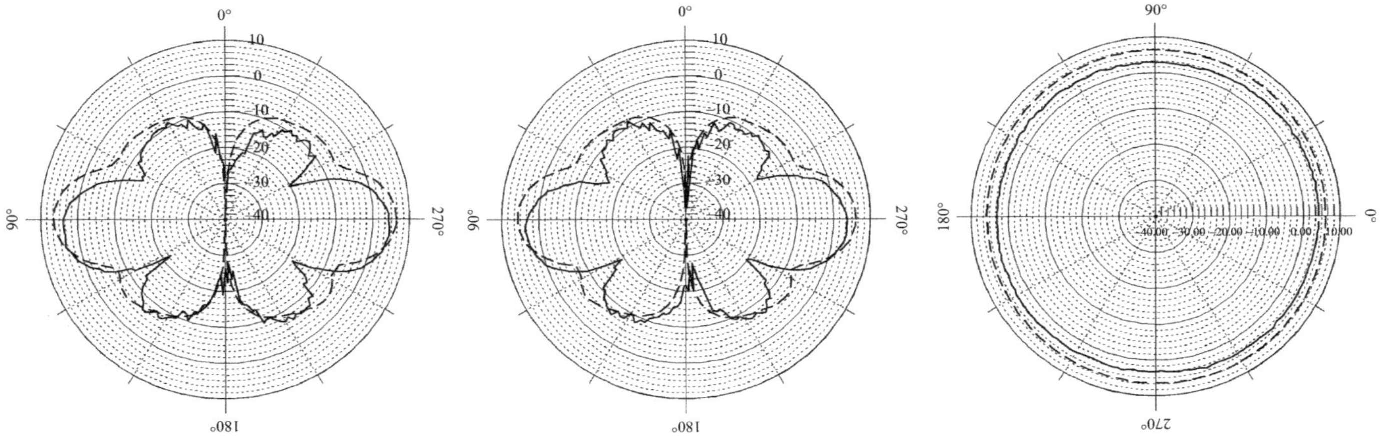

Figure 7.5 *y–z plane (left), x–y plane (center), and x–y plane (right) radiation patterns of OMA computed using FDTD analysis (dashed) and measured (solid). 2.586 GHz*

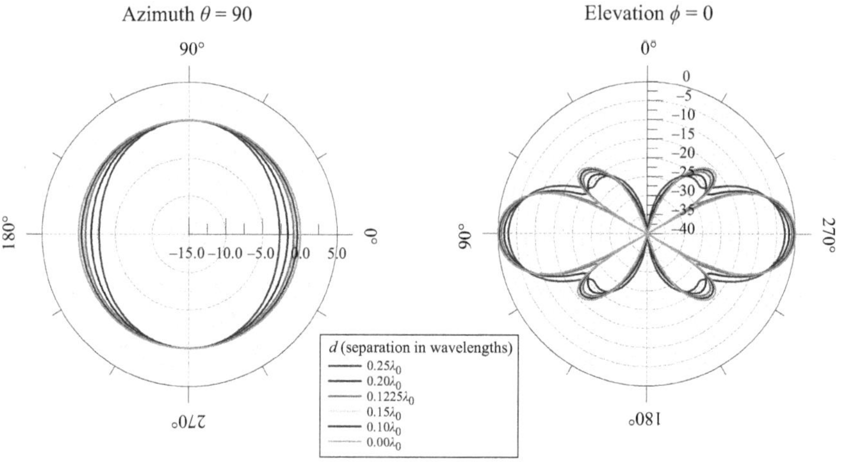

Figure 7.6 The effect of W_e on the pattern variation of an OMA in the azimuth (omni), and elevation plane, modeled with uniform amplitude sinusoids. The azimuth gain variation changes from 0.0 dB to 2.77 dB as W_e varies from $0.0\lambda_0$ to $0.25\lambda_0$

the phase relationship along the array and is the cause of this beam squint. The small cable used to feed the array was impractical to model with FDTD. The maximum gain was predicted to be 6.4 dBi versus 4.6 dBi measured at 2.586 GHz. The antenna sidelobes are approximately −11 dB below the main lobe.

The optimum match for the antenna is at 2.4 GHz, with a 371 MHz 2:1 VSWR impedance bandwidth. The normalized bandwidth is 15.45%, which is very good for a printed antenna. However, the pattern bandwidth is only 5%–6%. The driving point is unbalanced, and thus a balun is not required when feeding this antenna with a coaxial cable.

The radiation of the OMA originates from the currents at each edge of the rectangular elements. When W_e is small, the pair of currents become almost colinear, and the antenna pattern has very little variation in the omni-plane. As W_e becomes large, the two currents begin to array, and the pattern deviates significantly from a circle. One can use a set of uniform amplitude sinusoids (UAS) to model the radiation from an OMA. The pattern results of this analysis are presented in Figure 7.6. The predicted pattern variation correlates well with HFSS computations [11]. The predicted pattern variation value is 0.0 dB to 2.77 dB as W_e ranges from $0.0\lambda_0$ to $0.25\lambda_0$.

7.1.1 Low sidelobe omnidirectional MSA

The OMA presented in Section 7.1 has a uniform illumination along its length. A uniform amplitude distribution along an array produces sidelobes which are −13.2 dB below the main beam. The example uniform OMA has sidelobes which are as high as −11 dB. The uniform OMA has radiating elements of identical widths. We can control

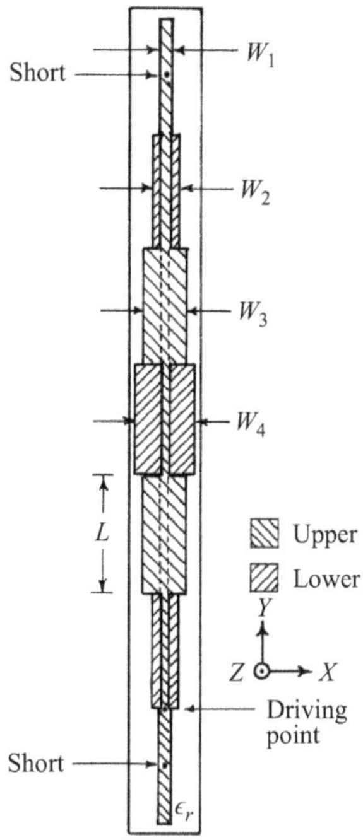

Figure 7.7 Seven section omnidirectional microstrip antenna with linear taper

the amount of radiation from each of the elements by varying their widths [12]. Figure 7.7 shows a seven section OMA which has elements of different widths. The relative width of each element corresponds with a −6 dB linear taper on a pedestal using (6.15).

FDTD analysis was used to vary the center width W_4 (with the other widths dependent upon W_4) until the desired distribution to produce a −22.5 dB SLL is obtained. The design uses a 0.762 mm thick dielectric substrate with $\epsilon_r =$ 2.6 and 0.0025 $\tan\delta$. The element widths are $W_1 = 3.0$ mm, $W_2 = 7.32$ mm, $W_3 =$ 11.66 mm and $W_4 = 16.0$ mm with 50 Ω interconnects of 2.03 mm width. The length of each element is 36.15 mm.

An antenna with dimensions from the previous FDTD analysis (Section 7.1) was fabricated with the altered element widths. The antenna patterns were optimum at 2.628 GHz, but the input impedance has a slight series inductive reactance, which produced an unacceptable mismatch (2.5:1 VSWR). A 1.0 pF capacitor was used as a via at the driving point to match the antenna with a return loss of better than 25 dB. The normalized impedance bandwidth of the matched −6 dB taper OMA is 3.8%, which is smaller than the 14.58% bandwidth of the uniform design.

The directivity predicted by FDTD is 5.39 dB. The measured gain of the fabricated antenna is 5.0 dBi. The measured and predicted radiation patterns are presented in Figure 7.8.

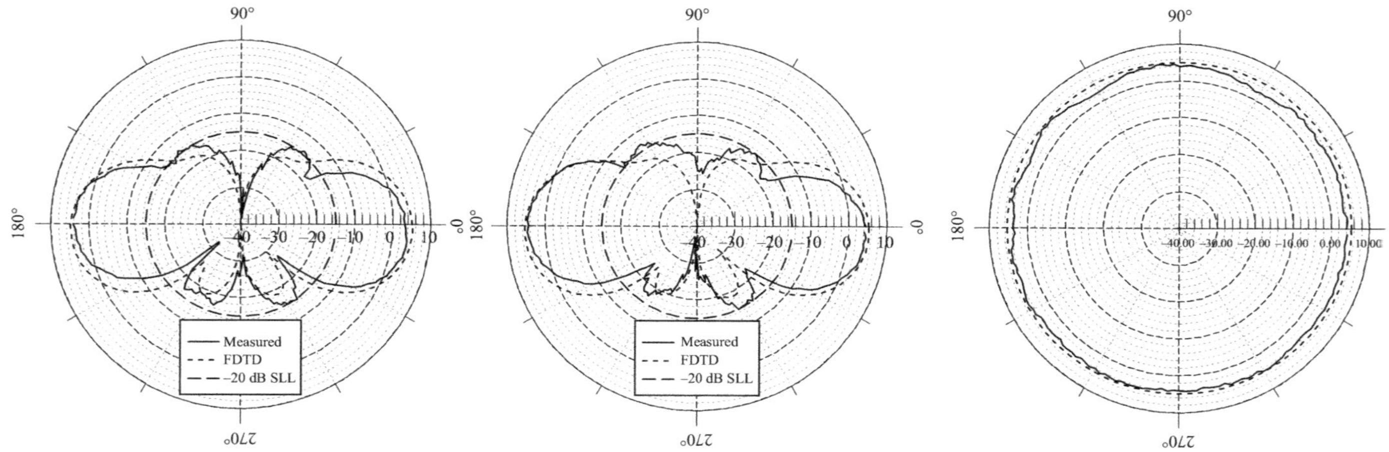

Figure 7.8 *Omni plane radiation patterns of seven section omnidirectional microstrip antenna with linear taper: y–z plane, x–z plane, and x–y plane. The dashed line is from FDTD analysis. The solid line is measured*

The FDTD patterns have a −22.5 dB SLL. The measured patterns are close to −20 dB SLL. This is approximately a 9 dB improvement over the −11 dB SLL of the uniform OMA design.

When designing an omnidirectional antenna, it is often useful to know the approximate directivity of an omnidirectional pattern versus the half-power beamwidth HPBW. Pozar developed a curve-fit equation based on a $\sin^N \theta$ pattern to relate HPBW of an omnidirectional pattern without sidelobes to its directivity [13]:

$$D = 10\log_{10}\left(191.0\sqrt{0.818 + \frac{1}{HPBW}} - 172.4\right) \text{dB} \tag{7.1}$$

Where:

HPBW is the elevation-plane half-power beamwidth in degrees.

This equation is valid up to a beamwidth of 140°.

When sidelobes are present with an assumed uniform current distribution, McDonald has developed a relationship which uses a $\sin(b\theta)/(b\theta)$ pattern as its basis [14]:

$$D = 10\log_{10}\left(\frac{101.5}{HPBW - 0.00272(HPBW)^2}\right) \text{dB} \tag{7.2}$$

7.1.2 Element shaping of OMA

The OMA's discussed thus far all have rectangular elements. The use of other shapes can provide some advantages in the design of an Omnidirectional Microstrip Antenna. In Figure 7.9, we have five combinations of circular, rectangular and elliptical elements.

We have seen previously with rectangular elements, that as the width of the element is increased, the efficiency of the antenna also increases. For a 4.9 GHz design, HFSS predicts that the efficiency of each design decreases from left to right in Figure 7.9. The efficiency of the circular OMA design (a) is 96.8%, (b) is 95.8%, (c) is 93.5%, (d) is 92.52% and (e) is 92.5% as predicted using HFSS. The efficiency change from the element shape changes is only 0.2 dB. The gain of the designs changes, decreasing from left to right in Figure 7.9. The gain of the circular OMA design (a) is 7.7 dBi, (b) is 6.7 dBi, (c) is 6.8 dBi, (d) is 6.7 dBi and (e) is 6.2 dBi.

In all five cases, the physical antenna length is approximately equal for equivalent operating frequency, but there appears to be approximately a 1.0–1.5 dB advantage in gain by using circular elements as opposed to all rectangular elements.

The SLLs predicted by HFSS indicate that the combination of ellipses and rectangles produces the lowest SLL, with elements of uniform width. The predicted SLLs for (a) is 11.8 dB, (b) is 11.3 dB, (c) is 13.1 dB, (d) is 14.5 dB and (e) is 11.8 dB.

The driving point impedance is proportional to the element widths as seen in the case of the rectangular designs. The driving point impedance at resonance is the lowest for the circular elements (a), and increases to a maximum value with the

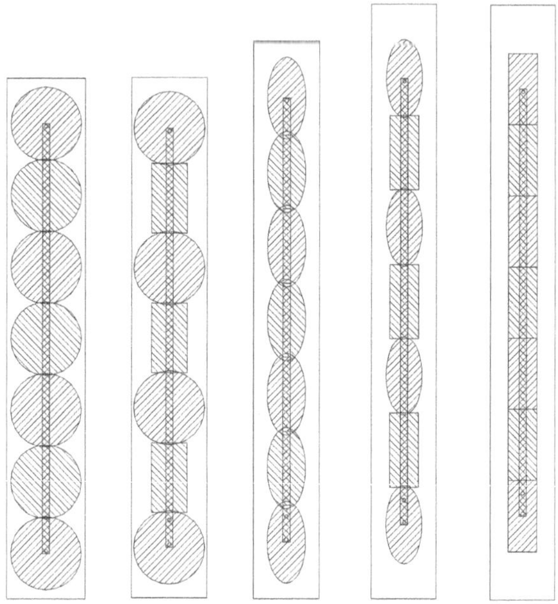

Figure 7.9 Seven section dual-short shaped element omnidirectional microstrip antenna designs (a) circular, (b) circular rectangular, (c) ellipse, (d) ellipse rectangular and (e) rectangular

rectangular elements (d). The impedance bandwidth of the omnidirectional mode is approximately the same for all the element widths for a dual-short design.

7.1.3 Single-short omnidirectional microstrip antenna

The dual-short OMA design, presented in Figure 7.1, has the advantage that one can directly connect a coaxial transmission line, and match the driving point impedance with the proper choice of element width and driving point location. A second option is to use a single short at the top of the antenna, and use the bottom section of the antenna as a platform for broadband impedance matching.

An illustration of a single-short OMA with circular elements, and a broadband impedance matching network, is shown in Figure 7.10. A number of useful impedance matching techniques are presented in Appendix E. This design used a theoretical driving point impedance with transmission line analysis software to design a broadband matching network. This network was input into a full wave analysis package (HFSS) and optimized. A prototype antenna was realized, Its measured VSWR, and the prediction made by HFSS, are plotted on the left side of Figure 7.11. This OMA has an impressive 25% 2:1 VSWR impedance bandwidth. This bandwidth covers many commercial frequency bands with a single antenna. The theoretical antenna gain ranges from 6.4 to 7.6 dBi. The elevation pattern of this antenna as predicted by HFSS is

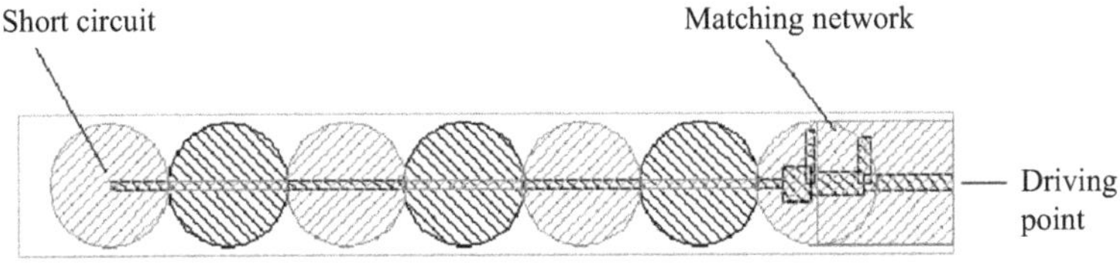

Figure 7.10 *Single-short omnidirectional antenna with broadband matching network*

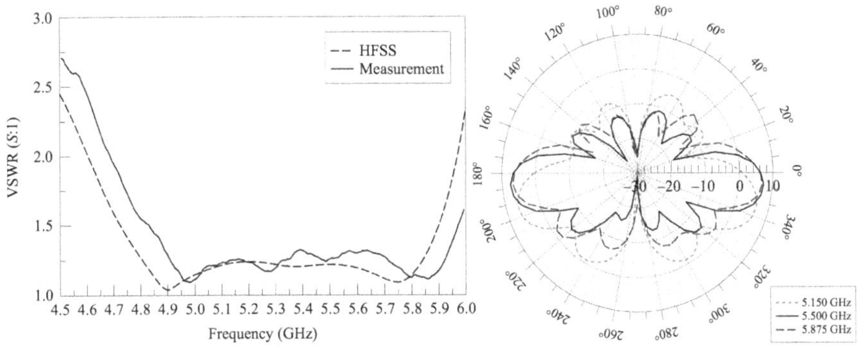

Figure 7.11 *VSWR (left) and elevation pattern (right) of single-short omnidirectional microstrip antenna with broadband matching network*

shown in the right image of Figure 7.11. The single-short OMA offers more design options than the dual-short antenna.

7.1.4 Corporate-fed omnidirectional microstrip antenna

The series-fed OMA design has the advantages of simplicity and maximum use of physical aperture. One disadvantage is the amount of frequency-dependent beam scanning. A common alternative omnidirectional antenna design uses a corporate-fed network. This type of antenna design is shown in Figure 7.12 and derived from a series-fed design presented by Wong *et al.* [15]. The antenna array is fed using a parallel plate transmission line (Appendix C) [16]. The end of the parallel plate transmission line corporate feed network is terminated using metal strips, which are routed in opposite directions. The current on the parallel-plate transmission line is differential, which suppresses feed radiation. The elements of the metal strips are in opposite directions, which produces unopposed current on each element (i.e. the current on all the radiating strip pairs is in the same direction and phase). All the elements have current in the same direction, producing omnidirectional radiation.

Planar antennas are generally fed using microstrip transmission lines, which is unbalanced. As shown in Figure 7.12, a balun should be used to transition the unbalanced microstrip transmission line mode to the balanced parallel-plate transmission

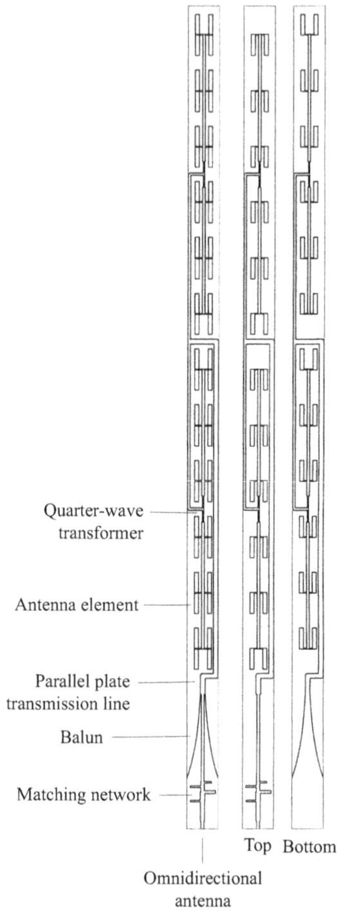

Figure 7.12 Corporate-fed omnidirectional microstrip antenna array using parallel-plate transmission lines

line mode. The 50 Ω parallel-plate line splits into a pair of 100 Ω lines. Quarter-wave transformers connect to each of the three element sub-arrays and may be used to match the antenna driving point impedance at a desired frequency. Some applications require as large an impedance bandwidth as possible. A matching network may be synthesized to produce this. Figure 7.13 shows the VSWR of the antenna, with an impedance matching network, designed to operate from 4.5 to 5.0 GHz (<2:1 VSWR).

The corporate feed decreases the beam scanning with frequency, but at the cost of extra losses from the corporate feed network. In the case of the design shown in Figure 7.12, a 1.5–2.0 dB loss is introduced by the feed network. The realized gain is approximately 10.0 dBi for this design. This produces an antenna which has a smaller realized gain than a series-type OMA.

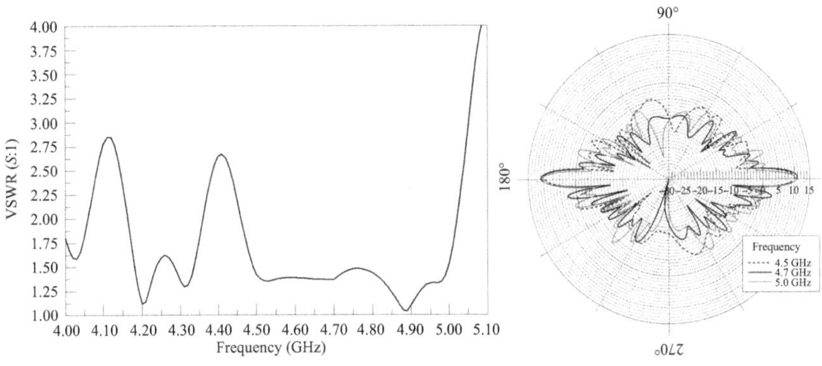

Figure 7.13 VSWR of corporate-fed microstrip antenna (left) elevation and azimuth radiation patterns of corporate-fed microstrip antenna (right) designed to operate from 4.5 to 5.0 GHz, using a parallel-plate transmission line, predicted with HFSS

The corporate feed can also introduce some scattering of the desired radiation and produce pattern distortion. The design of Figure 7.12 has dual radiators which will have a similar pattern distortion to that produced by the edges of the OMA. This is shown in Figure 7.6. The predicted patterns of the design of Figure 7.12 are shown in Figure 7.13. Other designs of this type often use single radiating elements rather than dual.

7.2 Tapered (Vivaldi) antenna

The Vivaldi antenna was introduced in 1979 by Gibson [17]. Equations presented by Shin and Schaubert offer more flexibility in the design of a taper for a Vivaldi antenna than the original equations of Gibson [18]. These may be written as

$$y = \pm C_1 e^{Rx} + C_2 \tag{7.3}$$

$$C_1 = \frac{y_2 - y_1}{e^{Rx_2} - e^{Rx_1}} \tag{7.4}$$

$$C_2 = \frac{y_1 e^{Rx_2} - y_2 e^{Rx_1}}{e^{Rx_2} - e^{Rx_1}} \tag{7.5}$$

The coordinate system is shown in Figure 7.15. The y values for a given point along the x-axis are given by (7.3). The point of origin for the taper is (x_1, y_1), and the terminating point of the curve is at (x_2, y_2).

The advantage of this set of equations is that the *taper rate* parameter R may be varied to change the rate at which the exponential taper diverges from the slotline width. When $R \to 0$, the taper is essentially linear. As the value of R increases, the slotline transition becomes more gradual, and most of the exponential taper is moved toward the aperture end of the antenna. The property is illustrated in Figure 7.14.

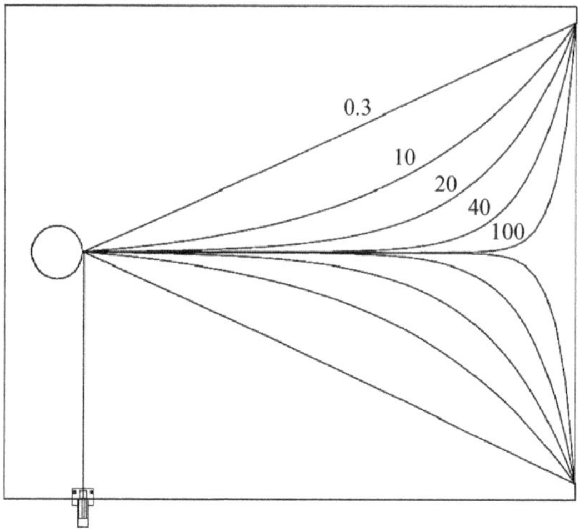

Figure 7.14 Antenna shape versus taper rate R

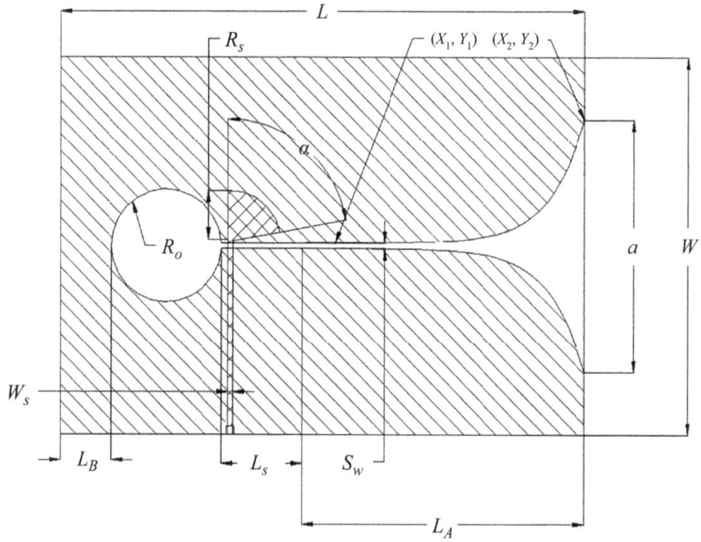

Figure 7.15 Geometry of typical Vivaldi (exponential taper slot antenna) which utilizes a stripline radial stub and circular bilateral slotline open circuit transition

According to Gibson, the Vivaldi antenna "... has theoretically unlimited instantaneous frequency bandwidth." A limitation to realizing this theoretical potential is the bandwidth of the transition and the maximum size of the antenna aperture.

The proverbial Vivaldi design, shown in Figure 7.15, which is generally found in the literature, has a stripline to bilateral slotline transition, utilizing a circular bilateral

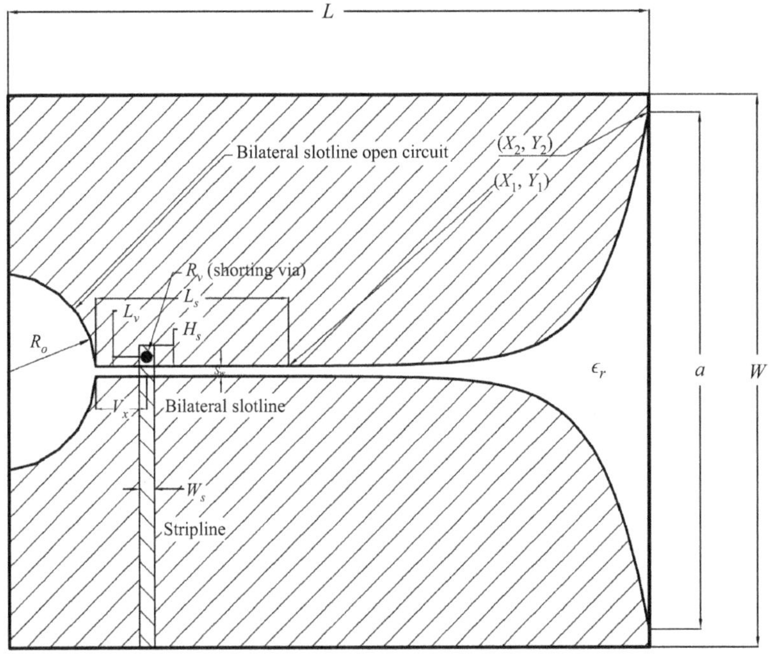

Figure 7.16 Geometry of revised Vivaldi (exponential taper slot antenna) design with a stripline to bilateral slotline transition, a bilateral slotline open circuit, and a shorting via

slotline open circuit stub and a stripline open circuit stub. The geometry of this stripline to bilateral slotline transition limits both the lower and upper frequency bandwidth of this antenna (Figure 7.15). When the frequency becomes small enough so the circular bilateral slotline open circuit is electrically small, it appears as a short circuit, rather than as an open. At higher frequencies, the open stripline stub becomes large with respect to a wavelength and produces a mismatch at the transition. This stripline to bilateral slotline transition bandwidth limitation, in turn, limits the bandwidth of the Vivaldi antenna design. Some researchers simply allow the slotline open circuit to become truly open at the back end of the design by cutting the circle in half [19,20]. Then, the Vivaldi antenna element length, width, and taper rate may be chosen to produce acceptable VSWR over the design bandwidth.

Full-wave analysis indicates that the use of a half-circle bilateral slotline open and a short-circuit via in place of the stripline radial stub provides a much larger impedance bandwidth (Figure 7.16). Some texts utilize microstrip and slotline rather than stripline and bilateral slotline. In general, bilateral slotline is chosen to drive Vivaldi antennas in this text because [21]

1. The use of a stripline to bilateral slotline decreases the amount of radiation from the feed network and shields the radial impedance stub. This is of considerable importance in array applications.

2. Bilateral slotline has less dispersion relative to single-sided slotline.
3. The characteristic impedance of bilateral slotline in many instances is approximately half of the characteristic impedance of single-sided slotline.
4. The design width of a slot for a given impedance value is larger for a bilateral slotline than a single-sided slotline. This allows for more realistic fabrication tolerances for bilateral slotline than single-sided slotline.

Metallization on either side of the total substrate thickness of H is removed to create the bilateral slotline transition and antenna element. A circular bilateral open circuit, with a radius of R_o, is connected to a slot of length L_s and width S_w. The exponential taper begins after this slot and extends to the end of the PCB, which is the length of the antenna element. The aperture of the element is a. The board length and width is L and W, respectively. The stripline feed is of width W_s and terminates in a short circuit via of radius R_v, which connects the stripline to the upper and lower conductors. The stripline extends past the upper slot by a distance of H_S. The short circuit via is at a distance of L_v from the bilateral slotline edge and V_x from the slotline open circuit.

The design of the bilateral slotline to stripline transition is critical to maximizing the impedance bandwidth of a Vivaldi antenna [22]. The radius R_s of the shorting via is also critical. The shorting via radius is chosen to produce the best overall bandwidth for the transition.

An optimized short-circuit transition was then used as a transition to design an optimized Vivaldi exponential curve antenna element. The slotline width is 0.25 mm. The stripline edge is 0.508 mm from the bilateral slotline open circuit. The stripline extends past the top of the slot 0.1875 mm and has a shorting via with a diameter of 0.125 mm. The radius of the BLS open circuit is 50 mm. The aperture of the Vivaldi is 262 mm, with a taper rate of 42.5. The board material is Taconic RF-43, which has a relative permittivity of 4.3 and a 0.0033 loss tangent. The total thickness of the board is 1.524 mm.

The VSWR performance of this design (Figure 7.16) is given in Figure 7.17. The predicted VSWR remains below 2:1 from 500 MHz to 6 GHz for a 12:1 impedance bandwidth. The measured VSWR remains below 2:1 VSWR over the same frequency range with the exception of 801–888 MHz, when it edges just above 2.0:1 VSWR with a 2.2:1 maximum. The computed E- and H-plane radiation patterns at selected frequencies are given in Figure 7.18.

7.3 Microstrip-fed slot antenna

A slot antenna fed with a microstrip transmission line is illustrated in Figure 7.19. The radiating slot has a width W_s and length L_S. The slot antenna may be viewed as a slot in the groundplane of a microstrip transmission line, which is shorted to the back side of the slot radiator. The microstrip transmission line is offset from the center of the slot by X_S. This is done to match the transmission line to the driving point impedance of the antenna.

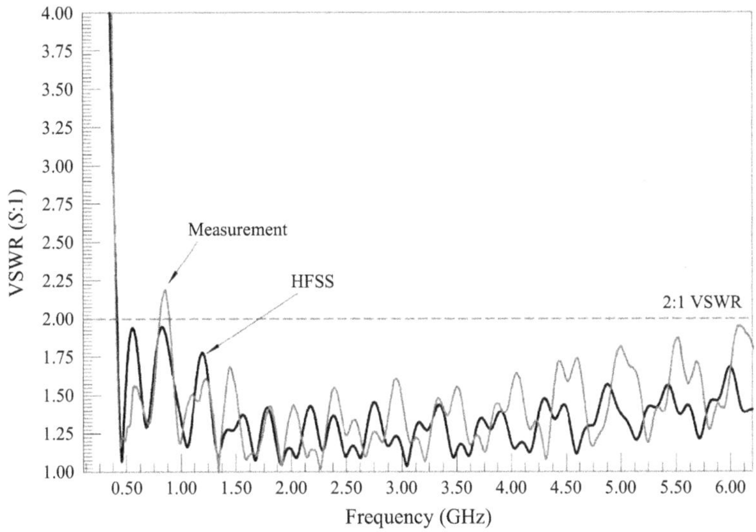

Figure 7.17 Reflection loss of Vivaldi antenna design [22]

A $\lambda/2$ slot antenna is complimentary to a $\lambda/2$ dipole antenna. The relationship between the input resistance of a slot antenna (Z_s) in an infinite groundplane and the input of a complimentary dipole antenna (Z_d) is

$$Z_d Z_s = \frac{\eta}{4} \tag{7.6}$$

The input impedance of round $\lambda/2$ dipoles has been extensively computed. The equivalent radius of a round wire dipole is about one-half of the strip width of the complimentary (flat) dipole. If we restrict ourselves to a thin flat dipole, $W_s \ll \lambda$, its resonant resistance is approximately 67 Ω. Using (7.6), we compute a typical value for the resonant resistance of a slot:

$$R_s = \frac{(376.73)^2}{4 \cdot 67} \approx 530\ \Omega \tag{7.7}$$

The driving point resistance is maximum at the center of the slot and decreases to zero as the driving point moves toward the end of the slot. The current increases in a sinusoidal manner as one moves from the center of the slot toward an edge. The voltage decreases in a sinusoidal manner as one moves from the center of the slot to an edge. This implies the driving point resistance (R_{drv}) will vary as [39]:

$$R_{drv} \approx R_s \sin^2 k\left[\frac{L_s}{2} - X_s\right] \tag{7.8}$$

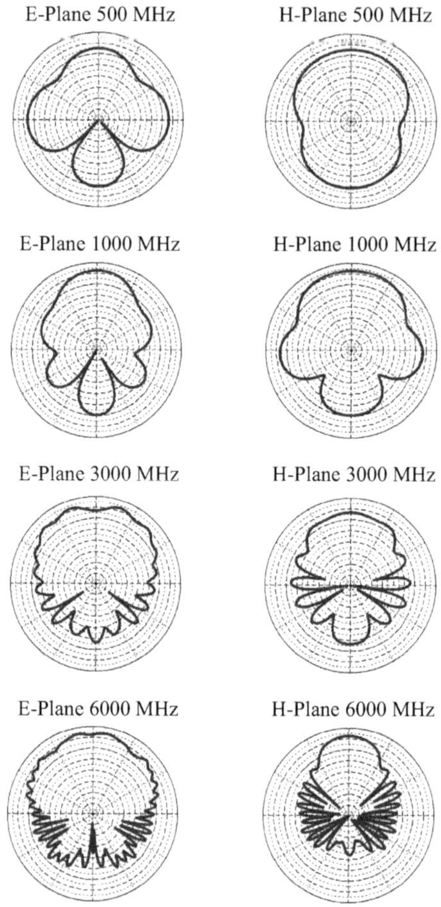

Figure 7.18 Radiation patterns of example Vivaldi design computed with HFSS. The plot range is −30 to 10 dBi

This allows us to compute the approximate position of a 50 Ω ($R_{drv} = 50\ \Omega$) driving point:

$$X_s = \frac{L_s}{2} - \frac{\lambda}{2\pi}\sin^{-1}\sqrt{\frac{R_{drv}}{R_s}} \tag{7.9}$$

For a slot which is at resonance, with $L_s = 0.46\lambda$, and $R_s = 530\ \Omega$, the 50 Ω driving point location X_s is

$$X_s = \frac{0.46\lambda}{2} - \frac{\lambda}{2\pi}\sin^{-1}\sqrt{\frac{50}{530}} = 0.180\lambda$$

which is 0.050λ from the end of the slot.

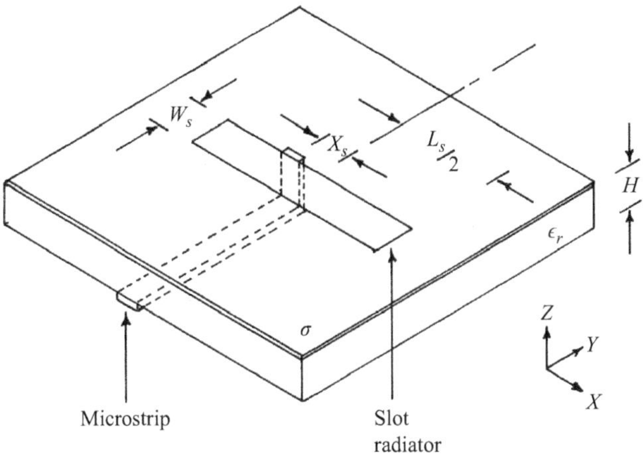

Figure 7.19 A $\lambda/2$ slot antenna of width W_s, and length L_s is fed with a microstrip transmission line, which is shorted to the back side of the slot. The position of the microstrip feed line is placed at a location X_s between the slot center and its edge, which provides a driving point resistance which matches the microstrip transmission line

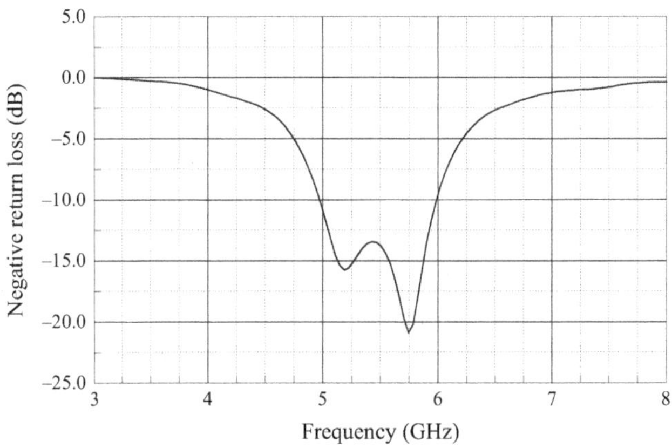

Figure 7.20 Negative return loss (dB) of example slot antenna

As an example rectangular slot antenna design, we use a substrate with $\epsilon_r = 3.5$ and a 0.0055 tan δ. The thickness of the substrate is $H = 0.5$ mm. The slot dimensions are $L_s = 22$ mm and $W_s = 1.0$ mm. The feedpoint is $X_s = 8.0$ mm. The microstrip line width is 1.0 mm.

The predicted negative return loss plot of the rectangular slot antenna is presented in Figure 7.20. The antenna is seen to have an 18.78% (2:1 VSWR) impedance bandwidth.

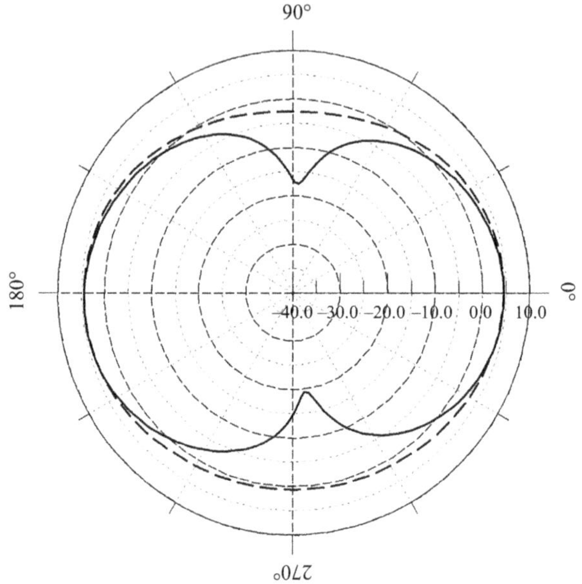

Figure 7.21 Radiation pattern of the example rectangular microstrip-fed slot antenna computed with FDTD analysis. The solid line is the E-plane pattern. The dashed line is the H-plane pattern

The current has a maximum value along the ends of the slot radiator. These currents are directed parallel with the $\hat{y}$-axis. The electric field of the radiator is across the narrow dimension of the slot, which produces $\hat{y}$ directed polarization. The maximum driving point impedance at the center of the slot antenna allows one to cut the slot radiator in half to produce a $\lambda/4$ version. A single $\hat{y}$ directed current then becomes the radiation source. Further reducing the groundplane surrounding the $\lambda/4$ slot produces a printed circuit version of an inverted F antenna (IFA) [40].

Figure 7.21 shows the E- and H-plane radiation patterns of the example microstrip-fed slot antenna computed using FDTD. The pattern directivity is 4.66 dB.

7.3.1 *Slot antenna "fictitious resonance"*

One may have noted the considerable bandwidth and a pair of negative return loss minimums of the previous microstrip slot antenna design example. It was noted by Behdad and Sarabandi that a slot fed in this manner can contain two resonances [41,42]. The first resonance is the expected normal resonance of the full slot, but a second "fictitious" resonance may also exist in some slot designs. The fictitious resonance occurs when the tangential electric field component of the feed conductor is at a position along the slot which cancels the electric field of the slot excited by the ground plane return current.

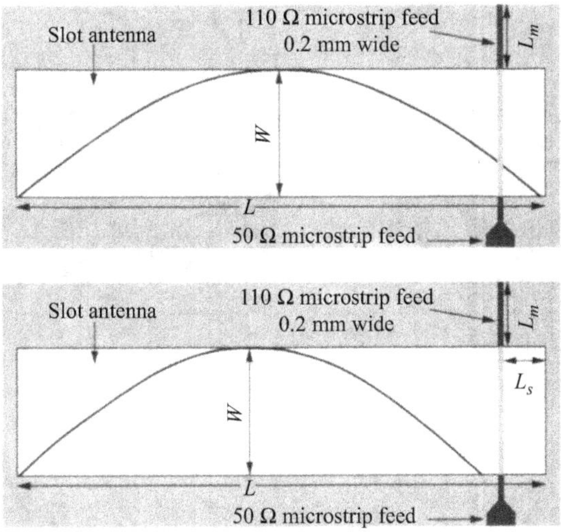

Figure 7.22 Magnetic current distribution of the normal (top) and "fictitious" (bottom) resonance of a slot antenna fed with a narrow line from a microstrip transmission line and impedance matched using an open microstrip line (from [42], copyright 2004, IEEE, reprinted with permission)

The width of the feed line, the slot width, and the distance from the edge of the slot to the center of the feedline (L_s) determines the existence and location of the fictitious resonance. The frequency of the fictitious resonance may be increased by increasing L_s, which increases the total impedance bandwidth or produces dual-band operation. The overall length of the slot L determines the frequency of the lower resonance. The length of the microstrip line L_m above the slot feed conductor allows for matching.

The magnetic current distribution, illustrated in Figure 7.22, is similar at both resonances, producing a relatively stable radiation pattern over the impedance bandwidth.

A parametric study of the design of this slot antenna is available in the literature [43].

In light of the results obtained by Behdad and Sarabandi, FDTD analysis of the design example slot antenna presented above (and in the First Edition) was undertaken at the two negative return loss minimums. A sinusoidal source was used at 5.189 and 5.744 GHz. The results are presented in Figure 7.23. Although the researchers indicated that a fictitious resonance was best driven by using a thin conductive strip across the slot, it appears in this case that the fictitious resonance was at work to produce a wideband slot antenna.

This design method was implemented to create a multi-band antenna for laptop use. A half slot is used for 802.11b (2.4–2.5 GHz) band and a full dual-band

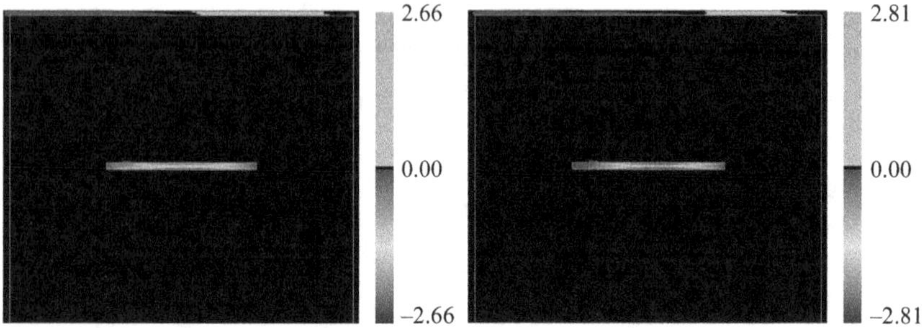

Figure 7.23 Electric field magnitude (linear) for 5.189 GHz resonance of example design. Electric field magnitude (linear) for 5.744 GHz resonance of example design (left). The fictitious resonance as described by Behdad and Sarabandi is clearly visible (right)

slot (with normal and "fictitious" resonances) which covers multiple high frequency bands (4.9–5.875 GHz). The use of a half slot for the low band removes the odd numbered resonances $F_1, F_3, \ldots$ ($F_0 = 2.45$ GHz) by removing the required short circuit boundary condition for these slot resonances. The removal of the F_1 resonance allows the bottom full dual-band slot antenna to operate from 4.9 to 5.875 GHz without interference from the upper slot's driving point impedance [44].

7.4 Stripline Series Slot Antenna

The first author to describe the design of a Stripline Series Slot Antenna (SSSA) was Milligan in 1985 [45]. The geometry of the SSSA is presented in Figure 7.24. The antenna is fed using a 50 Ω stripline transmission line. A slot is cut in the upper groundplane, across which, the stripline crosses the slot, and terminates with a shorting via.

The normalized conductance, G_s, presented by the slot, with respect to the stripline transmission line characteristic admittance, was given by Oliner in 1955 [46]. Milligan presented Oliner's equation with an assumed 50 Ω stripline characteristic impedance. This is reproduced as (7.10):

$$G_s = \frac{8\sqrt{\epsilon_r}}{45\pi^2}\left(\frac{L_s}{\lambda}\right)^2\left[1 - 0.374\left(\frac{L_s}{\lambda}\right)^2 + 0.130\left(\frac{L_s}{\lambda}\right)^4\right] \tag{7.10}$$

The length of the slot is L_s, and ϵ_r is the relative dielectric permittivity of the board material. The effective relative dielectric permittivity, ϵ_e, presented to the slot is

$$\epsilon_e = \frac{2\epsilon_r}{1 + \epsilon_r} \tag{7.11}$$

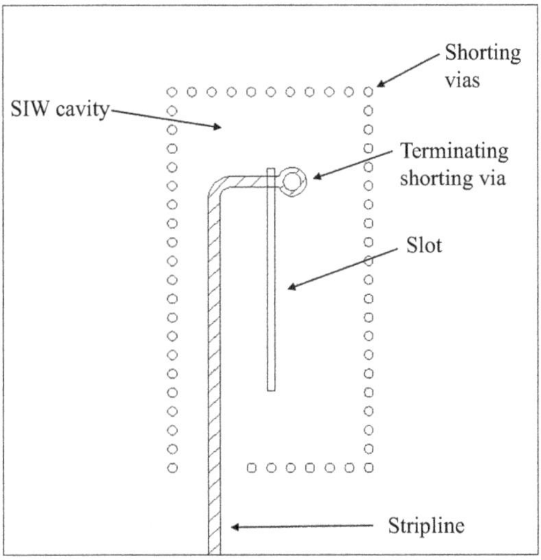

Figure 7.24 Stripline Series Slot Antenna (SSSA) uses stripline to feed a slot cut into the upper groundplane, centered within a Substrate Integrated Waveguide (SIW) cavity. The stripline is terminated with a via that is shorted to the upper ground plane which contains the slot

When $\epsilon_r = 3.0$ dielectric material (Rogers RO3003) is utilized for the stripline design, the effective permittivity is 1.5. The resonant length of a slot in free space is $L_s = 0.48\lambda_0$. The dielectric loading reduces this length to:

$$\frac{L_s}{\lambda} = \frac{0.48}{\sqrt{1.5}} = 0.392$$

Equation (7.10) is then used to obtain the conductance presented to the stripline as 4.533 mS or 220.6 Ω. An estimate of the 50 Ω driving point impedance location $X_{R_{in}}$ can be obtained using (7.12):

$$X_{R_{in}} = \frac{L_s}{2} - \frac{\lambda}{2\pi\sqrt{\epsilon_e}} \cdot \sin^{-1}\sqrt{\frac{R_{in}}{R_{center}}} \tag{7.12}$$

$$X_{50\Omega} = \frac{0.392\lambda}{2} - \frac{\lambda}{2\pi\sqrt{1.5}} \cdot \sin^{-1}\sqrt{\frac{50}{220.6}} = 0.131\lambda$$

At 10 GHz, $\lambda = 30$ mm, and $X_{50\Omega} = 3.93$ mm from the center of the slot. $L_s = 11.76$ mm.

The stripline is designed using two 500 μm thick layers of material for a total thickness of 1000 μm. The width of the 50 Ω stripline center conductor is 615 μm. The total stripline plate separation is 1000 μm. The conductor thickness is 17 μm. The approximate formulas presented, provide useful starting values, which can then be empirically adjusted for optimum performance. When optimized, the width of the

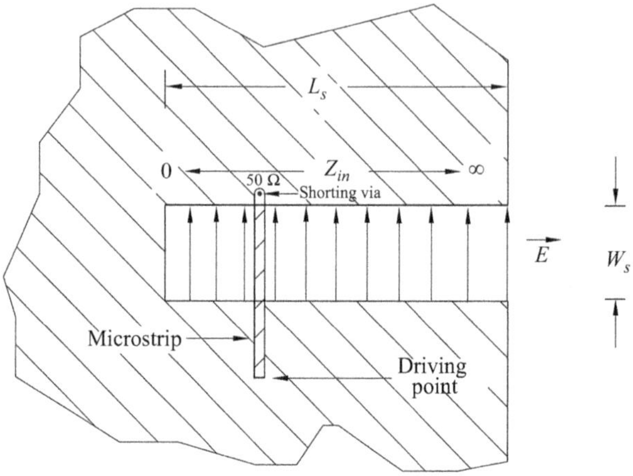

Figure 7.25 A half-wavelength slot antenna is cut through its high impedance center to form a quarter-wave slot antenna

slot is 0.4 mm with a length of 12.162 mm (400 μm × 12 162 μm). The 50 Ω driving point location, from the center of the slot, is 5.35 mm.

The shorting pins surrounding the slot enclose any parallel plate modes that are generated. Parallel plate modes have no lower frequency cutoff. The slot is centered above the cavity boundaries. Any TE_{10} mode that is driven in the SIW cavity has a null at its center, and therefore the slot does not couple to it. The dimensions of the via enclosure are 11 mm × 20.5 mm, with 0.5 mm via diameters. The diameter of the terminating via is 1 mm with a 1.6 mm pad diameter.

This 10 GHz example has a gain of 6.39 dBi. The radiation efficiency computed is 91.94%. The 2:1 VSWR bandwidth is 265 MHz, or about 2.65%. The bandwidth remains stable with increasing groundplane separation.

The SSSA allows one to fabricate a two element slot array with a free space wavelength separation of $\lambda_0/2$ using a high dielectric substrate. This is equivalent to the radiating edge separation of a rectangular microstrip antenna using an air dielectric which is the square patch design with maximum directivity. This allows the designer the option of using common solid dielectric material. The trade-off is increased design complexity.

7.5 Inverted F antenna

The inverted F antenna (IFA) is derived from a slot antenna. A slot antenna generally has a length of about one-half guide wavelength in the slot. It was shown in Section 7.3 that the driving point impedance varies from zero at each end of the slot to maximum at the center which can be approximated as an open circuit. The slot is cut in half at this open-circuit plane when forming an IFA. This is illustrated in Figure 7.25.

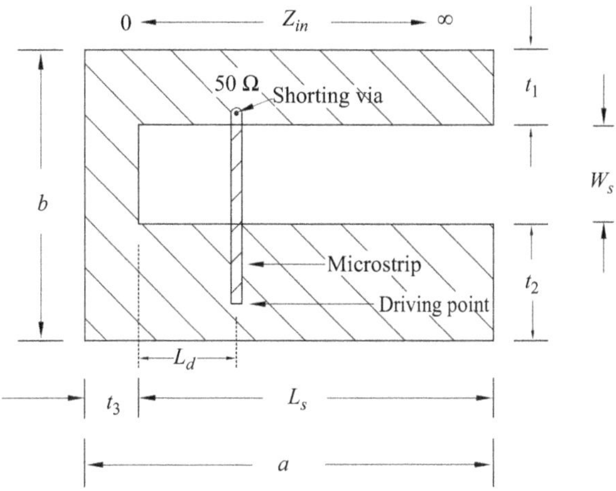

Figure 7.26 The quarter-wave slot antenna in Figure 7.25 has its infinite groundplane reduced on three sides to form an inverted F antenna

The infinite groundplane surrounding the left side of the quarter-wave slot antenna is truncated on all sides. The result is an antenna geometry that is often placed at the edge of a printed circuit board of a wireless device. The resulting antenna is shown in Figure 7.26. The overall footprint of the antenna is a by b. The length of the slot is L_s. The width of the slot is W_s. The driving point impedance varies from zero to open circuit, and a distance from the short to a 50 Ω driving point is selected. The equations of Section 7.3 may be used to estimate the location of the desired driving point impedance as a starting point for a full-wave analysis. If the values of t_1, t_2 and t_3 are large enough, the IFA is very close to the ideal case of a quarter-wave slot at the edge of an infinite groundplane.

This antenna can be placed along a radius, or curved, as long as the length of the slot maintains one-quarter of a guide wavelength. The versatility and ease of integration with many wireless devices has made the IFA on a PCB a very popular choice.

For thicker substrates, the microstrip feed line can become rather wide compared with the short distance L_d required to match the antenna. The width of the microstrip line can interfere with the matching for thick substrates at microwave frequencies. A choice of higher dielectric permittivity can both slightly decrease the slot length, and reduce the width of the microstrip feed line.

A common IFA design frequency is 2.45 GHz. As an example, an IFA was designed on common FR4 and analyzed with HFSS at this frequency. The results are presented in Table 7.1. The distance of the microstrip transmission line from the edge of the slot (L_d) increases as the slot width (W_s) is increased. As the width is decreased, it becomes harder to feed the antenna. It is clear that increasing the width of the slot has only a small affect on bandwidth and efficiency.

Table 7.1 IFA design: a = 25 mm, b = 30 mm, $\epsilon_r = 4.4$, $\tan\delta = 0.02$, $t_1 = 3$ mm, $t_2 = 20$ mm, $t_3 = 3.25$ via radius = 0.5 mm, substrate thickness = 762 μm (2.45 GHz)

W_s	IFA design example			
	L_s	**Bandwidth**	**Bandwidth**	**Efficiency**
4 mm	21.3 mm	188.4 MHz	7.69 %	90.3 %
5 mm	21.1 mm	190.9 MHz	7.79 %	91.5 %
6 mm	21.7 mm	211.7 MHz	8.64 %	92.5 %
7 mm	21.8 mm	209.1 MHz	8.53 %	93.4 %

7.6 Electrically small antennas

7.6.1 Electrically small antenna limitations

In some applications, an electrically small planar antenna can be an appropriate solution. Electrically Small Antennas (ESAs) have fundamental limitations to their bandwidth. The definition of an ESA is an antenna whose maximum dimension is less than $\lambda/2\pi$ [23]. This relation is often expressed as

$$ka < 1 \tag{7.13}$$

$k = \frac{2\pi}{\lambda}$
λ = free space wavelength
a = radius of sphere enclosing the maximum dimension of the antenna

An ESA has an inherent *minimum* value of Q. This places a limit on the attainable impedance bandwidth of an Electrically Small Antenna (ESA). The higher the antenna Q, the smaller the impedance bandwidth.

The efficiency of an ESA is determined by the amount of losses in the conductors, dielectrics and other materials out of which the antenna is constructed compared with the radiation loss. This can be expressed as

$$\eta_a = \frac{R_r}{R_r + R_m} \tag{7.14}$$

η_a = efficiency of ESA
R_r = radiation resistance (Ω)
R_m = material loss resistance (Ω)

The input impedance of an ESA is capacitive, and in order to provide the maximum transfer of power into the antenna's driving point, a matching network is required. The efficiency of the antenna and its matching network is expressed as

$$\eta_s = \eta_a \eta_m \tag{7.15}$$

η_s = efficiency of system (i.e. antenna and matching network)
η_m = efficiency of matching network

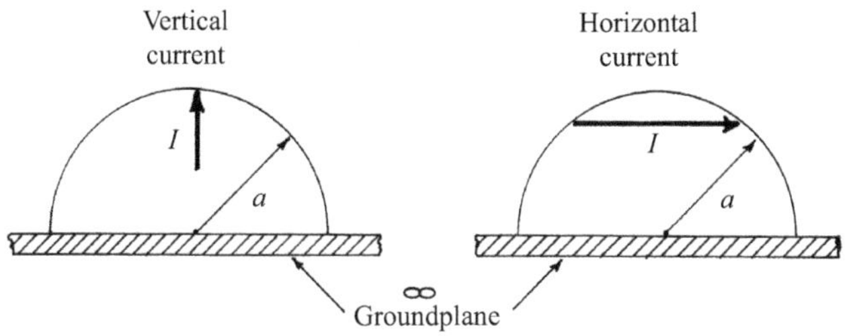

Figure 7.27 *Vertical and horizontal ESA's (current) over a large groundplane and their enclosing spheres*

Using common assumptions, the efficiency of the matching network is approximately:

$$\eta_m \approx \frac{\eta_a}{1 + \frac{Q_a}{Q_m}} \tag{7.16}$$

Q_a = Q of ESA
Q_m = Q of matching network

In 1996 McLean refined and corrected earlier work on the minimum Q of an ESA [24]. The minimum Q for an electrically small linear antenna in free space is expressed as

$$Q_L = \frac{1}{k^3a^3} + \frac{1}{ka} \tag{7.17}$$

The minimum Q for an ESA which is circularly polarized is

$$Q_{cp} = \frac{1}{2}\left[\frac{1}{k^3a^3} + \frac{2}{ka}\right] \tag{7.18}$$

Equations (7.17) and (7.18) assume a *perfect lossless* matching network.

The minimum Q relationship was originally derived for the case of an ESA in free space. In any realistic environment, an ESA is near some type of groundplane or other structure. In 2001, Sten *et al.* evaluated the limits on the fundamental Q of an ESA near a groundplane [25]. These relationships provide useful guidelines on theoretical limits to the development of an ESA with a desired impedance bandwidth.

The Q for the case of a horizontal current element, and a vertical current element, over a groundplane, are analyzed as illustrated in Figure 7.27. The formulas for the Q of both instances are found in Sten *et al.*

The approximate bandwidth for an RLC type circuit in terms of Q is

$$BW = \frac{S-1}{Q\sqrt{S}} \tag{7.19}$$

$S = S : 1$ VSWR
BW = normalized bandwidth

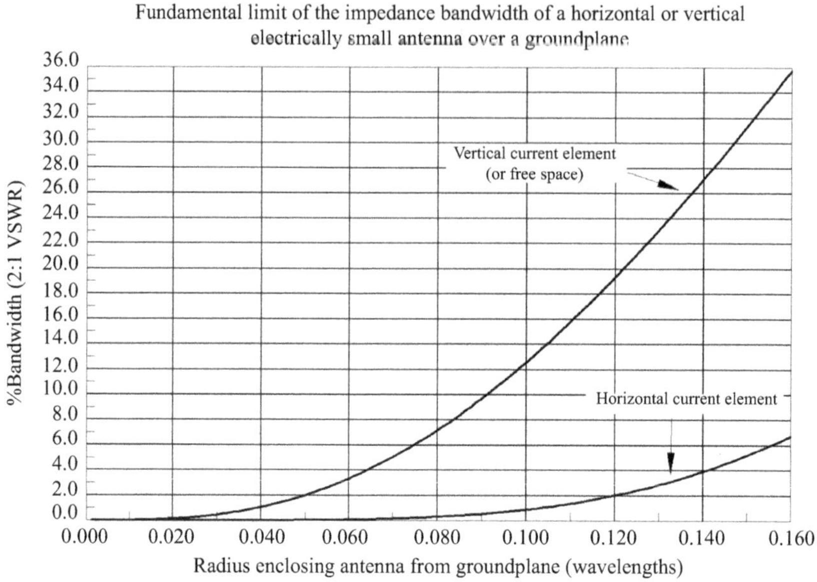

Figure 7.28 Fundamental bandwidth limits of a vertical and horizontal current

Figure 7.28 presents these impedance bandwidth results in a graphical form. The maximum (normalized) percent impedance bandwidth for the vertical and horizontal polarization cases, with respect to the radius of a sphere which encloses the ESA, may be determined with Figure 7.28. In the situation of a vertical ESA over a groundplane, we find its Q is equivalent to the free space case. When a horizontal current is over a groundplane, the radiation efficiency is reduced. The tangential electric field at the surface of a perfect conductor is zero. As a horizontal ESA is moved closer and closer to the surface of a conductor, the radiation decreases, the energy in the stored near fields increases, the Q becomes large, and the bandwidth becomes small. In many practical cases, the proximity of a groundplane will decrease the attainable bandwidth of an ESA. The ESA limit is fundamental, and has been validated extensively [57]. The difficulty of ESA design produces ample opportunities for self-deception [58].

7.6.2 Meanderline antenna

In Figure 7.29, we have the basic meanderline antenna geometry. The antenna itself is the rectangular meandering conductor trace which is of width W_c with separation W_s, and is of overall length L and W. An electrically small matching network, $(\lambda/10)$, using microstrip transmission line of length L_m and W_m is just below the ESA's driving point. The characteristic impedance of this transmission line section may be determined using computer optimization to provide enough series inductive reactance to cancel the large capacitive reactance of the meanderline ESA. This matching section is attached

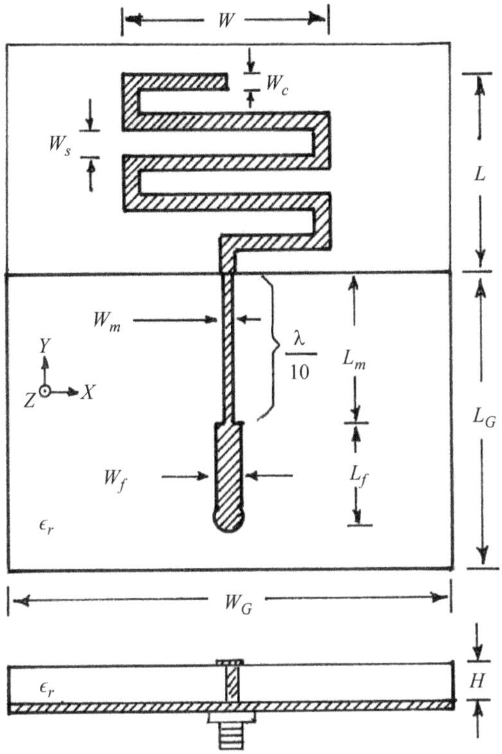

Figure 7.29 Electrically small meanderline antenna geometry

to a 50 Ω microstrip line which is fed with a coaxial probe. The antenna is above a groundplane of width W_G and length L_G.

The meanderline antenna operates in a manner analogous to a self-resonant inductor. The capacitance between the windings of an inductor and the parallel inductance of each coil winding form an equivalent resonant *LC* circuit. When an inductor is driven at its self-resonant frequency it radiates electromagnetic waves. The meanderline antenna element's vertical sections act like inductors which are in parallel with each pair of horizontal traces which act like capacitors. The current on each pair of horizontal traces flows in opposite directions, and cancel in the far field. The radiation is produced by each of the short vertical sections. The electric field polarization in the far field is along the *Y*-axis.

We will assume that even though the meanderline resonator and groundplane section are thin, that the minimum Q restrictions for a vertically polarized ESA over an infinite groundplane will approximately apply to this geometry.

We will now use a 1.575 GHz ($\lambda = 190.48$ mm) meanderline antenna to estimate the best case impedance bandwidth we can expect to obtain for this geometry. The substrate is $H = 0.762$ mm (0.030 inch) FR4 with $\epsilon_r = 3.9$ and a 0.02 $\tan\delta$.

The groundplane dimensions are $W_G = 16.7$ mm and $L_G = 39.0$ mm. The meanderline antenna dimensions are $W_c = 1.07$ mm and $W_s = 1.71$ mm ($L = 13.91$ mm and $W = 14.552$ mm). The matching network values are $W_m = 0.660$ mm and $L_m = 10.27$ mm. The 50 Ω microstrip transmission line dimensions are $W_f = 1.65$ mm and $L_f = 8.02$ mm.

The radius of a sphere which can enclose the meanderline antenna, assuming an infinite groundplane, is $a = 15.63$ mm. We calculate the free space wavelength, and wave number, which allows us to evaluate $ka = 0.515$.

We can see that ka is less than one, and this 1.575 GHz meanderline antenna is, by definition, an electrically small antenna. This antenna is known to be linear, and polarized vertical to the groundplane, so we easily calculate the radiation Q using (7.17) to be $Q_L = 9.22$.

We choose a 2:1 VSWR limit and compute the bandwidth:

$$BW = \frac{1}{Q_L\sqrt{2}} = 0.0291 = 7.66\%$$

Unfortunately, this does not match with the computed percent bandwidth of 17.4% from FDTD. At first glance, this antenna appears to violate a fundamental limit of ESAs. We can better understand the situation by computing the Q value which corresponds to a 17.4% (0.174) impedance bandwidth. We obtain $Q_L = 4.06$ for this bandwidth. We next determine what ka value is required to produce a 4.06 value for Q_L. The value is $ka \approx 0.72$, which is still electrically small and should be restricted by minimum Q. We know the value of k at 1.575 GHz. The value of enclosing radius is

$$a = 0.72/(32.987 \cdot 10^{-3} \text{ radians/mm}) = 21.83 \text{ mm}$$

In the case where we have an ESA with vertical polarization over a groundplane, the radius of the antenna appears to be expanded from 15.63 to 21.83 mm. The explanation for this is that the radiation of the meanderline structure includes about 6.2 mm of the groundplane. These extra currents are seen in the FDTD analysis results presented in Figure 7.30 on the upper left and upper right vertical edges of the groundplane. These patches of current are in phase with the four vertical high current radiating sections on the meander line. One can see the horizontal currents on the meander line sections cancel. The complement of currents on the groundplane cancel with the currents on the upper microstrip to form a transmission line.

If one increases the width of the meanderline antenna groundplane, the impedance bandwidth will decrease until it reaches a limit. When the bandwidth limit is reached, the dimensions of the groundplane have become large enough so that the vertical currents on the meanderline do not drive currents along the edges of the groundplane. FDTD analysis confirms this occurs. The FDTD results of Figure 7.31(a) and (b) show the meanderline ESA with its original groundplane and modified groundplane width. The thermal plots show the magnitude of the edge current which correlates to the direction shown in Figure 7.30. $W_G = 66.7$ mm. One can see by comparison with Figure 7.31(d), no significant currents exist on the edges when the groundplane is

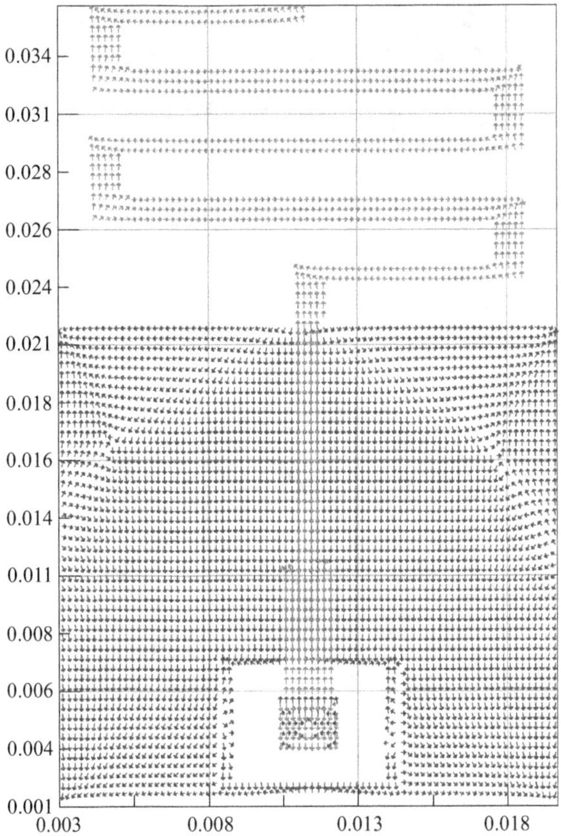

Figure 7.30 *Surface current on the meanderline element and its groundplane as computed by FDTD. The two sets of parasitically driven currents on either side of the narrow groundplane are clearly visible. These are responsible for the measurement of a larger bandwidth than expected for a meanderline on a narrow groundplane ESA. The square section is a square coaxial probe feed*

widened. The width of the electrically small matching section had to be increased in width ($W_m = 4.8$ mm) to cancel the increased capacitive reactance of the meanderline antenna driving point as the antenna's Q increased.

The main difference between the narrow and wide groundplane meanderline ESA antennas is the impedance bandwidth. When the groundplane width is increased to 66.7 mm, the bandwidth of the element decreases to 5.19% bandwidth. This value is in fair agreement with our computed estimate of 7.66%. Realizing the fundamental bandwidth limit in practice has proven elusive. Theoretical work by Thiele *et al.*

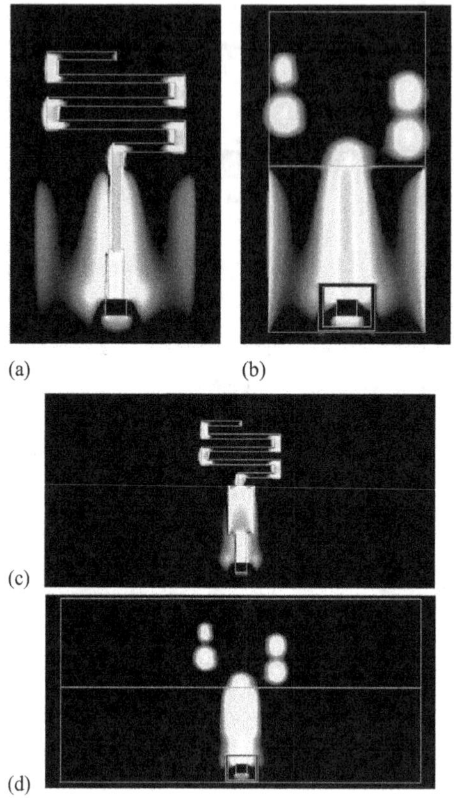

Figure 7.31 *Magnitude of surface current on (a) ESA conductor, (b) ESA groundplane, (c) ESA conductor + 25 mm GP and (d) ESA groundplane + 25 mm GP*

suggests that this theoretical limit is based on a current distribution which is not obtainable in practice [26].

Figure 7.32 shows the computed impedance bandwidth change for the baseline antenna groundplane width ($W_G = 16.7$ mm) and after 25 mm of extra groundplane are added to each side. The reduction in impedance bandwidth is clearly illustrated.

A pair of antennas were constructed using the dimensions obtained with FDTD. Figure 7.33 shows the measured impedance bandwidth change for the baseline antenna, and with 25 mm of extra groundplane. We note the measurements correlate very well with the predicted FDTD analysis. The measured antennas had a slightly higher resonant frequency than the analysis.

One must realize, that when an ESA is used in an application, the groundplane and environment have a considerable effect on the bandwidth, and what the electrical extent of the actual antenna is versus what one might assume.

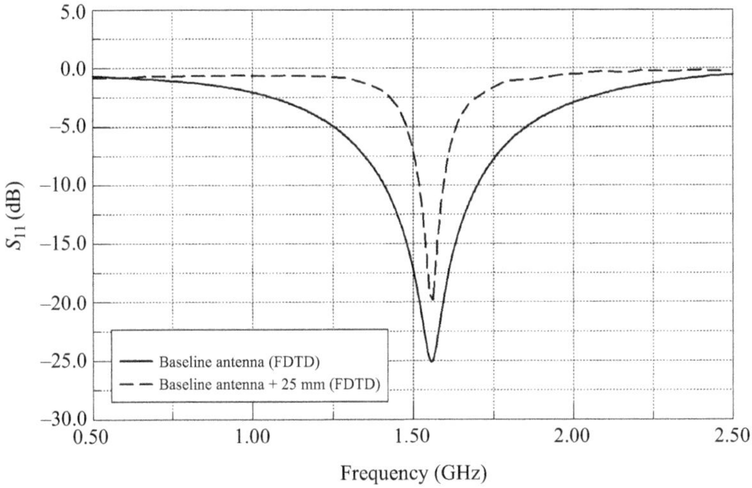

Figure 7.32 *Meanderline baseline antenna S_{11} dB with baseline groundplane width and with 25 mm added to each side predicted by FDTD analysis. The plot demonstrates that as the width of the groundplane is increased, the sides no longer contribute to the radiation, and the bandwidth reduces to that predicted by ESA fundamental limits*

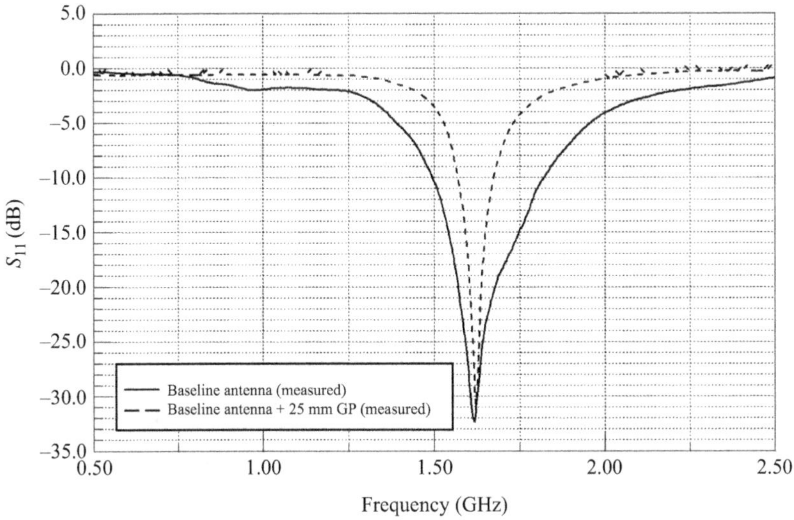

Figure 7.33 *Meanderline baseline antenna S_{11} dB with baseline groundplane width and with 25 mm added to each side measured on fabricated antenna. The bandwidth plot is consistent with Figure 7.32 which uses FDTD analysis*

7.6.3 Meanderline antenna radiation patterns

The antenna patterns computed by FDTD are nearly equivalent for the small and large groundplane (2.0 dB directivity) (Figure 7.35). Computed radiation patterns for the wide groundplane case are presented in Figure 7.34. The elevation plane radiation pattern in Figure 7.34 is similar to that of a dipole, and matches the direction of the radiating currents of Figure 7.30. The pattern is omnidirectional in the x–z plane, consistent with the computed radiating current.

The antenna patterns computed using FDTD, are essentially equivalent for the small and large groundplane (2.0 dB directivity). FDTD modeling allows for "perfect" feeding of the antenna, which minimizes perturbation from a coaxial feedline.

In practice, the gain of an ESA is bounded. This limitation has been expressed by Harrington [27] as

$$G = (ka)^2 + 2(ka) \tag{7.20}$$

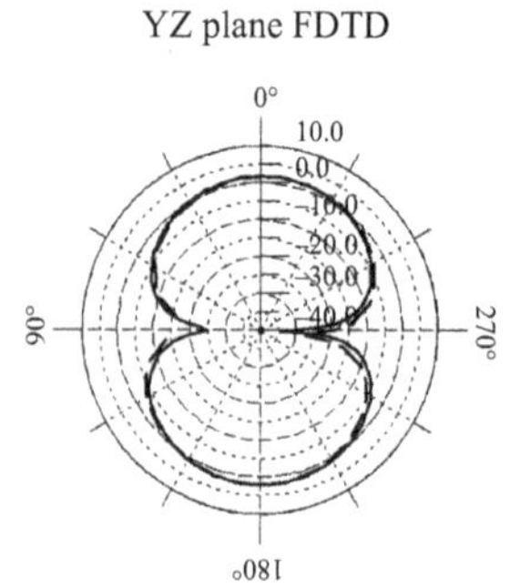

Figure 7.34 Radiation patterns of the narrow groundplane baseline (solid lines) meanderline ESA computed with FDTD, and the antenna with 25 mm wide groundplane width added (dashed lines)

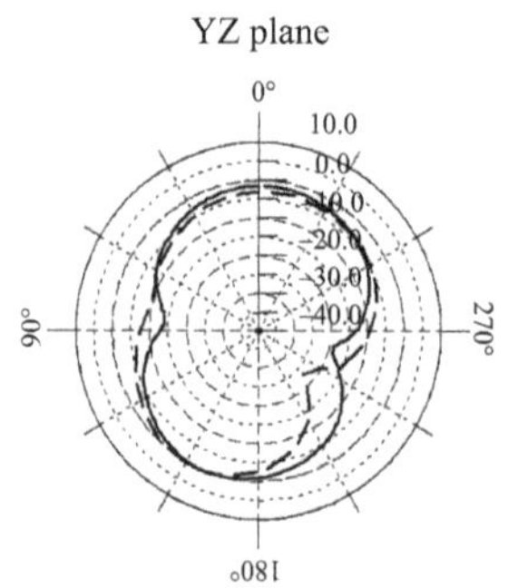

Figure 7.35 The measured radiation patterns of the baseline (narrow groundplane) meanderline ESA (dashed lines), and the antenna with 25 mm groundplane width added (solid lines)

When applied to the meanderline antenna, the maximum attainable gain for the antenna on a large groundplane ($a = 15.63$ mm) is 1.13 dBi, when the groundplane is reduced ($a = 21.83$ mm), we have a maximum *possible* gain of 2.9 dBi.

Meanderline antennas were fabricated and found to match in at 1.655 GHz (4.83% from 1.575 GHz). When measured, the maximum gain of the meanderline antenna with a large groundplane is 0.3 dBi. The measured gain value of the antenna with a smaller groundplane is 0.5 dBi. The smaller groundplane meanderline antenna generated more current along the coaxial cable which connects the antenna to the ESA than the wider antenna. This makes measuring the small groundplane antenna in isolation difficult and adds loss. This measurement problem has been noted and discussed by Staub *et al.* [28]. An ESA has a combination of balanced and unbalanced mode, which makes pattern measurement particularly problematic when using a coaxial (unbalanced) cable to feed the ESA.

7.6.4 Half patch with reduced SC plane (PIFA)

In 1987, Taga *et al.* introduced a modified version of a half-patch ($\lambda/4$) microstrip antenna, in which the width of the shorting plane was altered to reduce the resonant frequency of the antenna [33]. The geometry of a PIFA is presented in Figure 7.36. The antenna is fed along one edge at location F. The short is reduced in width with respect to the edge opposite that of the feed to width W. The element is of dimensions L_1 and L_2 and thickness H. The original PIFA uses air as its dielectric substrate [34].

When $W = L_1$, the antenna becomes a quarter wave microstrip (half-patch) antenna (Figure 2.11). As the width W becomes small enough to be equivalent to a shorting post, the antenna has the appearance of an IFA with a conductive plane attached to one side. The element was called a Planar Inverted F antenna (PIFA) from this view of the geometry.

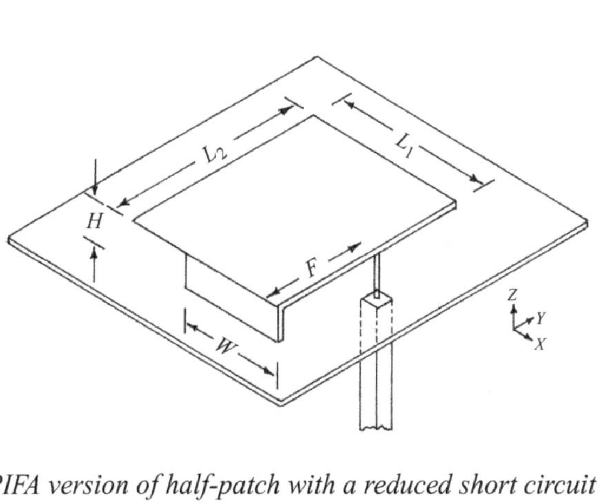

Figure 7.36 PIFA version of half-patch with a reduced short circuit plane. One patch edge has a probe feed. This same patch edge is terminated with a short circuit of width W

The resonant frequency of the antenna decreases as the value of W decreases. When $W/L_1 = 0.125$, it has been reported that the resonant frequency is about 40% of a half-patch, when $L_1/L_2 = 2.0$, for a ratio of $L_1/L_2 = 1.0$, the resonant frequency is about 60% of the half-patch, when $L_1/L_2 = 0.5$, it is lowered to about 70% of a normal quarter wavelength patch antenna.

Taga computed the input impedance for a PIFA, and then widened the short to create a quarter-wave microstrip antenna (QWMSA). In both cases the impedance was computed using an infinite groundplane. The antenna parameters are $L_1 = 30.0$ mm, $L_2 = 45.0$ mm, $H = 9.0$ mm. The PIFA (case A) has a feedpoint at $F = 3.0$ mm and a shorting plane width of $W = 6.0$ mm. The QWMSA has a feedpoint located at $F = 15.0$ mm and a shorting plane $W = L_1 = 30.0$ mm (case B).

FDTD was used to analyze these antennas. The resonant frequency of the QWMSA is 1.450 GHz, the PIFA has a resonant frequency of 1.020 GHz. The frequency of the PIFA is 29.7% lower than the QWMSA. The negative return loss plot for each case is presented in Figure 7.37. We note that the QWMSA has a normalized impedance bandwidth of 10.35%, and the PIFA has a 2.77% bandwidth. The reduction in resonant frequency comes at a considerable penalty in bandwidth.

One must keep in mind that these examples of a QWMSA and PIFA are both ESAs. The QWMSA has $ka = 0.865$ and the PIFA has $ka = 0.608$ which are both less than one. The impedance bandwidth of a PIFA can be increased by increasing H, which increases the volume of the antenna [33]. One can decrease the size of the groundplane upon which the PIFA is constructed, until the groundplane itself becomes part of the radiating structure, and increases the impedance bandwidth [35]. This was

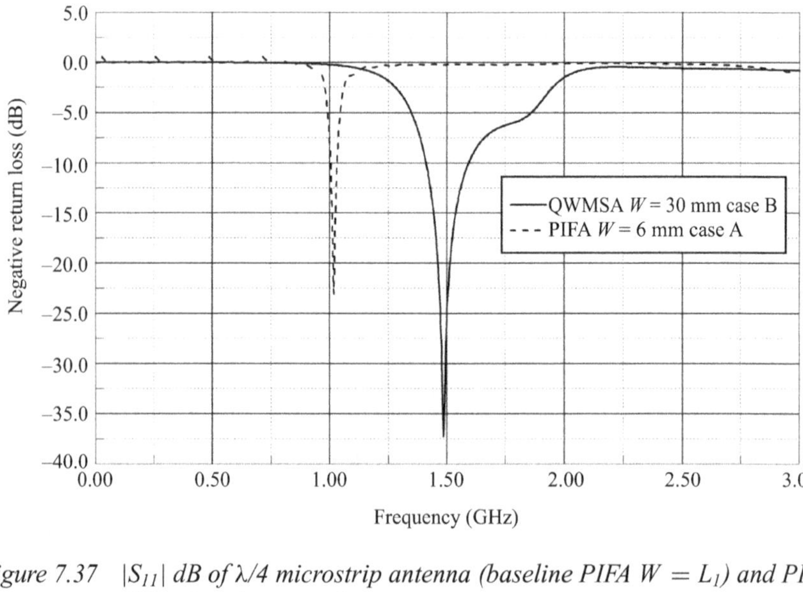

Figure 7.37 $|S_{11}|$ *dB of λ/4 microstrip antenna (baseline PIFA* $W = L_1$*) and PIFA (*$W = 6$ *mm) versus frequency*

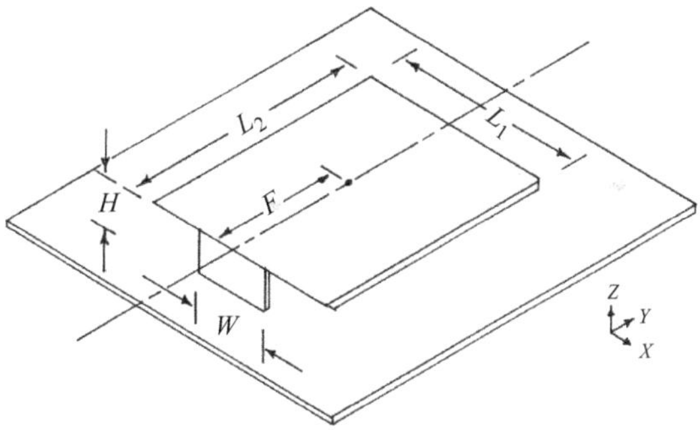

Figure 7.38 Centered short version of half-patch with a reduced short circuit plane. The probe feed is placed along the patch centerline. The short circuit plane is of width W

illustrated earlier with the Meanderline Antenna. One must trade-off the decrease in resonant frequency as W is decreased, with an associated decrease in impedance bandwidth for a given application.

One can create a PIFA type structure where the back shorting plane is centered, and the feed is positioned along the centerline (CPIFA). This is illustrated in Figure 7.38. If the width of the shorting plane is that of case A ($W = 6.0$ mm) with the CPIFA geometry, the resonant frequency is a slightly higher 1.112 GHz with 2.79% bandwidth.

When the shorting plane width is small, ($W \ll L_1$), the resonant frequency of a PIFA ($\epsilon_r = 1$) is approximately:

$$F_r = \frac{c}{4(L_1 + L_2)} \tag{7.21}$$

When the shorting plane width is large, correcting for the width produces a more accurate expression for resonant frequency:

$$F_r = \frac{c}{4(L_1 + L_2 - 0.978 \cdot W)} \tag{7.22}$$

A PIFA may be viewed as an LC circuit, where the upper conductor is a capacitor, and the shorting plane is an inductor. The inductance increases as the width of the shorting plane is decreased, which lowers the resonant frequency of the circuit. The PIFA is also analogous to a microstrip antenna which has an edge resistance (see Section 2.4). The edge resistance increases as the electrical width of the antenna decreases. The position of the 50 Ω location thus moves toward the shorting plane in a manner consistent with the half-patch transmission line model of Figure 2.12. Figure 7.39 illustrates this driving point location change (black dot) as the shorting plane is decreased, and the corresponding resonant frequency decreases (as computed with FDTD). The PIFA dimensions H, L_1 and L_2 are the same as Taga's, with a

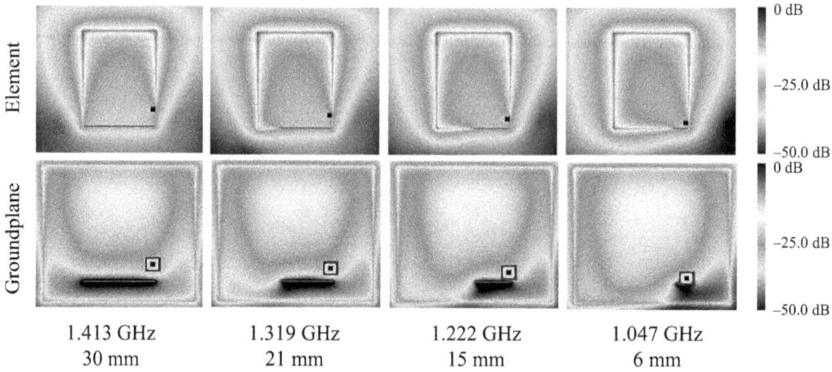

Figure 7.39 The operating frequency of a PIFA is decreased as the width of the shorting plane is decreased. The edge resistance of the patch increases because the electrical width of the antenna narrows. Because of this edge resistance increase, the location of a 50 Ω driving point moves toward the shorting plane. This is illustrated with electric field magnitude plots for a normal PIFA using FDTD above

63 mm × 63 mm groundplane. The impedance bandwidth of a PIFA with a finite-sized groundplane is generally smaller than that predicted for an infinite groundplane. This finding is consistent with the results for a centered shorting plane PIFA reported by Huynh and Stutzman [36].

The movement of the 50 Ω feed point location toward the shorting plane as W is decreased, for a centered shorting plane, is shown in Figure 7.40. The results are very similar to those for the standard PIFA geometry.

In practice, a PIFA is often realized using a shorting post rather than a shorting plane. FDTD analysis demonstrates that as the radius of the shorting post decreases so does the resonant frequency. As the resonant frequency decreases, the 50 Ω driving point position moves closer to the shorting post as seen previously with the shorting planes.

HFSS analysis illustrates that a contour of 50 Ω driving point locations exist for the case of a PIFA realized with a shorting post (Figure 7.41). The PIFA dimensions H, L_1 and L_2 are the same as Taga's with a 63 mm × 63 mm groundplane and 2 mm radius shorting post.

As the shorting plane width or shorting post radius is decreased, the location of a 50 Ω driving point can become so close to the short to be unrealizable.

In many practical applications, the groundplane and PIFA dimensions can produce a driving point anti-resonance which renders the antenna unrealizable. The currents on the feed center conductor, which have upward and downward currents which tend to cancel, can begin to couple to the PIFA shorting post/plate. When the feed probe is close enough, the coupling increases to the point where the current on the feed and shorting post/plate travel in phase (i.e. in the same direction), and in practice have produced an anti-resonance.

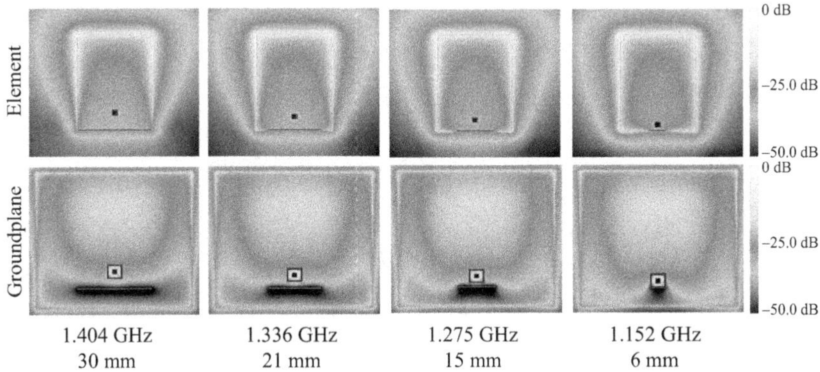

Figure 7.40 *The operating frequency of a PIFA is decreased as the width of the centered shorting plane is decreased. The edge resistance of the patch increases because the electrical width of the antenna narrows. Because of this edge resistance increase, the location of a 50 Ω driving point moves toward the shorting plane. This is illustrated with electric field magnitude plots for a normal PIFA using FDTD above*

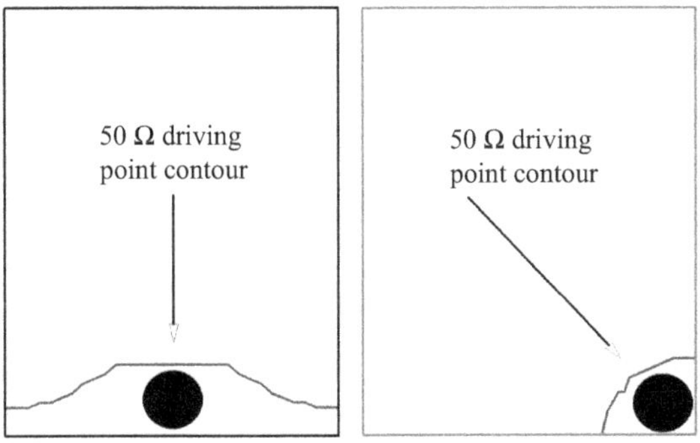

Figure 7.41 *50 Ω driving point locations for PIFA with centered (2 mm radius) shorting post (left) and offset shorting post (right). The 50 Ω contour is longer for the centered shorting post case but the resonant frequency is also higher. The resonant frequency for the centered shorting post is 1.267 GHz whereas the resonant frequency of the offset shorting post is 1.090 GHz*

The main advantage of using a PIFA is that it is a very compact ESA. The majority of the radiation is from the vertical current on the shorting plane (post) which is between two thin conductive plates. The impedance bandwidth and gain of a PIFA is consistent with the fundamental limits of an ESA with a vertical current, as described by (7.17) and (7.20). Other comparable printed antenna designs (such as a half slot) must be physically thicker than a PIFA for the same effective height because

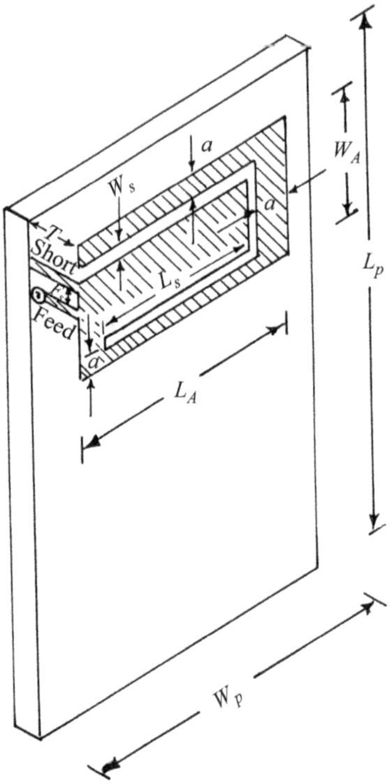

Figure 7.42 Geometry of dual-band PIFA antenna on a metal plate which represents a mobile device

the radiating length is less than the physical height. Alternative methods of decreasing the resonant frequency of a patch without the use of a short exist. Slots and rectangular apertures have been cut into rectangular microstrip antenna elements to reduce the resonant frequency. Details of these designs may be found in the literature [37].

7.6.5 Dual-band PIFA

The Planar Inverted F Antenna (PIFA) has proven popular as a dual-band antenna for mobile devices. Often a slot is cut into the planar conductor to create a dual-band element. A radiating and balanced mode analysis of a dual-band PIFA performed by Boyle and Ligthart reveals important insight into this design [38]. The dual-band PIFA design geometry is shown in Figure 7.42. The slot introduces an additional resonance, which is proportional to the length of the slot. As the slot length becomes approximately one-half wavelength, the currents on each side of the slot shift from co-phase to anti-phase. The natural resonant mode of the radiator is between the two modes. The proper choice of dimensions allows for dual-band operation. A long slot is required to create a dual-band PIFA with this geometry. The PIFA is analyzed on a circuit board which represents a mobile device. The half-wave resonances of the circuit board need to be understood, and taken into account, for a workable design.

The radiating/balanced mode analysis reveals there is an impedance bandwidth tradeoff between the two desired frequencies. As the impedance bandwidth of one resonance increases, it is at the expense of the second resonance bandwidth. The dual-band PIFA may be represented by a series (one resonance) and parallel (second resonance) pair of resonance circuits in series. Both bandwidths are maximized when the two circuits resonate at the geometric mean between the two desired design frequencies. If either resonant frequency differs from the geometric mean a bandwidth tradeoff occurs.

If a dual-band PIFA design has a short circuit (vertical current) which is so electrically small (i.e. T is very small) that the radiation from the horizontal (planar) current dominates over the vertical current at both frequencies, the radiation efficiency from the co-phase mode will be much larger than that of the anti-phase mode. For many systems this large decrease in radiation efficiency renders the antenna unusable at the anti-phase frequency in this case.

Boyle and Ligthart designed a dual-band PIFA to operate at 920 MHz (GSM) and 1800 MHz (DCS). The design parameters are $F_s = 2$ mm, $a = 4$ mm, $L_A = 30$ mm, $W_A = 20$ mm, $L_s = 23$ mm, $W_s = 1$ mm, $T = 8$ mm, $L_p = 100$ mm and $W_p = 40$ mm.

7.7 Tapered balun printed dipole

The printed dipole of Section 7.8 uses the groundplane as a reflector to increase the gain of the antenna. A more traditional printed dipole may be created using a microstrip line with a tapered groundplane which acts as a balun.

The geometry of the Microstrip-Fed Tapered Balun Dipole Antenna (MFTBA) is found in Figure 7.43. The taper chosen is generally exponential as expressed in (7.23):

$$\pm x = \pm W_0 e^{-ay} \tag{7.23}$$

$$a = \frac{-\ln (W_m / W_0)}{(L_0 + W_e)} \tag{7.24}$$

An example antenna was analyzed with HFSS and FDTD, and realized using Taconic TLC-32 substrate. The thickness H is 1.5748 mm (0.062 inch) with $\epsilon_r = 3.2$ and 0.003 $\tan \delta$. The antenna dimensions are $W = 62.97$ mm, $L = 40.96$ mm, $L_e = 27.0$ mm, $W_e = 1.794$ mm, $L_0 = 34.68$ mm, $W_0 = 15.536$ mm and $W_m = 1.794$ mm (width of microstrip transmission line). The design frequency is 2.20 GHz.

The negative return loss predicted by FDTD, HFSS, and measured is found in Figure 7.44. The best match was measured at 2.207 GHz, FDTD predicted 2.212 GHz and HFSS predicted 2.200 GHz. The measured value essentially splits the predicted ones. The impedance match was found to be better for increased values of W_0 for this design.

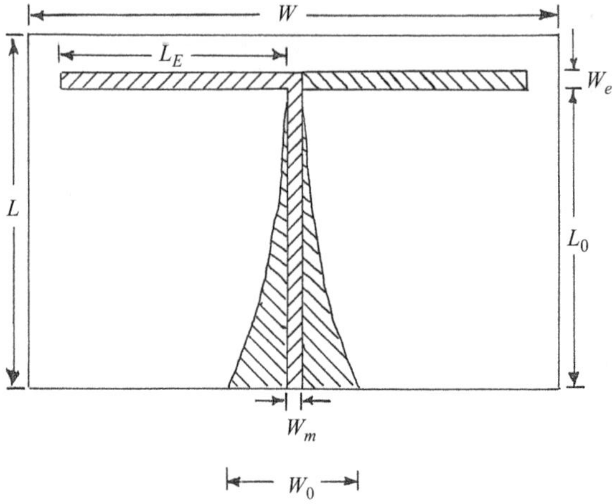

Figure 7.43 The geometry of a microstrip-fed tapered balun dipole antenna (MFTBA)

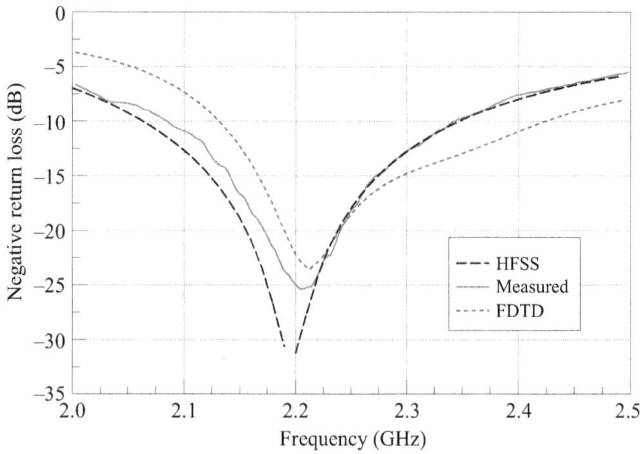

Figure 7.44 Negative return loss of printed dipole antenna with microstrip tapered balun as predicted by FDTD, HFSS and measured

The radiation patterns of the MFTBA are presented in Figure 7.45. The directivity predicted by FDTD is 2.54 dB; HFSS predicts 2.30 dB directivity and 2.15 dBi gain. The measured gain is 2.53 dBi. The FDTD patterns are closer to measurement than the HFSS predictions.

The tapered balun of the MFTBA produces a pattern with some beam squint, but still reduces the current on a coaxial connection line to small values. This dipole design is useful when a dipole is required which does not have as pure dipole patterns

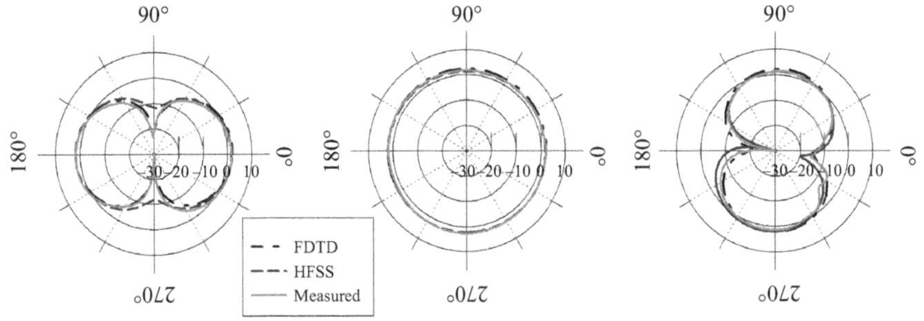

Figure 7.45 Radiation patterns of a printed dipole antenna with microstrip tapered balun, as predicted by FDTD, HFSS, and measured

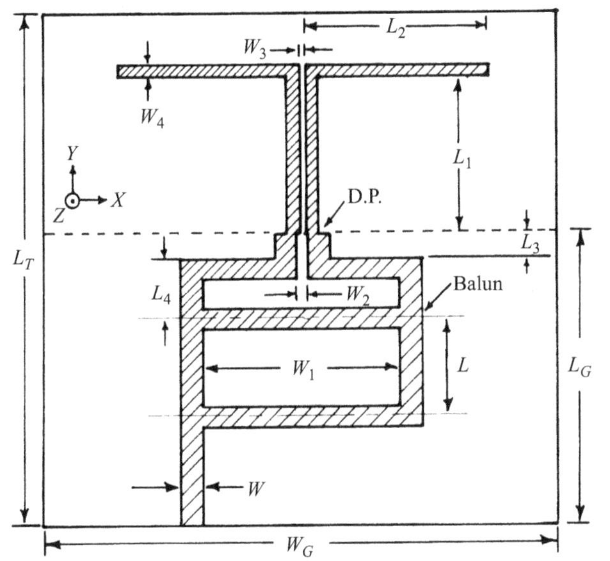

Figure 7.46 Geometry of printed dipole antenna with microstrip balun

as a Roberts Dipole, but is required for a test fixture or other purpose where the ease of fabrication may be of greater importance [49–51].

7.8 Log-periodic balun dipole

A printed dipole antenna fed using microstrip is illustrated in Figure 7.46 [47]. Microstrip is an unbalanced transmission line and requires a balun to produce the desired balanced transmission line mode for the printed twin lead transmission line, which in turn feeds the dipole. The microstrip input is at the bottom of the figure, and

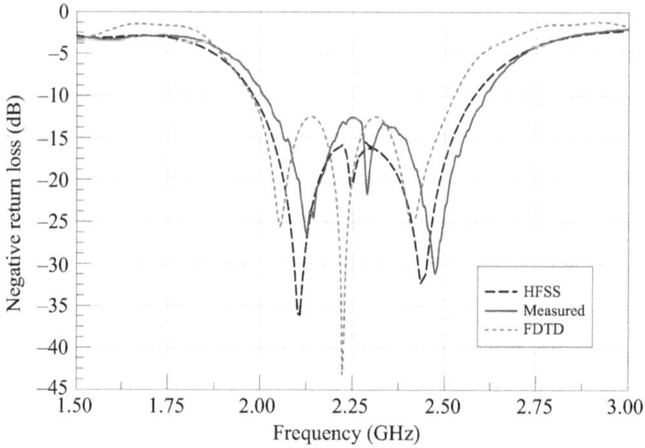

Figure 7.47 Negative return loss of printed dipole antenna with microstrip balun example predicted by HFSS and FDTD with measurement

connects to the input port of a two section log-periodic microstrip balun described by Al Basraoul and Shastry [48]. The two 50 Ω microstrip transmission lines are of opposite phase and equal magnitude as they emerge from the balun. They connect to flat balanced twin lead transmission line which feeds the printed dipole.

A planar dipole of this design can have as much as a 25% normalized impedance bandwidth. An example design, which operates with a center frequency of 2.22 GHz, has a substrate with $\epsilon_r = 3.2$ and a thickness of 1.524 mm (0.060 inch) is used for illustration. The 50 Ω microstrip line which forms the balun has a width of $W =$ 3.64 mm. The value of $L = 19.09$ mm which is a $\lambda/4$ guide wavelength at the center frequency. The width of the delay section of the balun is $W_1 = 33.30$ mm which is approximately $\lambda/2$ guide wavelength at the center frequency. The delay section produces a 180° phase shift. $L_1 = 30.61$ mm which is approximately $\lambda/4$ free space wavelength from the edge of the groundplane. Each dipole element is $L_2 = 30.77$ mm. $W_2 = 2.0$ mm, $W_3 = 1.374$ mm and $W_4 = 2.0$ mm. The last horizontal section of microstrip has an edge which is 5.0 mm below L_G. $L_G = 55.907$ mm. The width of the groundplane section is $W_G = 78.0$ mm. The total length of the antenna board is $L_T = 91.0$ mm. $L_3 = 5$ mm and $L_4 = 9.545$ mm.

In Figure 7.47, the negative return loss plot of the printed dipole with microstrip balun example is plotted. We can see the dipole example has (approximately) a 25% (2:1 VSWR) impedance bandwidth. The maximum directivity of the example dipole design varies from 5.52 dB to 4.82 dB over the 2:1 VSWR bandwidth. The wave radiated by the dipole in the direction of the groundplane edge is reflected back toward the dipole. The quarter-wave spacing of the dipole from the groundplane edge causes the wave to add in phase with the wave radiating outward away from the groundplane. This produces a directivity which is larger than that of a dipole in free

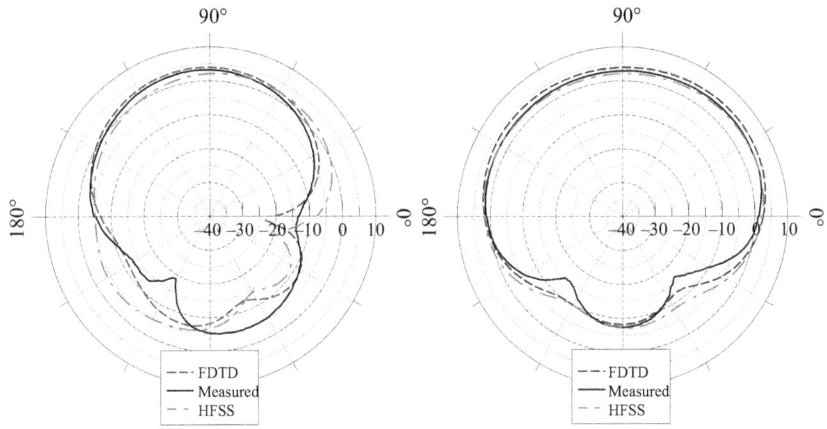

Figure 7.48 E-Plane (left) and H-plane (right) radiation patterns of printed dipole antenna with microstrip balun example design at 2.222 GHz measured and predicted by FDTD and HFSS.

space. In Figure 7.48, we find the antenna radiation patterns computed using FDTD, HFSS and measured at 2.222 GHz.

7.9 Loop antenna and coupler for RFID

Sensor designs for reading RFID tags at 915 MHz are often optimized to read either electrically small tags (which have very small values of radiation resistance) that are essentially directly coupled, or radiative tags which are not electrically small. A single antenna geometry has been developed which has both a section which acts as a direct RFID coupler and also radiates [52]. We will refer to this antenna as an LALC for Loop Antenna/Loop Coupler.

The resonant and anti-resonant dimensions of circular loop antennas have been discussed by Schekunoff and Friis [53]:

$$C = n \cdot \lambda \quad n = 1, 2, 3, \ldots \quad \text{(resonances)} \tag{7.25}$$

$$C = \frac{(2n+1)}{2}\lambda \quad n = 0, 2, 4, \ldots \quad \text{(antiresonances)} \tag{7.26}$$

Equation (7.25) gives the resonant circumferences (C) of a loop antenna in terms of wavelength λ, and (7.26) relates the anti-resonant circumferences.

When the circumference of a loop antenna is $\lambda/2$ or less, neither resonance nor anti-resonance exists. If this electrically small loop is fed in a balanced manner, the current is approximately uniform around the perimeter of the loop. The radiation from the current on one side of the loop will cancel with radiation from a current on the opposite side of the loop in the far field. When the small loop is fed in an unbalanced manner (common mode), the two sides of the loop have currents in the same direction and will radiate more efficiently.

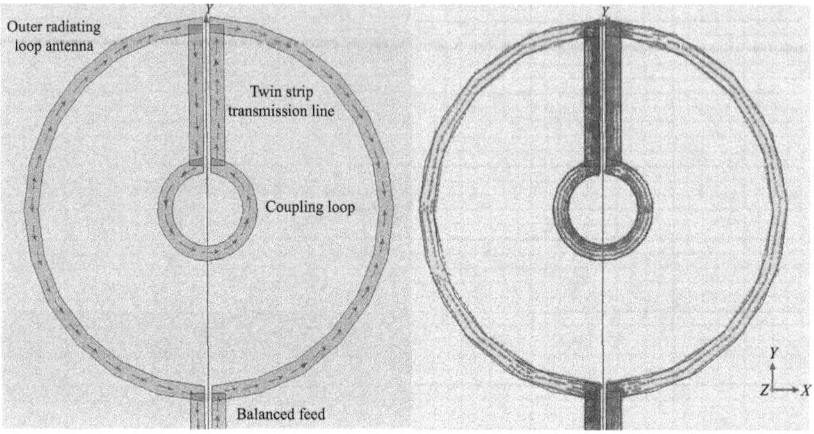

Figure 7.49 *(Left) Desired current distribution on integrated loop antenna and loop coupler geometry. (Right) Realized current distribution as predicted by HFSS (26 mm outer radius). This current distribution is very close to that of the desired distribution*

The first loop resonance occurs when the perimeter of the loop is one-wavelength ($n = 1$). The currents on either side of the outer loop are similar to a pair of parallel curved monopoles with in-phase current.

Through the introduction of a small loop coupler in the center of a larger loop, connected with "twin strip" transmission line, we will produce a current distribution on the larger loop which is similar to that of a one-wavelength loop antenna but with a much smaller dimension. The interior loop is electrically small, and will not radiate efficiently. The connecting twin strip transmission line also produces only a minute amount of radiation.

The desired current distribution is illustrated in Figure 7.49. The loop in the center is electrically small and has a uniform current distribution. For every current on the small loop there is a current on the opposite side which cancels radiation in the far field. The twin strips transmission line feeding the small loop has opposing currents which also cancel in the far field. The entire non-radiating geometry of the center acts as a phase shifter which brings the exterior currents in line which allows them to radiate.

The details of a complete LALC design are shown in Figure 7.50. The design procedure begins with obtaining the driving point impedance of the LALC element. This is done using HFSS. The antenna has a large capacitive reactance which may then be removed using a pair of inductors. This is done so the driving point now presents a pure resistance. This resistance value is then used to design a lattice balun which will both perform an impedance transformation from the antennas driving point resistance to a 50 Ω microstrip transmission line, but also transforms the field structure from an unbalanced to balanced mode. The design equations for a lattice balun are presented in Appendix F, Section F.2.

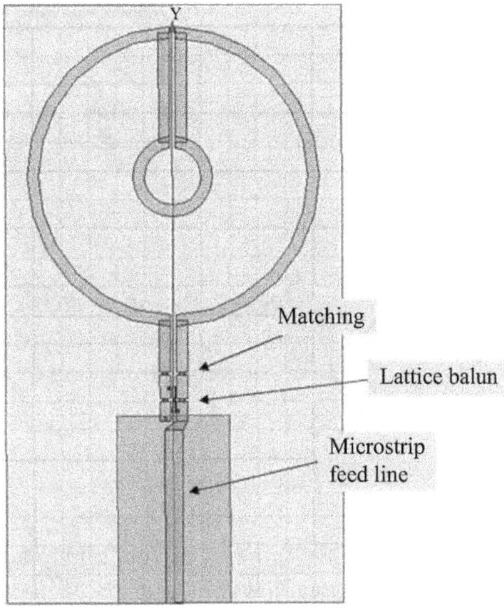

Figure 7.50 *Design details of an LALC. The antenna is first matched to produce a real input resistance. The antenna is a balanced structure and requires a balun to feed it. A lattice balun is used to both act as a transition from unbalanced microstrip to the balanced antenna, and as an impedance transformer*

An LALC was designed with a 26 mm radius at 915 MHz. 1.524 mm thick FR4 was used as a substrate. The width of the planar strip is 3.168 mm. The external radius of the center coupling loop is 7 mm with a 2 mm wide planar strip. The matching inductors are 14.88 nH and the components for the lattice balun are 4.78 pF and 6.32 nH. The predicted antenna gain is 1.1 dBi. Details of this design may be found in the literature.

7.10 CPW flexible monopole

There are design spaces that can require a physically flexible antenna. Thin dielectric materials, such as metalized Kapton, can be used to create an OMA like that of Section 7.1. Often this produces a very low efficiency antenna design because of the extremely thin substrate. A useful option is to use a thin single-sided flexible dielectric to create a coplanar waveguide (CPW) fed monopole antenna. There are instances where flexible antennas are useful for deployment within a compact form-factor, also in temporary biological implants.

The geometry of this antenna is given in Figure 7.51. The upper planar monopole element of width W and length L transitions from CPW at a small connection between the two, which has a gap of length L_N and a trace width of W_N. The terminating end

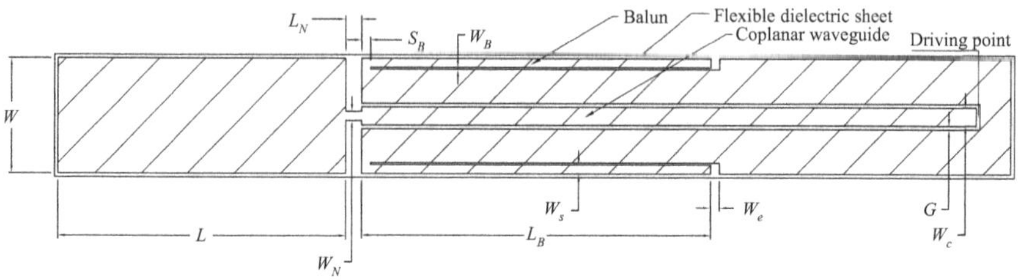

Figure 7.51 The flexible coplanar waveguide monopole consists of a radiating element of area L by W fed with a small trace connected to the CPW feedline. A pair of current traps on the side act as a balun to enforce that all radiating currents are on the monopole radiator

of the CPW that feeds the monopole requires a current trap or a balun structure. The currents will normally attempt to flow down the outside of the narrow groundplane opposite the radiating element. In a true monopole, a large groundplane acts to enforce the monopole mode, and functions as a balun. The length and width parameters for the balun are L_B and W_B. The length of L_B is approximately one-quarter wavelength. The antenna as shown in Figure 7.51 can be fed with a short coaxial cable, or the CPW can be extended to the right edge and fed directly by an edge launch connector or IC.

This antenna can be embedded in a cylinder of silicone for temporary implantation within a living organism, and remain flexible. When using Kapton as a dielectric, water absorption can occur, and change the losses and permittivity significantly. When embedded in a hydrophobic material, this can be mitigated. Some thin film dielectrics can also cold flow, and may not be appropriate in some applications. See Appendix A.

7.11 Characteristic mode antenna

The desire for broadband-printed monopole antennas has been addressed with the use of characteristic modes as an antenna design method by Cabedo-Fabrés *et al.* [54,55]. The idea is to study the eigenvectors that describe the currents on a rectangular plate. Rather than simply drive the plate as if it is a monopole [56], one should drive the plate in a manner that matches the natural mode on the metal plate. When this is done, it produces an antenna that is very broadband, and has a monopole pattern. The authors show a plate monopole over a perpendicular rectangular groundplane, fed at a single point in the center, and bifurcated either side of the centerline. The single feed antenna provides approximately 70% bandwidth. The bifurcated feed produces about 135% bandwidth. The author recast this design into a planar version, with both a bifurcated and trident feed.

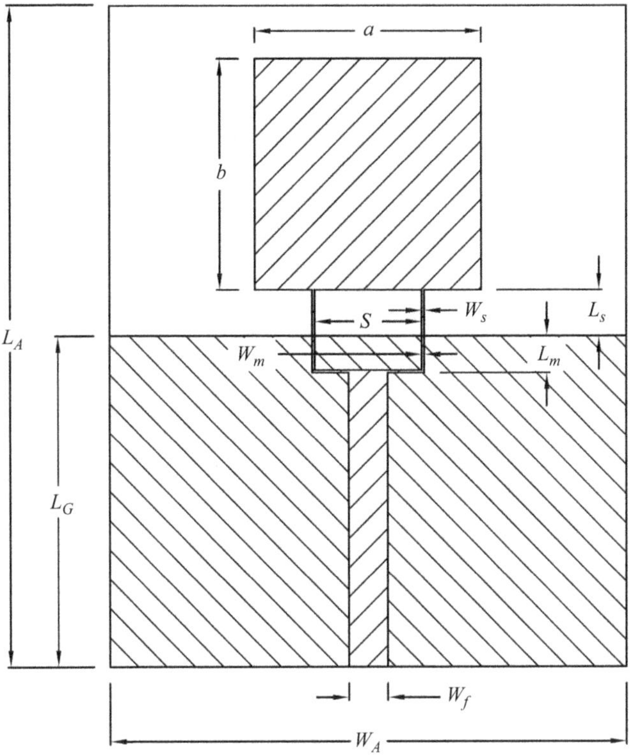

Figure 7.52 Geometry of characteristic mode printed antenna with bifurcated feed

The antenna, recast into a planar design, is shown in Figure 7.52. The antenna has a driven plate, which is generally square. The antenna is fed with a microstrip transmission line which splits into a pair of narrower microstrip lines. These lines then become skywires that feed the plate in its characteristic mode. This produces an antenna with an impedance bandwidth that is generally greater than an octave (2:1).

As an example, we use common FR4 with a thickness of 1.524 mm, and $\epsilon_r = 4.4$ with a 0.2 tan δ. The antenna substrate is 40 mm (L_A) by 33 mm (W_A). The lower microstrip groundplane length (L_G) is 20 mm. The bifurcated microstrip feed and skywires are both 0.158 mm wide ($W_m = W_s$). The length of each leg of the bifurcated microstrip transmission line (L_m) 2.06 mm. The length of each of the two skywires (L_s) is 2.8 mm. The separation (S) between the bifurcated microstrip, and skywire lines is 3.38 mm. The VSWR computed with HFSS is shown in Figure 7.53.

The VSWR is below 2:1 from 2.45 GHz to 6.90 GHz for a 2.8:1 bandwidth, well over an octave.

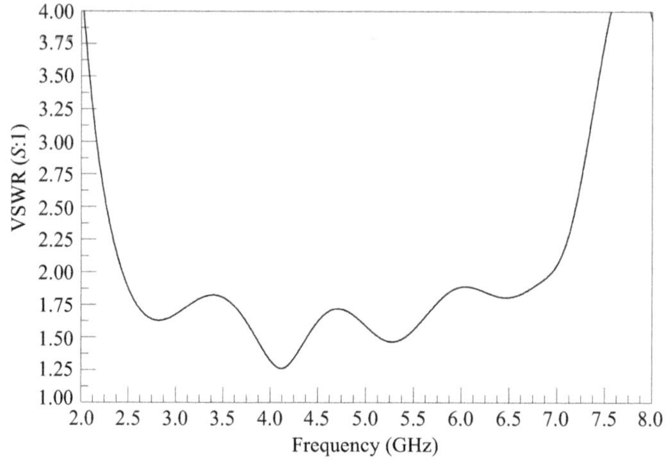

Figure 7.53 VSWR of example characteristic mode printed antenna

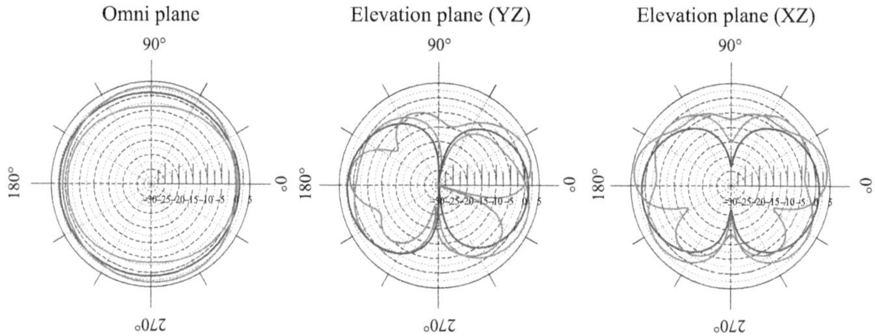

Figure 7.54 CMPA radiation patterns, computed at 2.45, 4.675 and 6.9 GHz

The antenna patterns computed at 2.45, 4.675 and 6.9 GHz are shown in Figure 7.54.

The radiation efficiency is very good over the bandwidth. It ranges from 91.6% at 2.45 GHz to 84.5% at 6.9 GHz. The directivity of the patterns span 2.6 dB at 2.5 GHz to 5.22 dB at 6.9 GHz.

References

[1] Nesic, A., and Nesic, D., "Omnidirectional Uniplanar Electromagnetically Coupled Antenna Array," *Microwave and Optical Technology Letters*, March 2004, Vol. 40, No. 6, pp. 516–518.

[2] Bancroft, R., and Bateman, B., "An Omnidirectional Microstrip Antenna," *IEEE Transactions on Antennas and Propagation*, November 2004, Vol. 52, No. 11, pp. 3151–3153.
[3] Bancroft, R., and Bateman, B., US Patent No. 7,518,554 B2, April 14, 2009.
[4] Bancroft, R., "Design Parameters of an Omnidirectional Planar Microstrip Antenna," *Microwave and Optical Technology Letters*, December 2005, Vol. 47, No. 5, pp. 414–418.
[5] Jasik, H., Myslicki, R.L., and Rudish, R.M., US Patent No. 3,757,342.
[6] Rudish, R.M., "Comments on 'An Omnidirectional Planar Microstrip Antenna,'" *IEEE Transactions on Antennas and Propagation*, November 2005, Vol. 53, No. 11, p. 3855.
[7] Hill, R., "A Twin Line Omni-Directional Aerial Configuration," *8th European Microwave Conference*, September 1978, pp. 307–311.
[8] Ono, M., Numazaki, T., and Mizusawa, M., "A High-Gain Omnidirectional Antenna Made of a Printed Element," *Transactions of the IECE of Japan*, January 1980, Vol. E 63, No. 1, pp. 58–60.
[9] Wei, K.P., Zhang, Z.J., and Feng, Z.H., "Design of a Dualband Omnidirectional Planar Microstrip Antenna Array," *Progress in Electromagnetics Research*, March 2012, Vol. 126, pp. 101–120.
[10] Yee, K.S., "Numerical Solution of Initial Boundary Value Problems Involving Maxwell's Equations in Isotropic Media," *IEEE Transactions on Antennas and Propagation*, 1966, Vol. 14, pp. 302–307.
[11] Bancroft, R., "Radiation Properties of an Omnidirectional Planar Microstrip Antenna," *Microwave and Optical Technology Letters*, January 2008, Vol. 50, No. 1, pp. 55–58.
[12] Bancroft, R., and Bateman, B., "An Omnidirectional Microstrip Antenna with Low Sidelobes," *Microwave and Optical Technology Letters*, July 2004, Vol. 42, No. 1, pp. 68–69.
[13] Pozar, D. (Antenna Designer's Notebook, Hal Shrank ed.), "Directivity of Omnidirectional Antennas," *IEEE Transactions on Antennas and Propagation Magazine*, October 1993, Vol. 35, No. 5, pp. 50–51.
[14] McDonald, N., "Omnidirectional Pattern Directivity in the Presence of Minor Lobes: Revisited," *IEEE Transactions on Antennas and Propagation Magazine*, April 1999, Vol. 41, No. 2, pp. 63–65.
[15] Wong, H., and Chiou, T. "Omnidirectional Planar Dipole Array Antenna," *IEEE Transactions on Antennas and Propagation*, February 2004, Vol. 52, No. 2, pp. 624–628.
[16] Duffley, M., and Mikavika, A., "A Wide-Band Printed Double-Sided Dipole Array," *IEEE Transactions on Antennas and Propagation*, February 2004, Vol. 52, No. 2, pp. 628–631.
[17] Gibson, P.J., "The Vivaldi Aerial," *Proceedings of the 9th European Microwave Conference*, 1979, pp. 101–109.
[18] Shin, J., and Schaubert, D.H., "A Parameter Study of Stripline-Fed Vivaldi Notch-Antenna Arrays," *IEEE Transactions on Antennas and Propagation*, May 1999, Vol. 47, No. 5, pp. 879–886.

[19] Deng, C., and Xie, Y., "Design of Resistive Loading Vivaldi Antenna," *IEEE Antennas and Wireless Propagation Letters*, 2009, Vol. 8, pp. 240–243.
[20] Cerny, C., Nevrly, J., and Mazanek, M., "Optimization of Tapered Slot Vivaldi Antenna for UWB Application," *19th International Conference on Applied Electromagnetics and Communications, 2007, ICECom 2007*, October 2017.
[21] Janaswamy, R., "Even Mode Characteristics of the Bilateral Slotline," *IEEE Transactions on Microwave Theory and Techniques*, June 1990, Vol. 38, No. 6, pp. 760–764.
[22] Bancroft, R., and Chou, R.-C., "Vivaldi Antenna Impedance Bandwidth Dependence on Stripline to Bilateral Slotline Transition," *Microwave and Optical Technology Letters*, April 2013, Vol. 5, No. 54, pp. 937–941.
[23] Wheeler, H.A., "Fundamental Limits of Small Antennas," *Proceedings of The I.R.E. (IEEE)*, December 1947, pp. 1479–1484.
[24] McLean, J.S., "A Re-Examination of the Fundamental Limits on The Radiation Q of Electrically Small Antennas," *IEEE Transactions on Antennas and Propagation*, May 1996, Vol. 44, No. 5, pp. 672–675.
[25] Sten, J.C.-E., Hujanen, A., and Koivisto, P.K., "Quality Factor of an Electrically Small Antenna Radiating Close to a Conducting Plane," *IEEE Transactions on Antennas and Propagation*, May 2001, Vol. 49, No. 5, pp. 829–837.
[26] Thiele, G.A., Detweiler, P.L., and Penno, R.P., "On the Lower Bound of the Radiation Q for Electrically Small Antennas," *IEEE Transactions on Antennas and Propagation*, June 2003, Vol. 51, No. 6, pp. 1263–1268.
[27] Harrington, R.F., "Effect of Antenna Size on Gain, Bandwidth, and Efficiency," *Journal of Research of the National Bureau of Standards–D, Radio Propagation*, January–February 1960, Vol. 64D, No. 1, pp. 1–12.
[28] Staub, O., Zürcher, J.F., and Skrivervlk, A., "Some Considerations on the Correct Measurement of the Gain and Bandwidth of Electrically Small Antennas," *Microwave and Optical Technology Letters*, February 1998, Vol. 17, No. 3, pp. 156–160.
[29] Ju, X.-D., Fu, D.-M., and Mao, N.-H., "VHF TV Full Channel Loaded Circular Loop Antenna," *IEEE Transactions on Antennas and Propagation*, April 1984, Vol. AP-32, No. 4, pp. 425–428.
[30] Harrington, R., *Field Computation by Moment Methods*, Robert E. Krieger Publishing Company, 1968, pp. 85–87.
[31] Gibson, J.J., and Wilson, R.M., "The Mini-State—A Small Television Antenna," *IEEE Transactions on Consumer Electronics*, Vol. CE-22, No. 2, May 1976, pp. 159–175.
[32] Beverage, H.H., U.S. Patent No. 2,247,743 1938-12-10.
[33] Taga, T., Tsunekawa, K., and Sasaki, A., "Antennas for Detachable Mobile Radio Units," *Review of the ECL, NTT, Japan*, January 1987, Vol. 35, No. 1, pp. 59–65.
[34] Taga, T., "Analysis of Planar Inverted-F Antennas and Antenna Design for Portable Radio Equipment," *Analysis, Design, and Measurement of Small*

and Low-Profile Antennas, Hirasawa, K., and Haneishi, M. (eds.), Artech House, 1992, pp. 160–180.

[35] Zürcher, J.F., Staub, O., and Skrivervlk, A.K., "SMILA: A Compact and Efficient Antenna for Mobile Communications," *Microwave and Optical Technology Letters*, November 2000, Vol. 27, No. 3, pp. 155–157.

[36] Huynh, M.C., and Stutzman, W., "Ground plane effects on planar inverted-F antenna (PIFA) performance," *IEE Proceedings on Microwave, Antennas and Propagation*, August 2003, Vol. 150, pp. 209–213.

[37] Deshmukh, A.A., and Kumar, G., "Formulation of Resonant Frequency for Compact Rectangular Microstrip Antennas," *Microwave and Optical Technology Letters*, February 2007, Vol. 49, No. 2, pp. 498–501.

[38] Boyle, K.R., and Ligthart, L.P., "Radiating and Balanced Mode Analysis of PIFA Antennas," *IEEE Transactions on Antennas and Propagation*, January 2006, Vol. 54, No. 1, pp. 231–237.

[39] Milligan, T.A., *Modern Antenna Design*, McGraw Hill, 1985, pp. 83–84.

[40] Wong, K.-L., *Planar Antennas for Wireless Communications*, John Wiley and Sons, 2003, p. 10.

[41] Behdad, N., and Sarabandi, K., "A Novel Approach for Bandwidth Enhancement of Slot Antennas," *Proceedings of the 2003 Antenna Appl. Symposium*, Allerton Park, Monticello, IL, September 17–19, 2003, pp. 176–188.

[42] Behdad, N., and Sarabandi, K., "Dual Resonator Slot Antennas for Wireless Applications," *IEEE AP-S International Symposium Digest*, Monterey, CA, June 20–25, 2004.

[43] Behdad, N., and Sarabandi, K., "A Wide-Band Slot Antenna Design Employing A Fictitious Short Circuit Concept," *IEEE Transactions on Antennas and Propagation*, January 2005, Vol. 53, No. 1, pp. 475–482.

[44] Bancroft, R., "A Commercial Perspective on the Development and Integration of an 802.11a/b/g HiperLan/WLAN Antenna into Laptop Computers," *IEEE Antennas and Propagation Magazine*, August 2006, Vol. 48, No. 4, pp. 12–18.

[45] Milligan, T.A., *Modern Antenna Design*, McGraw-Hill, 1985, pp. 96–99.

[46] Oliner, A.A., "Equivalent Circuits for Discontinuities in Balanced Strip Transmission Line," *IRE Transactions on Microwave Theory and Techniques*, Vol. MTT-3, March 1955, pp. 134–143.

[47] U.S. Patent No. 7,098,863.

[48] Al Basraoui, M., and Shastry, P.N., "Wideband Planar Log-Periodic Balun," *International Journal RF and Microwave Computer Aided Design*, 2001, Vol. 11, pp. 343–353.

[49] Roberts, W.K., "A New Wide-Band Balun," *Proceedings ofThe I.R.E. (IEEE)*, December 1957, pp. 1628–1631.

[50] Fitzgerrell, R.C., "Standard Linear Antennas, 30 to 1000 MHz," *IEEE Transactions on Antennas and Propagation*, December 1986, Vol. AP-34, No. 12, p. 131.

[51] McLean, J., and Sutton, R., "The Minimization of Spurious Radiation from the Baluns and Feed Regions of Roberts Dipole Antennas," *Antenna*

Measurement Techniques Association 23rd Annual Meeting & Symposium, Denver, CO, October 21–26, 2001, pp. 217–223.

[52] Bancroft, R., "Design of an Integrated Loop Coupler and Loop Antenna for RFID Applications," *Microwave and Optical Technology Letters*, August 2009, Vol. 51, No. 8, pp. 1830–1833.

[53] Schekunoff, S.A., and Friis, H.T., *Antennas Theory and Practice*, John Wiley & Sons, 1952, pp. 269.

[54] Cabedo-Fabres, M., Antoniono-Daviu, E., Valero-Nogueria, A., and Bataller, M.F., "The Theory of Characteristic Modes Revisited: A Contribution to the Design of Antennas for Modern Applications," *IEEE Antennas and Propagation Magazine*, October 2007, Vol. 49, No. 5, pp. 52–68.

[55] Antonino-Daviu, C.M., Ferrando-Bataller, M., and Valero-Nogueira, A., "Wideband Double-Fed Planar Monopole Antenna," *Electronics Letters*, November 2003, Vol. 39, No. 23, pp. 1635–1636.

[56] Ammann, M.J., "Impedance Bandwidth of the Square Planar Monopole," *Microwave and Optical Technology Letters*, Vol. 24, No. 3, February 2000, pp. 185–187.

[57] Sievenpiper, D.F., Dawson, D.C., Jacob, M.M., *et al.*, "Experimental Validation of Performance Limits and Design Guidelines for Small Antennas," *IEEE Transactions on Antennas and Propagation*, January 2012, Vol. 60, No. 1, pp. 8–17.

[58] Hansen, R.C., *Electrically Small, Superdirective, and Superconducting Antennas*, John Wiley & Sons, 2006, Section 1.

Chapter 8
Millimeter wave microstrip antennas

8.1 General millimeter wave design considerations

Currently, commercial millimeter wave frequencies of interest are from around 20 to 80 GHz. Automotive collision avoidance radars operate at 24 and 77 GHz. The frequency spectrum between 30 and 300 GHz is referred to as the extremely high frequency (EHF) band.

At EHF frequencies, the electrical thickness of a substrate can be overlooked as a source of microstrip transmission line radiation. Radiation from microstrip transmission lines with a thick substrate was discussed in Section 6.4.6. A "rule of thumb" to prevent significant microstrip transmission line radiation is to keep the substrate less than about $0.020\lambda_0$ (free space wavelengths). In Table 8.1, a list of free

Table 8.1 Wavelengths and frequencies with maximum allowable microstrip transmission line non-radiation thickness and maximum available standard thickness

Extended EHF frequencies Microstrip radiation limits			
		Substrate	
Wavelength λ_0	**Frequency**	**Non-radiation thickness**	**Standard thickness**
15 mm	20 GHz	300 μm	254 μm
14 mm	21 GHz	280 μm	254 μm
13 mm	23 GHz	260 μm	254 μm
12 mm	25 GHz	240 μm	175 μm
11 mm	27 GHz	220 μm	175 μm
10 mm	30 GHz	200 μm	175 μm
9 mm	33 GHz	180 μm	175 μm
8 mm	38 GHz	160 μm	127 μm
7 mm	43 GHz	140 μm	127 μm
6 mm	50 GHz	120 μm	100 μm
5 mm	60 GHz	100 μm	100 μm
4 mm	75 GHz	80 μm	50 μm
3 mm	100 GHz	60 μm	50 μm
2 mm	150 GHz	40 μm	25 μm
1 mm	300 GHz	20 μm	25 μm

Table 8.2 Wavelengths and frequencies, maximum available standard thickness, and typical microstrip transmission line design widths, $\epsilon_r = 3.0$

Extended EHF frequencies
Microstrip transmission line widths

Wavelength λ_0	Frequency	Substrate Standard thickness	MS width 125 Ω	MS width 50 Ω
15 mm	20 GHz	254 μm	76 μm	615 μm
14 mm	21 GHz	254 μm	76 μm	615 μm
13 mm	23 GHz	254 μm	76 μm	615 μm
12 mm	25 GHz	175 μm	49 μm	420 μm
11 mm	27 GHz	175 μm	49 μm	420 μm
10 mm	30 GHz	175 μm	49 μm	420 μm
9 mm	33 GHz	175 μm	49 μm	420 μm
8 mm	38 GHz	127 μm	32 μm	301 μm
7 mm	43 GHz	127 μm	32 μm	301 μm
6 mm	50 GHz	100 μm	23 μm	235 μm
5 mm	60 GHz	100 μm	23 μm	235 μm
4 mm	75 GHz	50 μm	7 μm	112 μm
3 mm	100 GHz	50 μm	7 μm	112 μm
2 mm	150 GHz	25 μm	3 μm	52 μm
1 mm	300 GHz	25 μm	3 μm	52 μm

space wavelengths, their approximate frequencies, and a "rule of thumb" estimate of the maximum thickness of the substrate to prevent radiation from interconnecting microstrip transmission lines is given.

In general, most printed circuit board (PCB) houses have a lower limit etching dimension of about 150 μm. The edge resistance of a square microstrip antenna can be as high as 300 Ω. Often a quarter-wave length of microstrip transmission line is introduced to transform the edge resistance to 50 Ω. The characteristic impedance of this line is often around 125 Ω. Microstrip line widths for 125 and 50 Ω transmission lines for $\epsilon_r = 3$ are computed in Table 8.2. The widths of 125 Ω microstrip transmission lines are all too small for consistent etching. There are several options we will investigate. In many cases, 100 Ω microstrip lines are used in corporate-fed arrays. One should implement low impedance microstrip transmission lines to produce widths which can be realized by a PCB manufacturer. Designs at 77 GHz produce 50 Ω microstrip lines that are narrower than may be reliably etched.

As discussed previously, in the case of single layer PCB, this will limit the thickness of a microstrip antenna element, and in turn limit its impedance bandwidth. When one has the option to use a two layer PCB at millimeter wave frequencies, a coupled patch antenna is an option if more bandwidth is desired than a single layer PCB design can produce.

A misconception about millimeter wave microstrip antenna design is that transmission line losses are large. This concern was investigated as early as 1983 by James

Table 8.3 $H = 50\ \mu m$. Transmission line width = 126 μm (Rogers Ultralam 3850), $\epsilon_r = 2.9$, $\tan\delta = 0.0025$

Gradient model microstrip line loss Equivalent surface impedance (HFSS)			
R_q (RMS)	**10 GHz (dB/100 mm)**	**28 GHz (dB/100 mm)**	**60 GHz (dB/100 mm)**
250 nm	3.67	7.71	14.62
500 nm	4.65	10.75	21.58
1000 nm	6.73	16.82	30.59

Table 8.4 $H = 50\ \mu m$. Transmission line width = 126 μm (Rogers Ultralam 3850), $\epsilon_r = 2.9$, $\tan\delta = 0.0025$

Traditional closed-form model microstrip line loss		
10 GHz (dB/100 mm)	**28 GHz (dB/100 mm)**	**60 GHz (dB/100 mm)**
2.58	4.62	7.35

and Hall [1]. They concluded that "…the loss of microstrip at millimeter wavelengths is not excessive compared to microwave frequencies …"

Much of the concern often revolves around the surface roughness of the copper conductors. A surface roughness model based on triangular protrusions was proposed in the 1970s [2]. This model's simplicity and accessibility caused it to dominate general engineering usage. Recently, Gold has noted the inadequate nature of this model, and proposed a more physics based alternative [3–5]. The model uses a root mean square statistical roughness measurement, R_q,* to predict the amount of loss expected for a given surface roughness. Gold's gradient model is not readily implemented in a computer-aided design, but the information provided allows one to compute an equivalent surface impedance, which may be used at discrete frequencies with a full-wave electromagnetic program such as HFSS.

Gradient model results are presented in Table 8.3. It is interesting to note that even at 10 GHz, with rather smooth metal ($R_q = 250$ nm), it appears the surface roughness contributes to loss.

For comparison, Table 8.4 has results using a traditional closed form expression. Compared with this solution, even moderate surface roughness can contribute considerably to the transmission line losses. While using dB/100 mm is useful for noting

*DIN EN ISO 4287.

Table 8.5 $H = 50\ \mu m$. Transmission line width $= 126\ \mu m$ (Rogers Ultralam 3850), $\epsilon_r = 2.9$, $\tan\delta = 0.0025$

Gradient model microstrip line loss
Microstrip loss per guide wavelength (HFSS)

R_q (RMS)	10 GHz (dB/λ_{ϵ_e})	28 GHz (dB/λ_{ϵ_e})	60 GHz (dB/λ_{ϵ_e})
250 nm	0.721	0.541	0.476
500 nm	0.913	0.754	0.704
1000 nm	1.320	1.181	0.997

comparative differences, the loss per guide wavelength is useful for microstrip array, and matching network design.

Although surface roughness will change the guide wavelength compared with the traditional analysis, it is instructive to use it for values of guide wavelengths, for comparison of losses predicted by the gradient model.

The values for loss per microstrip guide wavelength, as predicted by the gradient model, are presented in Table 8.5. The loss per wavelength *decreases* with frequency. The losses per guide wavelength were calculated using the guide wavelength predicted by a traditional model with dispersion included. Gold and Helmreich indicate their model changes the guide wavelength in response to the surface roughness.

The notion that large losses in microstrip occur at millimeter wave frequencies may arise from the fact that a physical length of microstrip transmission line in dB/100 mm, increases rapidly with frequency. This is seen in Table 8.3. When written in terms of guide wavelength, λ_{ϵ_e}, the losses for elements such as a quarter-wave or half-wave transformer remain stable. This is seen when the losses are written in terms of the microstrip transmission line guide wavelength.

8.2 Corporate-fed patch arrays

8.2.1 28 GHz example

In theory, one can scale the corporate-fed arrays designed at lower frequencies to millimeter wave frequencies. In practice, this will produce microstrip transmission line widths which are narrower than 150 μm, which is a rule-of-thumb limit for PCB house etching. Generally, as one increases frequency, fabrication considerations become paramount in importance, especially for commercial designs.

One can redesign the corporate feed network to produce line widths larger than 150 μm throughout the array. This is shown in Figure 8.1. The thickness of the substrate is 254 μm, which is the thickest possible value not expected to produce significant radiation from microstrip feed lines at 28 GHz. The relative permittivity of the substrate is 3.0 (Rogers RO3003).

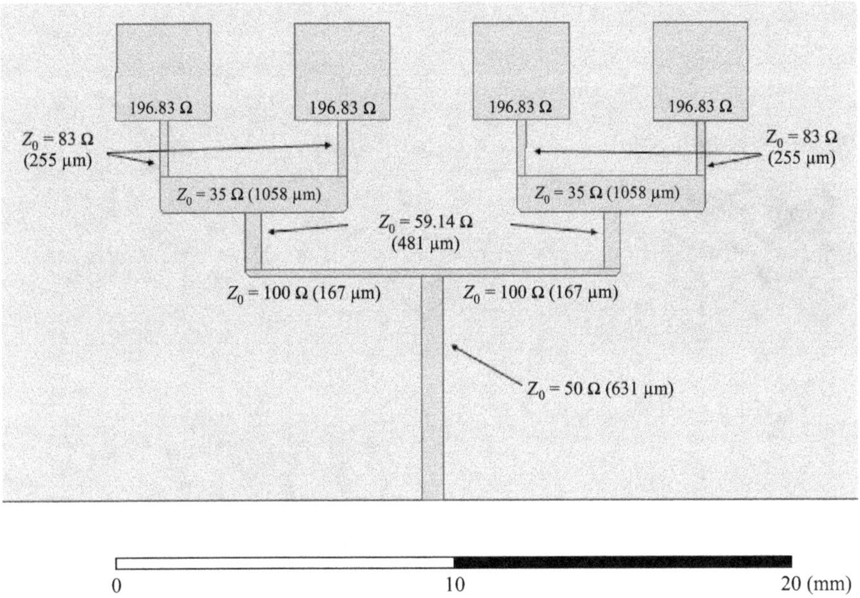

Figure 8.1 Corporate-fed microstrip array design for 28 GHz. The edge impedances of each patch are transformed to 35 Ω and then to 100 Ω so that all the interconnecting transmission lines have widths greater than 150 μm, which is an often cited lower limit for etching

The edge resistance of a square microstrip antenna at resonance is generally about 200 Ω for common substrates and thicknesses. Often this is transformed to 100 Ω, but in this case, a 141 Ω line for a quarter-wave transformer would be too narrow at 62 μm for an average PCB house to etch. What was done instead, was to transform the edge resistance to 35 Ω, which requires a quarter-wave transmission line with a characteristic impedance of only 83 Ω, and a width of 255 μm. The 35 Ω lines add in parallel to produce 17.5 Ω at the next junction. This impedance is then transformed to 100 Ω using a 59.14 Ω quarter-wave transformer with a width of 481 μm. The two 100 Ω lines add in parallel at the final junction to form a 50 Ω impedance where a 50 Ω feedline is attached to complete the design.

The efficiency of the antenna array is 90.23%. The directivity is 11.91 dB, and the gain is 11.46 dBi. The 2:1 VSWR bandwidth is 931 MHz or 3.33%.

8.2.2 60 GHz example

The essential limitation for the use of thin substrates at millimeter wave frequencies is etching. Generally, most PCB houses have a hard limit at around 150 μm. The accuracy of etching dimensions will be compromised below that value. A 50 Ω transmission line on 50 μm thick Rogers Ultralam 3000 has a width of about 126 μm. This

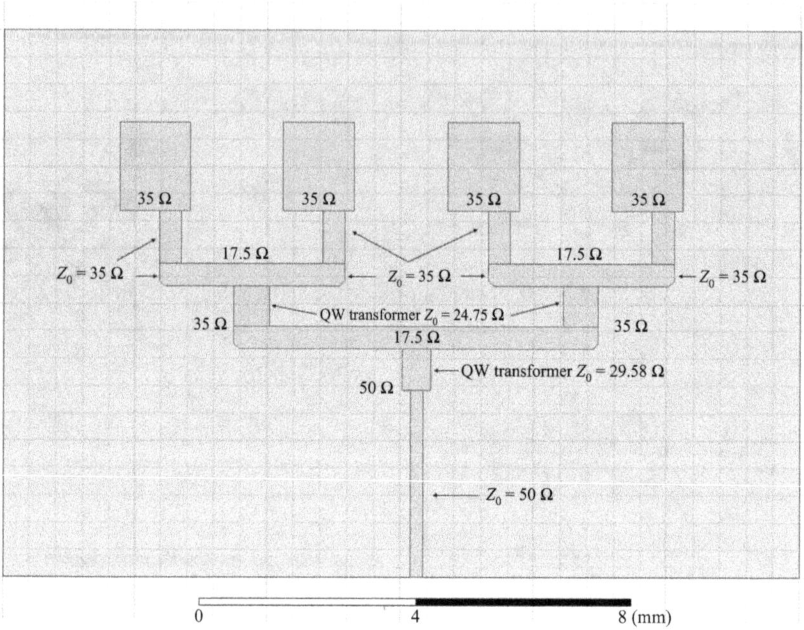

Figure 8.2 Edge-fed corporate array — all microstrip line impedances are less than 50 Ω. A 50 Ω line is 240 μm wide.

is below the common etching limit. The choice to use 100 μm thick material was to increase this 50 Ω width to about 240 μm.

In most corporate-fed arrays, patches are fed at the center of a radiating edge. The edge resistance at resonance is about 200–250 Ω or so, and thus a quarter-wave transformer of about 112 Ω would be required to transform to 50 Ω. The width of this transformer would need to be smaller than the 150 μm etching limit. Generally impedance transformers need to have characteristic impedances greater than 50 Ω, which precludes their feasibility for etching.

Here we introduce a different method of corporate feeding [6]. The patches are fed along their non-radiating edges (like a probe feed) at a low impedance feed point. Low impedance microstrip transmission lines are then used to create a corporate feed which are all wider than 50 Ω traces. This allows for the realization of transmission lines within the constraints of current PCB house limits for line widths.

The implemented array design is shown in Figure 8.2. The patches are rectangular, and fed along a non-radiating edge. The aspect ratio separates the TM_{01} and TM_{10} modes to preserve a pure patch resonance when it is fed along a non-radiating edge. The driving point resistance along the edge of a patch is chosen as 35 Ω. A pair of 35 Ω microstrip transmission lines ($W = 411$ μm) lead from a pair of patches on either side of the center. The two microstrip lines arrive at a junction where they add in parallel to form a local resistance of 17.5 Ω. A quarter-wave transformer is then introduced to step the impedance from 17.5 Ω back to 35 Ω. The characteristic impedance of the transformer is 24.75 Ω with a width of 531 μm. A pair of quarter-wave transformers are then connected with a 35 Ω line. At the center they form a 17.5 Ω local resistance. A 29.58 Ω quarter-wave transformer is then used to transform the impedance from 17.5 Ω to 50 Ω. A 50 Ω feed line provides a path to the edge of the PCB where a connector can be added.

References

[1] James, J.R., and Hall, C.M., "Investigation of New Concepts for Designing Millimetre-Wave Antennas," Final Technical Report, Royal Military College of Sciences, AD-A137 350, September 1983.

[2] Hammerstad, E.O., and Bekkadal, F., "Microstrip Handbook," ELAB Report STF 44 A74169, University of Trondheim, Norwegian Institute of Technology, 1975.

[3] Gold, G., and Helmreich, K., "A Physical Surface Roughness Model and Its Applications," *IEEE Transactions on Microwave Theory and Techniques*, October 2017, Vol. 65, No. 10, pp. 3720–3732.

[4] Bailer-Jones, D.M., *Scientific Models in Philosophy of Science*, University of Pittsburgh Press, 2009.

[5] Gold, G., "A Physical Model for Skin Effect in Rough Surfaces," *Proceedings of the 42nd European Microwave Conference*, 2012.

[6] Mahabub, A., Rifat A.A., and Toufikul, I., "Gain Improvement of Micro Strip Antenna Using Dual Patch Array Micro Strip Antenna," *Journal of Emerging Trends in Computing and Information Sciences*, December 2012, Vol. 3, No. 12, pp. 1642–1648.

Appendix A

Microstrip antenna substrates

A.1 Microstrip antenna/transmission line substrates

The selection of a substrate material is a balance between the required electrical, mechanical, and environmental performance required by a design versus economic constraints. Generally, if one has the available design volume to use air as a substrate for a microstrip antenna, this is a good choice. The antenna efficiency is high and the gain is maximized as is the impedance bandwidth of a conventional microstrip antenna. The surface wave loss when air is used as a substrate is minimal.

When a dielectric substrate is selected, one is often interested in a material with the lowest loss tangent ($\tan\delta$) available. The loss tangent is a metric of the quantity of electrical energy which is converted to heat by a dielectric. The lowest possible loss tangent maximizes the antenna efficiency (decreases the losses) and is expanded upon in Section 2.6 of Chapter 2.

The relative dielectric permittivity, ϵ_r, of the substrate determines the physical size of a patch antenna. The larger the dielectric permittivity, the smaller the element size, but also the smaller the impedance bandwidth and directivity. The surface wave loss also increases. The use of substrates with higher dielectric permittivity also tightens fabrication tolerances [1].

The tolerance of the dielectric permittivity value is also of significant importance in manufacturing yield. A Monte Carlo-type analysis using the cavity model is a good method of estimating antenna manufacturing yield for a rectangular microstrip antenna when etching tolerance, substrate thickness tolerance, feed point location tolerance, and dielectric tolerances are known. Substrate electrical and physical parameters also vary with temperature. Work by Kabacik and Bialkowski indicates that Teflon/fiberglass substrates can have a significant variation of dielectric permittivity for many airborne and space-borne applications [2]. The dielectric permittivity and loss tangent of Teflon/fiberglass often differed from what was quoted by manufacturers in their data sheets when compared with measurements and were valid over a much narrower temperature range than encountered in many aerospace applications. The performance variations are due to changes in the material dielectric properties—thermal expansion had a minor effect on microstrip antenna performance.

Commercially available substrates and laminates have different requirements for processing. Details of fabrication issues and methods may be found in the literature and directly from manufacturers [3]. Other fabrication options such as screen printing conductive inks directly on substrates have also been investigated [4]. In recent years, the use of inkjet printing methods has been explored to create antennas. This may alter the spectrum of substrates one may be able to use to design microstrip antennas [5].

Generally, relative dielectric permittivity ϵ_r and loss tangent $\tan\delta$ increase with temperature. In space applications, moisture outgassing produces a lower dielectric permittivity and loss tangent.

A.2 Metal cladding

Generally, the metal cladding attached to a dielectric substrate material is copper. Two types of copper foil are used as cladding, *rolled foil* and *electrodeposited foil*. Rolled foil is passed through a rolling mill a number of times until the desired physical dimensions are obtained and is then bonded to the substrate. Rolled copper has a polished mirror-like appearance. It is generally used for flexible circuitry. Electrodeposited foil is created by electrodeposition of copper onto an inert form, which is generally a rolling titanium or steel drum. A thin layer of copper is continuously removed from the form and then bonded to a substrate [6].

Oxidation of copper traces is the primary cause of copper bond failure over time. The copper traces oxidize on both sides (top and bottom) of the trace. The diffusion of oxygen in the substrate material causes the bottom side of the trace to oxidize. Continuous ambient temperatures above 250 °F will ultimately degrade the copper bonds on the surface of the board. The most effective way to mitigate this oxidation is to use an effective high temperature conformal coating. This coating will provide a barrier between the air and the surface of the resin and copper traces [7].

The computation of characteristic impedance and losses of a microstrip transmission line (see Appendix C) depend on the copper foil thickness. The copper cladding is not generally presented as a thickness, but is indirectly derived from a provided mass over a given area. In Table A.1, metric values in grams per square meter are given. The thickness is then derived using the density of copper (8.94 mg/mm^2). In the USA, copper cladding is archaically described in terms of weight per square foot. The thickness of the cladding may then be derived in inches and is listed in Table A.2 [8]. Thinner copper traces have smaller etching undercut than thicker traces. This allows for fine (narrow) traces as compared to thicker copper. Thicker copper has

Table A.1 Metric copper foil mass and foil thickness

Foil mass/area	Foil thickness	Tolerance
150 g/m^2	18 μm	$\pm$4.9 μm
300 g/m^2	34 μm	$\pm$4.9 μm
600 g/m^2	70 μm	$\pm$7.5 μm
1200 g/m^2	104 μm	$\pm$10.0 μm

higher power handling capability. This is important where the small cross-section of a trace could produce heat build-up and eventual failure.

The surface roughness of copper can produce considerable loss at higher frequencies. The effect becomes significant when the skin depth of copper becomes less than or equal to the surface roughness of the copper cladding. Often a rough surface is used to improve the adhesion of the copper cladding to a dielectric substrate. The surface roughness of commonly available copper cladding is given in Table A.3 [9].

The skin depth of copper for selected frequencies is given in Table A.4. The surface roughness of high-profile electrodeposited copper is just larger than a skin depth at 1 GHz. At 10 GHz, one must use low-profile electrodeposited copper to prevent surface roughness from introducing loss. The last frequency where one can choose a copper foil type which is on the order of a skin depth is 50 GHz. Above

Table A.2 US copper foil weight and foil thickness

Foil weight/area	Foil thickness	Tolerance
1/2 oz/ft^2	0.0007 in	±0.0002 in
1 oz/ft^2	0.0014 in	±0.0002 in
2 oz/ft^2	0.0028 in	±0.0003 in
4 oz/ft^2	0.0056 in	±0.0004 in

Table A.3 Copper foil surface roughness

Type of cladding	Copper surface roughness (RMS)
High-profile ED copper	2400 nm
Standard ED copper	1200 nm
Low-profile ED copper	600 nm
Rolled annealed copper	300 nm

Table A.4 Skin depth for copper with frequency

Frequency (GHz)	Skin depth δ (nanometers)
1	2063 nm
10	652 nm
50	292 nm
77	235 nm
110	197 nm

Table A.5 Common dielectric substrate material properties

Material	ϵ_r	tan δ
Teflon (PTFE)	2.1	0.0005
Rexolite 1422	2.55	0.0007
Noryl	2.6	0.0011
FR-4	4.1	0.02
Alumina (99.5 %)	9.8	0.0003

that, surface roughness will begin to have a significant contribution to microstrip transmission line loss. Once the surface roughness is larger than a skin depth, the amount of extra losses encountered will be proportional to the magnitude of the surface roughness; so, for minimum loss, copper with a minimum surface roughness should be chosen [10].

A.3 Dielectric materials

Dielectric substrates used to construct microstrip antennas (and radomes) may be grouped into plastics like fluoropolymers, such as Teflon [polytetrafluoroethylene (PTFE)] and cross-linked polystyrene (Rexolite 1422), ceramics, such as Al_2O_3 (alumina) or silicon nitride Si_3N_4, and glasses, such as E-Glass, which are used in most Teflon–fiberglass microwave substrates.

In some commercial applications, a planar antenna may be enclosed within an injection molded radome. The dielectric properties of the resins used are a very important consideration in an injection molded radome design. The resin should have known moisture absorption characteristics. Low moisture absorption is important in maintaining a stable resonance frequency and loss characteristics in environments which have humidity changes.

Materials are generally described as *hydrophobic* and *hydrophilic*. A hydrophobic material excludes water and incorporates only very minute quantities. The beading of water droplets on wax paper demonstrates the hydrophobic nature of wax. Hydrophilic substances have an affinity for water and absorb it.

Water has properties that are problematic for microstrip antenna designers. The relative dielectric permittivity (ϵ_r) of water varies from $\epsilon_r \approx 90$ at 1.5 °C to $\epsilon_r \approx 80$ at 25 °C. This is true from approximately 5–10 GHz. At 10 GHz, the dielectric permittivity of water rapidly decreases and the dielectric loss approaches a peak value before decreasing. The largest loss, at 10 GHz, occurs at just above freezing. When the temperature is decreased further, and water freezes, the dielectric permittivity changes to about three, and the dielectric loss becomes very small [11].

Table A.5 lists some common substrate materials used in the design of microstrip antennas and typical values of relative dielectric permittivity and loss tangent.

A.3.1 Plastics

Teflon (PTFE) has very desirable electrical qualities but is not recommended for many space applications. An extensive discussion of PTFE substrates and their fabrication may be found in the literature [12]. Teflon is soft and subject to creep (aka "cold flow") which allows it to deform from a stress below its yield point.

Rexolite 1422 (cross-linked polystyrene) is among the earliest materials used to create planar transmission line [13]. Rexolite 1422 is a very good material for space applications and has many desirable mechanical properties [14]. Rexolite 1422 is easily machined, and its dielectric permittivity remains stable up to 100 GHz. Rexolite 2200 is a fiberglass reinforced version which has similar properties to unfilled Rexolite 1422, but is more rigid and dimensionally stable.

Noryl (modified polyphenylene oxide) is suitable for many commercial microwave applications. It has a much lower loss than FR-4 and is relatively cost effective, but it is soft and melts at a relatively low temperature, which can create soldering complications and sometimes has unsuitable mechanical properties for some applications.

A.3.2 Ceramics

Alumina (Al_2O_3) has desirable microwave properties for applications which require a relatively high dielectric permittivity $\epsilon_r \approx 10.0$ and a low loss tangent. Its drawbacks are the difficulty involved in machining it and its brittleness. Alumina has good thermal conductivity, and in some aerospace applications, it more readily dissipates heat and remains cooler than other common microwave substrates. In some missile applications where high temperatures may compromise solder joints, alumina is a viable option for the dissipation of heat. Alumina's dielectric permittivity is very sensitive to the processing used to produce the alumina.

Alumina exists in a crystalline form called *corundum* or α-aluminum oxide. The crystal form has an anisotropic dielectric permittivity which can vary from 9.3 to 11.5 which makes it of limited use as a microwave dielectric substrate. Corundum may be ground up into a powder. Each particle of the powder is a tiny crystal, and when randomized in this fashion, it can produce a material with an isotropic dielectric permittivity.

The alumina powder may then be heated to a temperature below its melting point for a period of time (sintering) to form a *polycrystalline* material. The tiny crystals are called *grains*, and the void spaces between the grains are called *pores*. The ratio of pores to the desired material is called the *porosity* (P) of the material. The value of porosity varies from 0 to 1 or is sometimes stated as a percentage from 0% to 100%. It is well known that the effective dielectric permittivity value of two materials is related to the fraction of each material in a mixture; therefore, the porosity of a sintered alumina material will affect the dielectric permittivity of that material.

The porosity of sintered alumina directly affects dielectric permittivity; so, an understanding of the sintering process becomes important in understanding its electrical properties [15]. The sintering process is very complex. For illustration, we will assume that the alumina powder consists of uniform spherical particles. These particles (grains) slowly fuse as their contacting surfaces become larger and larger.

As sintering proceeds, the size of the grains in the material increases as the size of the pores decreases. Two types of grain growth are generally identified as *normal* and *abnormal* grain growth. When normal grain growth occurs, the size of the average grain increases, but the sizes and shapes of the grains remain within a narrow range. The distribution of the grain size remains relatively constant over time for normal grain growth. The grain size distribution shows self-similarity as the grain size increases. When abnormal grain growth occurs, a small number of large grains grow at a faster rate than the surrounding smaller grains. The grain size distribution can split into two peaks. The invariance of the distribution of grain sizes over time is lost. In some cases, the grain distribution can revert to a normal distribution over time. It has been reported that grain size can have a considerable effect on the loss tangent value of high density (98.1% $\pm$ 0.5%) alumina [16]. Smaller grain sizes (1–3 μm) have lower loss tangents than larger grain sizes (5–7 μm).

In some cases, small amounts of impurities are introduced to alumina to aid in sintering and suppress abnormal grain growth. These materials are often called a *solute* or *dopant*. Even though they are present in small amounts, they can have a considerable effect on the sintering rate and microstructure development. In the case of alumina (Al_2O_3), magnesium oxide (MgO) is sometimes used as a dopant to suppress abnormal grain growth. It has been reported that doping alumina with titanium dioxide can lower the loss tangent from 0.000027 to 0.000020 at 10 GHz for samples near theoretical density [17]. A consistent sintering process, without abnormal grain growth, is paramount for a consistent relative permittivity.

A.3.3 Glass transition temperature (T_g)

Polymer materials used for microstrip antenna substrates and radomes include plastics, rubber, and fibers. While rubber is a pliable material, plastic is still pliable but much stiffer. Rubber is also more properly called an *elastomer*. The parameter which generally delineates a plastic material from a rubber material is the position of its *glass transition temperature* (T_g). When T_g is above room temperature, a material is a plastic, if T_g is below room temperature, it is a rubber. Materials of this type at low temperatures may be thought of as a glassy solid [18].

Generally, carbon black is introduced into rubber to reinforce it for increased structural integrity. An exception is silicone rubber. Silicone rubber has minute fumed silica particles [19].

A.3.4 Composite dielectric substrates

When two or more dielectric materials are combined, and each retains their separate dielectric properties, we will refer to this as a composite dielectric material. A large number of commercially available microwave substrates combine two or more dielectric materials. A common combination is Teflon and fiberglass. Another is epoxy and fiberglass. Many manufacturers offer materials in their pure form or with a percentage of fiberglass added. For instance, Rexolite 1422 is pure cross-linked polystyrene, but

Figure A.1 Cross-section of 12.7 mm (1/2 inch) thick FR-4 which shows the resin (dark) and glass cloth (lighter) layers

is also offered with fiberglass reinforcement as Rexolite 2200. Ultem (Polyetherimide) may be obtained in pure form or with 10%, 20%, or 30% fiberglass reinforcement.

A.3.5 FR-4

FR-4 is an inexpensive epoxy fiberglass material which finds use in many commercial applications (usually below 1 GHz).* FR-4 is a combination of thermoset epoxy and woven fiberglass sheets. The epoxy used in FR-4 has a relative dielectric permittivity of $\epsilon_r \approx 3.2$, and E-Glass is $\epsilon_r \approx 6.2$. The ratio of woven fiberglass to epoxy determines the dielectric permittivity. The material can be used for some wireless applications, but great care must be taken to budget and minimize the losses when it is used as a substrate above 1 GHz. The relative dielectric permittivity of FR-4 formulations varies considerably from 3.9 to 4.6 [21].

Plastic materials are commonly of two types: thermoset and thermoplastic. When exposed to heat, thermoset materials undergo a chemical change which is irreversible. When re-heated, thermoset materials do not melt but generally begin charring. FR-4 is an example of a substrate made from a thermoset material. When exposed to heat, thermoplastic materials melt, but do not undergo an irreversible chemical change. They may be re-melted after molding. Thermoplastic materials can be reprocessed a number of times. The reprocessed material is often mixed with a virgin material to create *regrind*. Regrind is not generally recommended for use as a microwave dielectric material. The polymer will degrade after a large number of reprocessing cycles. Noryl is an example of a thermoplastic substrate.

*FR-4 is a flame-retardant version of G-10. FR-4 can generally be used in applications which call for G-10, but G-10 should not be used when FR-4 is specified. G-10 and FR-4 are rated at 285 °F. FR-4 does not melt above its rated temperature but instead begins charring.

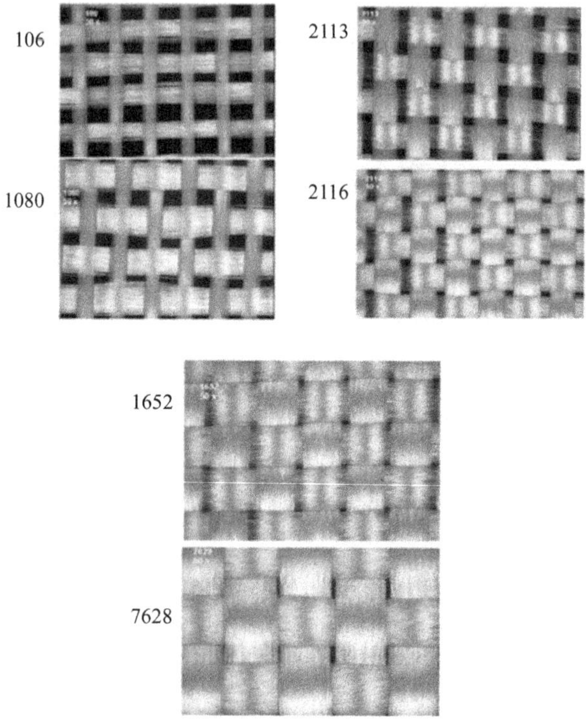

Figure A.2 Different types of fiberglass weaves

A.3.6 Fiberglass

When glass is drawn into thin fibers (which can be 1/100th the width of a human hair), it becomes flexible enough to combine into a yarn [22] (Figure A.1). These yarns may then be woven into a fiberglass cloth. A common glass cloth is Type 7628 in which each filament is specified to have 9.40 μm (0.37 mil) diameter. Each yarn has 408 filaments. When woven into cloth Type 7628 is 44 yarns/inch (warp) by 32 yarns/inch (weft). Therefore, this glass cloth has 17 952 filaments/inch (warp) by 13 056 filaments/inch. The typical thickness of Type 7628 is 150–200 μm (0.006–0.0078 inches). An image of the weave of Type 7628 is shown in Figure A.2. Ten layers of 7628 material may be impregnated with a resin and then stacked and laminated to create a 1575 μm (0.062 inches) thick woven fiberglass material.

Type 7628 is created using E-glass which is an electrical grade glass [23]. This is considered a general-purpose glass fiber. The relative dielectric permittivity is $\epsilon_r =$ 6.1 at 10 GHz, with a 0.0026 loss tangent [24]. The effective dielectric permittivity of the material will be dependent on the ratio of the fiberglass to the material impregnated into the glass cloth. In the case where Teflon (PTFE) is filled to produce a relative dielectric permittivity of 6.0, the laminate is almost isotropic. This is because E-Glass has a 6.1 dielectric permittivity, and the material consists of mostly glass fibers [25].

When fiberglass is being drawn, air bubbles may be present in the glass, which end up as capillaries in the glass fibers. These are often referred to as "hollow fibers." These hollow fibers can provide a conductive path, which can create a conductive filament and possibly lead to board failure [26,27].

One method used by industry to produce a composite substrate with a relative dielectric permittivity near 10 is to combine Teflon ($\epsilon_r = 2.1$), fiberglass ($\epsilon_r = 6.1$), and powdered titanium dioxide (TiO_2). Titanium dioxide has a very high dielectric permittivity ($\epsilon_r \approx 100$), which will bring the effective dielectric permittivity of the composite material up to the desired value of 10.

Microwave substrates with woven fiberglass are inherently anisotropic. The spacing of the weave may be electrically small over most of the common microwave region (1–10 GHz), but as one encounters millimeter wave frequencies (30–100 GHz), the periodic change in permittivity becomes significant. When microstrip transmission lines are constructed from impregnated woven fiberglass substrates, at high frequencies they produce a non-uniform transmission line (Appendix F). The periodic change in the transmission line can produce resonances, which an associated increases in insertion loss [28].

A.3.7 Dielectric foam

Structural foams are a composite dielectric material, with air as one dielectric and a plastic material as the other. The plastic materials used to produce structural foams are polyvinyl chloride (PVC), polyurethane, polystyrene, styrene acrylonitrile, polyetherimide (PEI), and polymethacrylimide (PMI) [29]. These foams are available in prescribed densities and a variety of thicknesses. Table A.6 shows the densities and permittivities of a number of commercial structural foams.

Foam materials are described as either *open cell* or *closed cell*. An open cell foam has interconnected chambers which allow it to absorb water or other liquids. A closed cell foam has chambers that are not interconnected. This type of foam generally repels water. If a foam is to be used in an elevated temperature environment, it needs to have been heat stabilized. In some cases, the foam can shrink up to 10% if not heat stabilized.

Caution must be exercised when using urethane foams. They are very flammable and produce extremely poisonous gases while burning. One should not use hot wire cutting to produce desired shapes from this foam. This type of foam should never be incinerated due to the toxic nature of the gases generated, even in an open area. Urethane foams are also rather fragile below a density of 60 Kg/m^3, this should be the lower limit value for any structural design [30].

PVC foams are stronger at lower densities (40 Kg/m^3) and slightly outperform urethane foams to 96 Kg/m^3. Above 96 Kg/m^3, higher quality urethane foams have higher strength than PVC foams.

Another option is PMI foam which is manufactured under the name Rohacell. The mechanical properties of Rohacell are superior to urethane or PVC foams at all densities. PMI foam resists nearly all solvents and chemicals and retains its properties up to at least 121 °C.

Table A.6 Structural foam materials

Manufacturer	Product	ϵ_r	tan δ	Density
Styrofoam	–	1.02	–	32 Kg/m^3
Divinycell	H45	1.065	0.0008	45 Kg/m^3
Divinycell	H60	1.075	0.0010	60 Kg/m^3
Divinycell	H80	1.10	0.0013	80 Kg/m^3
Divinycell	H100	1.10	0.0016	100 Kg/m^3
Divinycell	H130	1.15	0.0021	130 Kg/m^3
Divinycell	H200	1.23	0.0032	200 Kg/m^3
Divinycell	H250	1.28	0.0040	250 Kg/m^3
Klegecell	R100	1.08	–	100 Kg/m^3
Last-a-Foam	FR-3703	1.05	–	48 Kg/m^3
Last-a-Foam	FR-3704	1.05	–	67 Kg/m^3
Last-a-Foam	FR-3704.5	1.06	–	72 Kg/m^3
Last-a-Foam	FR-3705	1.07	–	80 Kg/m^3
Last-a-Foam	FR-6706	1.08	–	96 Kg/m^3
Last-a-Foam	FR-6707	1.12	–	112 Kg/m^3

Table A.7 Common radome materials

Material	ϵ_r	tan δ
Polypropylene (PP)	2.26	0.0028
Nylon 66	3.00	0.0066
Acrylonitrile butadiene styrene (ABS)	2.74	0.0051
Polycarbonate (PC) Lexan	2.76	0.0050
Polyetherimide (PEI) Ultem 2300	3.15	0.0072
Polyetheretherketone (PEEK)	3.16	0.0006
Acetal (POM)	2.79	0.0081

Syntactic foam was used as the dielectric substrate for the design of the rectangular microstrip antennas that make up the main mission antennas (phased array) on iridium satellites [20]. Syntactic foams are fabricated by filling plastic materials with small plastic, glass, or ceramic spheres which can be solid or hollow. Syntactic foams, with a relative permittivity range from 1.2 to 2.5, provide dielectric values that lie between structural foam materials and solid dielectric materials.

A.4 Radome materials

There are a number of plastics and composite dielectric materials which are used as radomes to cover microstrip and printed antennas (see Table A.7). In some cases, a microstrip antenna may be encapsulated (over molded) using injection molding methods. In other cases, a "clamshell" dielectric cover may be used to protect the antenna from its environment. The increased use of inkjet printing methods to create printed antennas may make some of these materials viable as microstrip antenna substrates in the future.

The addition of coloration to plastics can have a small effect on dielectric permittivity. Black pigment often increases the dielectric permittivity. In the case of polypropylene (PP), it has been noted that natural PP has $\epsilon_r = 2.26$, but that black pigment can increase this value to about 2.42. A disadvantage of polypropylene is that it becomes brittle at around 0 °C.

Materials such as acetal can be degraded by ultraviolet light. Carbon black may be introduced by manufacturers to reduce this degradation, which has the potential to affect the dielectric properties of the material. Acetal has a low coefficient of friction, which makes it good for bearings and moving parts.

Polycarbonate (Lexan) is one of the most useful engineering thermoplastics. The polymer was first created in 1898. It has very good mechanical properties, but is susceptible to a number of solvents such as benzene and toluene. Care must be taken not to design parts with tight radii as polycarbonate is susceptible to stress cracking.

Polyetherimide (Ultem) is a relatively new material which has become a popular choice for radome designs in aerospace applications. It has properties similar to polyimides (Kapton) which makes it heat resistant [31]. Ultem is available with glass-fiber and carbon-fiber reinforcement.

There are a number of structural thermoplastics available for use as structural parts of antenna designs, many are heat resistant. Two examples are polyetheretherketone (PEEK) and polyethylsulphones (Udel). PEEK is able to withstand temperatures required for lead-free soldering operations. The melting point of unfilled PEEK is 370 °C. These plastics are also combined with fillers to produce mixed materials which resist creep at high temperatures.

Often combinations of plastics are used with injection molding to produce parts. The ratio of the plastics can affect dielectric permittivity. In the case of high Q electrically small antenna designs, this has the potential to produce significant production variation.

A.5 Water absorption

Water absorption is the amount of weight gain in percent (%) experienced by a polymer, after immersion in water for a specific length of time, under a controlled environment. Water has a very high dielectric permittivity ($\epsilon_r \approx 81$) which will have a large effect on the dielectric permittivity of a material.

Nylons (polyamides) are well known to absorb water, which often makes them unacceptable for many applications. Water absorption causes nylon to change its dimensions, which can preclude its use for parts with tight dimensional tolerances. Water absorption also produces dielectric changes.

Polycarbonate cannot stand long-term exposure to hot water. It can absorb 0.2% of its weight in water when exposed to an environment of 33 °C with 65% relative humidity [32].

A.6 Dielectric films

In some applications, a printed antenna must be formed around a given plastic part. This can be the case in some RFID applications. The preferred material for this is generally Kapton film. Kapton is a polyimide which has excellent physical properties that allow it to be used in harsh environments. Kapton is prized for its high melting point. Kapton may be used in applications where a continuous temperature of up to 260 °C is required. This allows Kapton, which has a copper cladding, to be directly soldered without melting [33].

A.7 Passive intermodulation

In general, antennas are linear passive devices and not thought to contribute to intermodulation [34]. Intermodulation occurs when a device has a non-linear response to a signal. This non-linear response produces harmonics that are not just multiples of frequencies, but also sum and differences of frequencies. Diodes are passive devices that can produce intermodulation. Intermodulation in antennas generally occurs from corrosion of antenna fasteners, but can also be present in printed antennas in particular circumstances.

Although PIM problems in antennas are generally associated with high power transmitters in the range of 100 W, the author has seen significant third-order intermodulation in a system that transmitted only 500 mW. The radiated third harmonic generated was significant enough to preclude the system from passing its regulatory emissions tests. The antenna had been designed with a version of FR-4 ($\epsilon_r \approx 4.4$) and failed unintentional radiation levels at the third harmonic. When the antenna design was implemented with a microwave substrate engineered to produce low PIM values (Taconic TLX-7-0600-CL1/CL1 $\epsilon_r = 2.6$) the spurious third-order radiation was reduced by 25–30 dB. The product then passed its regulatory emissions tests. The design is a printed dipole type of antenna and was readily redesigned from FR-4 to TLX-7.

There is no clear understanding of the mechanisms which produce PIM in antennas, but there are a number of contributing factors which when controlled are known to reduce PIM:

1. Dissimilar metals. When zinc is present in copper, it can cause intermodulation in antennas. Zinc is added to copper so that it adheres to the dielectric material better than does pure copper.
2. Copper dielectric interface. Small air gaps between the bottom of the copper traces on a microwave PCB and the dielectric substrate are known to also increase

PIM. Bonding materials which inherently reduce the number of air gaps decrease this contributor. Bonding materials used to adhere the copper to the dielectric, often called *adhesion promotion*, as well as surface treatment can also contribute to PIM [35].

3. The surface roughness of copper. The greater the surface roughness of copper, the larger the generated PIM. The use of reverse treat copper decreases the surface roughness. Reverse treat copper is electrodeposited copper which can be smoothed on one side by controlling the depositing rate and other factors. If one needs a smoother copper surface than reverse treat can provide, one may use rolled copper instead [36].
4. Dielectric material. Different dielectric materials are also known to contribute to PIM [37].

All these factors need to be controlled to decrease the PIM produced by a microwave board material. The solder used to connect the antenna can also be a contributor and must be taken into consideration. The amount of PIM is also proportional to the length of a microstrip transmission line and inversely proportional to the width (i.e. wider lines have less PIM) [38]. Thinner antenna PCBs also contribute to PIM because the power density is concentrated in a smaller conductor region. The under-etching of copper traces can produce dendrites along its edges. This edge roughness can contribute to PIM [39]. It has been suggested that ion beam milling can produce more consistent trace widths and much sharper trace edges than chemical etching. Using ion beam milling may be a viable option for low-PIM applications [40].

A.8 Solder mask and conformal coatings

At frequencies below one Gigahertz, the effects of solder mask are generally minimal. Solder mask chemistry can vary considerably, which in turn produces different values of relative dielectric permittivity and loss tangent. Some solder masks are acrylic and others epoxy based. The introduction of solder mask increases dispersion in microstrip lines and also loss per unit length [41]. In high-frequency applications, solder mask may be used only on the low-frequency sections of a PCB. On the high-frequency sections, often electroless nickel immersion gold (ENIG) is employed. Other available options are silver or tin and solder.

ENIG can compromise insertion loss performance. ENIG plating consists of a nickel alloy plating on top of a copper trace, followed with a thin layer of immersion gold, which serves to protect the nickel layer from oxidation. The nickel thickness is generally around 1910–7620 nm, whereas the gold layer is only 80–200 nm. The nickel alloy varies from manufacturer to manufacturer, but is much less conductive than copper. A common estimate is the nickel alloy is about one-third that of copper. Nickel has a high relative permeability (μ_r) from 100 to 600, whereas copper has a value near one. The nickel formulation is varied by ENIG suppliers so that μ_r is reduced so the ENIG plating is near copper.

Microwave Substrate Manufacturers

Arlon
1100 Governor Lea Road
Bear, DE 19701
(302)-834-2100
www.arlonmed.com

CoorsTek
17750 W. 32nd Avenue
Golden, Colorado 80401
(303)-277-4779
www.coorsceramics.com

Polyflon Company
One Willard Road
Norwalk, CT 06851
(203)-840-7555
www.polyflon.com

Rogers Corporation
One Technology Drive
P.O. Box 188
Rogers, CT 06263-0188
(860)-779-9605
www.rogerscorporation.com

Sheldahl
1150 Sheldahl Road
Northfield, MN 55057
(507)-663-8000
www.sheldahl.com

Taconic Advanced Dielectric Division
136 Coonbrook Road
Petersburgh, NY 12138
(518)-658-3202
www.taconic-add.com

Park Electrochemical Corporation (ParkNelco)
5 Dakota Drive
Lake Success, NY 11042
(516)-354-4100
www.parknelco.com

Structural Foam Manufacturers

DIAB Group
www.diabgroup.com

Elliott Company
9200 Zionsville Road

Indianapolis, Indiana 46268
(317) 291-1219
www.elliottfoam.com

Evonik
www.rohacell.com

Gurit
www.gurit.com

General Plastics Manufacturing Company
4910 Burlington Way
P.O. Box 9097
Tacoma, Washington 98409
(253) 473-5000
www.generalplastics.com

References

[1] James, J.R., Henderson, A., and Hall P.S., "Microstrip Antenna Performance is determined by Substrate Constraints," *Microwave System News (MSN)*, August 1982, pp. 73–84.

[2] Kabacik, P., and Bialkowski, M.E., "The Temperature Dependence of Substrate Parameters and Their Effect on Microstrip Antenna Performance," *IEEE Transactions on Antennas and Propagation*, June 1999, Vol. 47, No. 6, pp. 1042–1049.

[3] Laverghetta, T.S., *Microwave Materials and Fabrication Techniques*, Third Edition, Artech House, 2000.

[4] Bancroft, R., "Conductive Ink a Match for Copper Antenna," *Microwaves & RF*, February 1987, Vol. 26, No. 2, pp. 87–90.

[5] Vyas, R., Lakafosis, V., Rida, A., *et al.*, "Paper-Based RFID-Enabled Wireless Platforms for Sensing Applications," *IEEE Transactions on Microwave Theory and Techniques*, May 2009, Vol. 57, No. 5, pp. 1370–1382.

[6] James, J.R., and Hall P.S., eds., *Handbook of Microstrip Antennas Volume 2*, Peter Peregrinus Ltd., London, 1989, Chapter 15.

[7] Guiles, C.L., "Everything You Ever Wanted To Know About Laminates...But Were Afraid To Ask," Eighth Edition, Version 3.0, Arlon Inc., 2000.

[8] Coombs, C.F., *Printed Circuits Handbook*, Second Edition, McGraw Hill, 1979, pp. 2–6.

[9] Coonrod, J., "Choosing Circuit Materials for Millimeter Wave Applications," *High Frequency Electronics*, July 2013, pp. 22–30.

[10] Horn, A.F, Reynolds, J.W., and Rautio, J.C., "Conductor Profile Effects on the Propagation Constant of Microstrip Transmission Lines," *IEEE MTT-S International Microwave Symposium*, 2010, pp. 868–871.

[11] Cruickshank, D.B., *Microwave Materials for Wireless Applications*, Artech House, 2011, pp. 44–46.

[12] Traut, G.R., "Advances in Substrate Technology," in James, J.R., and Hall P.S., eds., *Handbook of Microstrip Antennas Volume 2*, Peter Peregrinus Ltd., London, 1989, Chapter 15.
[13] Howe, H., Jr., "Dielectric Material Development," *Microwave Journal*, November 1978, pp. 39–40.
[14] Bouquet, F.L., Price, W.E., and Newell, D.M., "Designer's Guide to Radiation Effects on Materials for use on Jupiter Fly-Bys and Orbiters," *IEEE Transactions on Nuclear Science*, August 1979, Vol. NS-26, No. 4, pp. 4660–4669
[15] Rahaman, M.N., *Sintering of Ceramics*, CRC Press, 2008.
[16] Penn, S.J., Alford, N.M., Templeton, A., *et al.*, "Effect of Porosity and Grain Size on the Microwave Dielectric Properties of Sintered Alumina," *Journal of American Ceramic Society*, Vol. 80, No. 7, pp. 1885–1888.
[17] Alford, N., and Penn S.J., "Sintered Alumina with Low Dielectric Loss," *Journal of Applied Physics*, November 1996, Vol. 80, No. 7, pp. 5895–5898.
[18] Harper, C.A., ed., *Handbook of Plastics Technologies*, McGraw Hill, 2006, pp. 1.4–1.5.
[19] Harper, C.A., ed., *Handbook of Plastics Technologies*, McGraw Hill, 2006, pp. 3.41.
[20] Rohwer, A.B., Desrosiers, D.H., Bach, W., Estavillo, H., Makridakis, P., and Hrusovsky, R., "Iridium Main Mission Antennas—A Phased Array Success Story and Mission Update," *2010 IEEE International Symposium on Phased Array Systems and Technology*, Waltham, MA, October 12–15, 2010, pp. 504–511.
[21] Wang, C., "Determining Dielectric Constant and Loss Tangent in FR-4," *UMR EMC Laboratory Technical Report: TR00-1-041*, University of Missouri, Rolla, March 2000.
[22] Lamm, M., "The Fiberglass Story," *Invention and Technology*, Spring 2007, pp. 8–16.
[23] Wallenberger, F.T., Watson, J.C., and Li, H., "Glass Fibers," *ASM Handbook, Vol. 21: Composites (#06781G)*, www.asminternational.org.
[24] Hartman, D., Greenwood, M., and Miller, D., "High Strength Glass Fibers," *AGY Inc Technical Paper*, 1996 (Table 3).
[25] Olyphant, M., and Nowicki, T.E., "MIC SUBSTRATES—A REVIEW," Paper prepared for presentation at *ELECTRO-80*, May 13, 1980, Boston, MA, p. 5.
[26] Rogers, K., Van Den Driessche, P., and Pecht, M., "Do You Know That Your Laminates May Contain Hollow Fibers?," *Printed Circuit Fabrication*, April 1999, Vol. 22, No. 4, pp. 34–38.
[27] Rogers, K., Hillman, C., and Pecht, M., "Hollow Fibers Can Accelerate Conductive Filament Formation," *ASM International Practical Failure Analysis*, August 2001, Vol. 1, No. 4, pp. 57–60.
[28] Torres-Torres, R., Romo, G., Schauer, M., Nwachukwu, C., and Baek, S., "Modeling Resonances in Transmission Lines Fabricated Over Woven Fiber Substrates," *IEEE Transactions on Microwave Theory and Techniques*, July 2013, Vol. 61, No. 7, pp. 2558–2565.

[29] Black, S., "Getting To The Core Of Composite Laminates," *Composites Technology*, October 2003.

[30] Lee, S.M., ed., *Handbook of Composite Reinforcements*, John Wiley and Sons, 1992, pp. 173–176.

[31] Harper, C.A., ed., *Handbook of Plastics Technologies*, McGraw Hill, 2006, p. 2.26.

[32] Harper, C.A., ed., *Handbook of Plastics Technologies*, McGraw Hill, 2006, p. 2.20.

[33] Harper, C.A., ed., *Handbook of Plastics Technologies*, McGraw Hill, 2006, p. 2.35.

[34] Coonrod, J., "Materials Make the Difference in Low-PIM PCB Antennas," *Microwaves & RF*, October 2016, pp. 51–54.

[35] Shitvov, A.P., Olsson, T., Francey, J., Zelenchuk, D.E., Schuchinsky, A.G., and Banna, B.E., "Effects of Interface Conditions and Long-Term Stability of Passive Intermodulation Response in Printed Lines," *IET Microwaves, Antennas & Propagation*, 2011, Vol. 5, No. 1, pp. 68–76.

[36] Schuchinsky, A.G., Francey, J., and Fusco, V.F., "Distributed Sources of Passive Intermodulation on Printed Lines," *Proceedings of the IEEE AP-S Symposium*, Washington, DC, July 2005, pp. 447–450.

[37] Zelenchuk, D.E., Shitvov, A.P., and Schuchinsky, A.G., "Effect of Laminate Properties on Passive Intermodulation Generation," *Proceedings of LAPC 2007*, Loughborough, April 2007, pp. 169–172.

[38] Shitvov, D.E., Zelenchuk, D.E., Schuchindky, A.G., and Fusco, V.F., "Passive Intermodulation in Printed Lines: Effects of Trace Dimensions and Substrate," *IET Microwaves, Antennas & Propagation*, 2009, Vol. 3, No. 2, pp. 260–268.

[39] Coonrod, J., "Choosing Circuit Materials for Low-PIM Antennas," *Microwaves & RF*, December 2017, pp. 54–59.

[40] Barrett, J., and Perkins, T., "Performance and Consistency Advantage of Ion Beam over Chemical Etching," *High Frequency Electronics*, December 2017, pp. 28–37.

[41] Coonrod, J., "Ambiguous Influences Affecting Insertion Loss of Microwave Printed Circuit Boards," *IEEE Microwave Magazine*, Vol. 13, No. 5, July/August 2012, pp. 66–75.

Appendix B

Numerical methods

B.1 Numerical integration

The integrations presented in Chapter 2 to compute the directivity of a rectangular microstrip antenna using (2.47) and the integrations in Chapter 3 to compute the radiation conductance and radiation Q of a circular microstrip antenna (3.9) have no known closed-form solution and are numerically evaluated.

When an integrand is sufficiently smooth and contains no singularities, a very efficient method of numerical integration is Gaussian Quadrature. Gaussian Quadrature integrates a polynomial which curve fits the integrand. The integrations presented in this book have well-behaved integrands [except for (C.21)] and may be integrated using this method. The integrations of interest are of this form [1]:

$$\int_a^b f(x)dx \quad \text{or} \quad \int_a^\infty f(x)dx \tag{B.1}$$

$$\int_a^b f(x)dx = \frac{b-a}{2}\sum_{i=1}^{N} W_i f\left(\frac{Z_i(b-a)+b+a}{2}\right) \tag{B.2}$$

$$\int_a^\infty f(x)dx = 2\sum_{i=1}^{N} \frac{W_i}{(1+Z_i)^2} f\left(\frac{2}{1+Z_i}+a-1\right) \tag{B.3}$$

where N is the number of Gaussian quadrature points, W_i are the weights, and $\pm Z_i$ are the abscissas.

For a 16-point Gaussian quadrature, the sums are evaluated as

$$\begin{aligned}\int_a^b f(x)dx = \frac{b-a}{2}\Bigg(&\sum_{i=1}^{8} W_i f\left(\frac{Z_i(b-a)+b+a}{2}\right)\\ &+\sum_{i=1}^{8} W_i f\left(\frac{-Z_i(b-a)+b+a}{2}\right)\Bigg)\end{aligned} \tag{B.4}$$

and

$$\int_a^\infty f(x)dx = 2\sum_{i=1}^{8} \frac{W_i}{(1+Z_i)^2} f\left(\frac{2}{1+Z_i} + a - 1\right)$$
$$+2\sum_{i=1}^{8} \frac{W_i}{(1-Z_i)^2} f\left(\frac{2}{1-Z_i} + a - 1\right) \quad \text{(B.5)}$$

The Gaussian weights W_i and abscissas $\pm Z_i$ for a 16-point Gaussian quadrature are presented below:

$\pm Z_i$	W_i
0.09501 25098 37637 440185	0.18945 06104 55068 496285
0.28160 35507 79258 913230	0.18260 34150 44923 588867
0.45801 67776 57227 386342	0.16915 65193 95002 538189
0.61787 62444 02643 748447	0.14959 59888 16576 732081
0.75540 44083 55003 033895	0.12462 89712 55533 872052
0.86563 12023 87831 743880	0.09515 85116 82492 784810
0.94457 50230 73232 576078	0.06225 35239 38647 892863
0.98940 09349 91649 932596	0.02715 24594 11754 094852

The integrations in this text are all calculated using a 96-point Gaussian quadrature [2].

B.2 Evaluation of sums

A common metric used to estimate the relative error when summing a series is [3]:

$$\frac{|p - p^*|}{|p|} < 5 \times 10^{-t} \quad \text{(B.6)}$$

where p is not equal to zero.

The value p^* is said to approximate p to t significant digits (or figures) if t is the largest nonnegative integer for which the relationship is satisfied. The assumption is that we know p *a priori* as the final value we are attempting to evaluate. This relationship is useful as it provides a continuous measure of relative error. Unfortunately, many summations in practice converge slowly enough to satisfy this relationship before it is sufficiently converged to an accurate approximation. For example, calculating the driving point impedance of a rectangular microstrip antenna using the cavity model entails computing this expression:

$$Z_{drv} = \sum_{m=0}^{\infty}\sum_{n=0}^{\infty} \frac{j\omega\alpha_{mn}}{\omega_{mn}^2 - (1 - j\delta_{eff})\omega^2} \quad \text{(B.7)}$$

One should account for the relative error of the real and imaginary parts of the driving point impedance separately. The contribution of the higher order modes to the driving point impedance can be slow in converging. It is often best to start with a fairly large number of modes, say, 100 (0–99), then compute nearly double that value, to 199, then calculate a relative error, and compute the sum a third time nearly doubling the number of modes again to 399. The relative error may be checked after each doubling inside a summation loop. Generally, this method produces convergent results even for slowly converging Green's functions expressed as a summation. Generally, testing 4–5 significant places of relative error is sufficient.

The summations are

$$Z^i_{drv} = \sum_{m=100(i-1)}^{(100i)-1} \sum_{n=100(i-1)}^{(100i)-1} \frac{j\omega\alpha_{mn}}{\omega^2_{mn} - (1 - j\delta_{eff})\omega^2} \tag{B.8}$$

where

$$i = 1, 2, 3, \ldots \frac{|Z^i_{drv} - Z^{(i+1)}_{drv}|}{|Z^i_{drv}|} < 5 \times 10^{-t}$$

where $t = 4$.

B.3 Fixed-point iteration

A solution to the equation $x = f(x)$ is said to be a fixed point of the function f. In the case of computing the resonant dimension L $(=W)$ of a square patch, we can use fixed-point iteration on (2.88), (2.6), and (2.7) to find the solution for an equal patch length and width. It is also useful for the computation of the radius of a circular patch antenna from (3.5).

The fixed-point algorithm presented in pseudocode is given below [4]:

```
***************************
** FIXED POINT ALGORITHM **
***************************

A solution for $x=f(x)$ may be found
given an initial approximation $x_0$
for the solution of this equation:

The INPUT is an initial approximation of $x_0$,
a desired relative accuracy TOL
and a maximum number of iterations $N_0$.

The OUTPUT is an approximate solution $x$
or failure to locate a fixed point.

STEP 1 Set $i = 1$.

STEP 2 WHILE $i \le N_0$ DO STEPS 3 to 6

   STEP 3 SET $x_i = f(x_{(i-1)})$
```

```
    STEP 4 IF $\mid x_{(i-1)}-x_{i}\mid / \mid x_{i} \mid$ <
              TOL THEN
           OUTPUT SOLUTION: $x_{i}$
           STOP

    STEP 5 SET $i=i+1$

    STEP 6 SET $x_{i-1}=x_i$

STEP 7 OUTPUT ''No fixed point after,'' $N_0$,''iterations''
       STOP
```

B.4 Bisection algorithm

Often it is useful to locate the root of an equation [i.e. the value of x, where $f(x) = 0$]. The transmission line model of a microstrip antenna can be used to predict the driving point impedance of a rectangular microstrip antenna. If the impedance function is represented by $Z(x)$, and we wish to find the value of x at which the real part of the impedance is R_0, we define $f(x) = RE(Z(x)) - R_0$ and locate the root. $RE(Z)$ is a function that returns the real part of a complex number Z.

A rectangular microstrip antenna in its lowest order mode has a maximum driving point impedance at one radiating edge and decreases to zero at its center. This provides boundaries that may be used with the bisection algorithm to determine the driving point location which provides a desired driving point resistance. Pseudocode for the bisection algorithm is

```
Given a continuous function $f$ on the interval [$a$, $b$],
where $f(a)$ and $f(b)$ have opposite signs.
A solution to $f(x)=0$ may be found with this procedure:

*****************************
** THE BISECTION ALGORITHM **
*****************************

STEP 1 Set $a_1 = a$ and $b_1 = b$

STEP 2 Set $i = 1$

STEP 3 Set $p_i = {1\over 2}(a_i + b_i)$

STEP 4 IF $p_i$ is an acceptable approximation, GOTO STEP 10
       IF $p_i$ is not an acceptable approximation, GOTO STEP 5

STEP 5 IF $f(p_i)\cdot f(a_i) > 0$  GOTO STEP 6
       IF $f(p_i)\cdot f(a_i) < 0$  GOTO STEP 8

STEP 6 Set $a_{i+1} = p_i$ and $b_{i+1} = b_i$
```

```
STEP 7 Add 1 to $i$ and GOTO STEP 3

STEP 8 Set $a_{i+1} = a_i$ and $b_{i+1} = p_i$

STEP 9 Add 1 to $i$ and GOTO Step 3

STEP 10 The procedure is completed
```

B.5 MSA Q-efficiency calculation

In this section, a method that uses HFSS full-wave analysis data with a cavity model to compute the loss constituents of a patch antenna is detailed.

The cavity model analysis relates the total cavity Q, Q_T, and its constituent Q's with this relation:

$$\frac{1}{Q_T} = \frac{1}{Q_c} + \frac{1}{Q_d} + \frac{1}{Q_R} + \frac{1}{Q_S} \tag{B.9}$$

where Q_c is the conductor Q, Q_d is the dielectric Q, Q_R is the radiation Q, and Q_S is the surface wave Q.

With HFSS, it is possible to eliminate terms in (B.9) to produce three independent equations. We initially re-write (B.9) ignoring the surface wave loss:

$$\frac{1}{Q_T} = \frac{1}{Q_c} + \frac{1}{Q_d} + \frac{1}{Q_R} \tag{B.10}$$

We first design a square patch antenna with a probe feed, which is matched at a desired operating frequency. We then assume that the Q of the HFSS S_{11} frequency plot is equal to Q_{T_1} which can be inverted for use with (B.11).

We next replace the copper conductors (patch and groundplane) with a perfect electrical conductor in the HFSS model. This will produce an infinite conductor Q (i.e. Q_c), which causes the first term on the right-hand side of (B.10) to become zero. The HFSS S_{11} frequency plot can be used to compute Q_{T_2} which can be inverted for use with (B.12).

Finally, we restore the copper conductors and set the loss tangent of the dielectric substrate to zero. This will effectively eliminate the second term of (B.10) producing (B.13). The Q of this HFSS S_{11} frequency plot is equal to Q_{T_3} which can be inverted for use with (B.13):

$$\frac{1}{Q_{T_1}} = \frac{1}{Q_c} + \frac{1}{Q_d} + \frac{1}{Q_R} \tag{B.11}$$

$$\frac{1}{Q_{T_2}} = \frac{1}{Q_d} + \frac{1}{Q_R} \tag{B.12}$$

$$\frac{1}{Q_{T_3}} = \frac{1}{Q_c} + \frac{1}{Q_R} \tag{B.13}$$

This allows us to express these equations in matrix form

$$\begin{vmatrix} 1 & 1 & 1 \\ 0 & 1 & 1 \\ 1 & 0 & 1 \end{vmatrix} \begin{vmatrix} A_1 \\ A_2 \\ A_3 \end{vmatrix} = \begin{vmatrix} B_1 \\ B_2 \\ B_3 \end{vmatrix}$$

where

$$A_1 = \frac{1}{Q_c} \qquad A_2 = \frac{1}{Q_d} \qquad A_3 = \frac{1}{Q_R}$$

$$B_1 = \frac{1}{Q_{T_1}} \qquad B_2 = \frac{1}{Q_{T_2}} \qquad B_3 = \frac{1}{Q_{T_3}}$$

HFSS is used to obtain the values for B_1, B_2, and B_3, and this allows one to compute A_1, A_2, and A_3, which provide us with values for Q_c, Q_d, and Q_R. The conductor efficiency is found using:

$$\eta_c = \frac{Q_d Q_R}{Q_d Q_R + Q_c Q_R + Q_c Q_d} \tag{B.14}$$

The dielectric efficiency is found with

$$\eta_d = \frac{Q_c Q_R}{Q_d Q_R + Q_c Q_R + Q_c Q_d} \tag{B.15}$$

The amount of loss computed using the radiation efficiency equation is actually a combination of the loss from radiation and surface waves:

$$\eta_{SR} = \frac{Q_c Q_d}{Q_d Q_R + Q_c Q_R + Q_c Q_d} \tag{B.16}$$

The radiation efficiency computed by HFSS for the model with conductor and dielectric losses present is the true radiation efficiency, η_r, without the surface wave loss included. HFSS states: "The radiation efficiency is the ratio of the radiated power to the accepted power. . . ." We compute the surface wave efficiency η_{sw} by subtracting the radiation efficiency computed by HFSS from the efficiency computed using (B.16):

$$\eta_{sw} = \eta_{SR} - \eta_r \tag{B.17}$$

This relationship is expressed as (B.17).

References

[1] Hewlett-Packard, *HP-41 User's Library Solutions High Level Math*, pp. 35–41.

[2] Abramowitz, M., and Stegun, I.A., *Handbook of Mathematical Functions*, Dover, 1972, pp. 916–917.

[3] Burden, R.L., and Faires, J.D., *Numerical Analysis*, Fourth Edition, PWS-Kent Publishing Company, Boston, 1989, pp. 12–14.

[4] Burden, R.L., and Faires, J.D., *Numerical Analysis*, Fourth Edition, PWS-Kent Publishing Company, Boston, 1989, p. 41.

Appendix C
Planar transmission lines

C.1 Microstrip transmission line design

There are many approaches for developing microstrip transmission line design and analysis equations [1–7]. Relationships presented by Kajfez and Tew have proven to be very flexible for the analysis of the microstrip transmission line [8,9]. The characteristic impedance of a microstrip line is given by

$$Z_0 = \frac{42.4}{\sqrt{\epsilon_r + 1}} \ln\left\{1 + \left(\frac{4h}{\acute{w}}\right)\left[b + \sqrt{b^2 + a\pi^2}\right]\right\} \tag{C.1}$$

where Z_0 is the characteristic impedance of the microstrip transmission line:

$$b = \left(\frac{14 + 8/\epsilon_r}{11}\right)\left(\frac{4h}{\acute{w}}\right) \tag{C.2}$$

$$\acute{w} = w + a \cdot \Delta w \tag{C.3}$$

where $\acute{w}$ is the corrected transmission line width and t is the thickness of the transmission line conductor (foil thickness) see Appendix A:

$$\Delta w = \frac{t}{\pi}\left\{1 + \ln\left(\frac{4}{\sqrt{\left(\frac{t}{h}\right)^2 + \left[\frac{1}{\pi\left(\frac{w}{t} + 1.1\right)}\right]^2}}\right)\right\} \tag{C.4}$$

where Δw is the width correction factor:

$$a = \frac{1 + 1/\epsilon_r}{2} \tag{C.5}$$

The use of a root finding method such as the bisection method (Appendix B) allows one to also use this set of equations for synthesis.

Dispersion is a very important property of microstrip lines, which must be accounted for, in order to expect a successful transmission line design, realized using microstrip. The guide wavelength of microstrip transmission line varies with frequency:

$$\lambda_g = \frac{\lambda_0}{\sqrt{\epsilon_{eff}}} \tag{C.6}$$

where λ_0 is the free space wavelength at a given frequency.

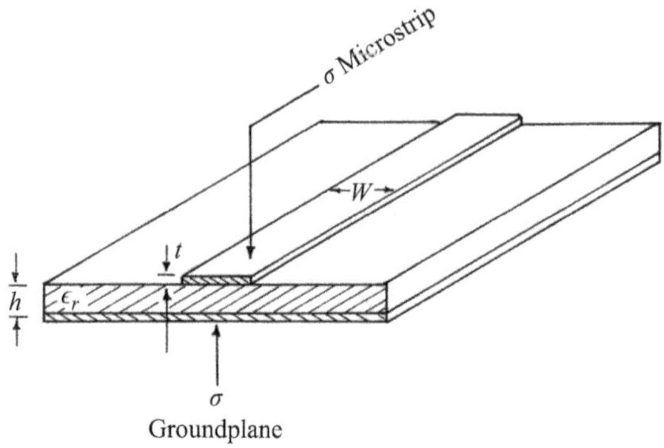

Figure C.1 Microstrip transmission line configuration

The relative effective dielectric permittivity computation is undertaken with

$$\epsilon_{eff} = \epsilon_r - \frac{\epsilon_r - \epsilon_{eff0}}{1 + G(f/f_p)^2} \tag{C.7}$$

$$G = 0.6 + 0.009Z_0 \tag{C.8}$$

$$f_p\,(\text{GHz}) = \frac{Z_0}{0.8\pi h\,(\text{mm})} \tag{C.9}$$

$$\epsilon_{eff0} = \left(\frac{Z_0^1}{Z_0}\right)^2 \tag{C.10}$$

where Z_0^1 is Z_0 computed with dielectric removed and Z_0 is computed with dielectric present.

The computation of microstrip transmission line loss using the following relationships has the advantage that different microstrip and groundplane conductivities may be used:

$$\alpha\,(\text{dB/m}) = 90.96\sqrt{\epsilon_{eff}} \cdot \frac{f\,(\text{GHz})}{Q} \tag{C.11}$$

$$Q = \frac{1}{1/Q_c + 1/Q_d} \tag{C.12}$$

$$\frac{1}{Q_c} = \frac{Z_0^2 - Z_0^1 + Z_0^3 - Z_0^1}{Z_0^1} \tag{C.13}$$

$$\frac{1}{Q_d} = \frac{1 - 1/\epsilon_{eff}}{1 - 1/\epsilon_r}\tan\delta \tag{C.14}$$

Table C.1 Computation of Z_0 values ($\epsilon_r = 1$)

Impedance	Width	Substrate	Metalization
Z_0^1	w	h	t
Z_0^2	w	$h+\delta_p/2$	t
Z_0^3	$w-\delta_s$	$h+\delta_s/2$	$t-\delta_s$

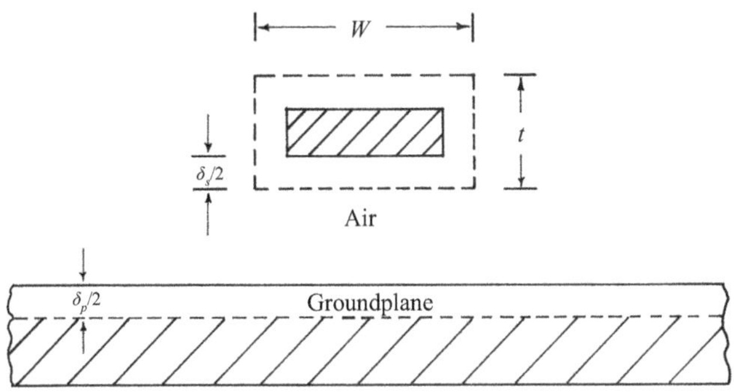

Figure C.2 Microstrip transmission line loss figure

$$\delta\ (\text{mm}) = \frac{1}{20\pi\sqrt{f(\text{GHz})\cdot 58\cdot \delta_r}}\ (\text{copper}) \tag{C.15}$$

The losses are evaluated using three computations of characteristic impedance with dielectric removed, Z_0^1, Z_0^2, and Z_0^3, as listed in Table C.1. The first calculation, Z_0^1, is the characteristic impedance of the transmission line with perfect electrical conductors (zero skin depth). We use h, w, and t. The second calculation, Z_0^2, is evaluated with a groundplane that has been decreased by half of a skin depth of the material from which it is constructed. The last calculation, Z_0^3, is computed for a microstrip transmission line in which all the sides of the microstrip conductor have been recessed by one-half of the skin depth of the material from which the conductor has been fabricated (Figure C.2). The conductor Q (Q_c) may then be computed. Next, the dielectric Q (Q_d) is obtained. The microstrip losses may then be obtained using the total Q.

C.2 Discontinuity compensation

When microwave devices such as a branchline hybrid are realized in the microstrip transmission line, they deviate from the ideal transmission line model representation

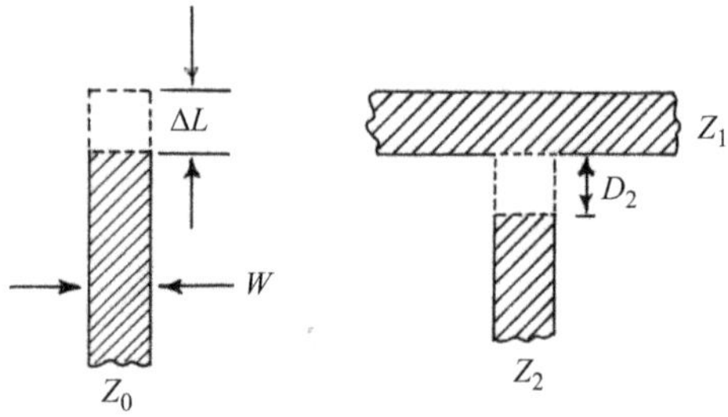

Figure C.3 (Left) Open-circuit end correction in microstrip and (right) shortening of the shunt arm of a microstrip tee junction

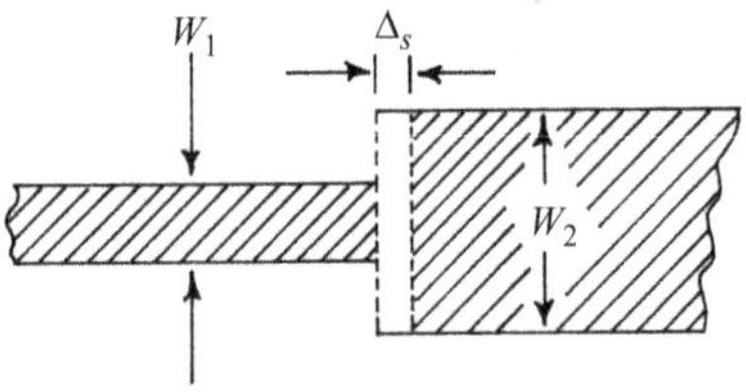

Figure C.4 Microstrip line step discontinuity

because of discontinuities. In order to design these devices as accurately as possible, it is necessary to compensate for discontinuities. In high-frequency devices, it is also imperative to include the effects of dispersion.

In Figure C.3, we have two common discontinuities encountered when realizing microstrip transmission line designs. The first case is an open-circuited transmission line. This is encountered when using open-circuited matching stubs. The fringing capacitance at the end of the transmission line causes it to be electrically longer than its physical length. We compensate for this effect by reducing the line length by ΔL [10]:

$$\Delta L = 0.412 \left(\frac{\epsilon_{re} + 0.3}{\epsilon_{re} - 0.258} \right) \left(\frac{W/h + 0.262}{W/h - 0.813} \right) h \tag{C.16}$$

where h is the substrate thickness, W is the microstrip line width, and ϵ_{re} is the effective dielectric permittivity of the microstrip transmission line.

The second case is tee junction compensation. The electrical length of the shunt arm of the tee junction must be reduced by length D_2. The currents take a shortened path close to the corner which requires compensation. The formula for the shunt arm reduction is

$$D_2 = \frac{120\pi}{Z_1\sqrt{\epsilon_{re}^1}} \left(0.5 - 0.16\frac{Z_1}{Z_2}\,[1 - 2\ln(Z_1/Z_2)] \right) h \tag{C.17}$$

where ϵ_{re}^1 is the relative effective dielectric permittivity of the through arm.

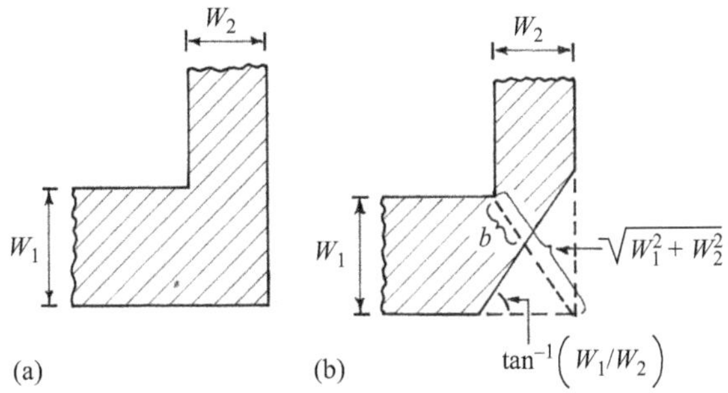

Figure C.5 (a) Microstrip transmission line with 90° bend and width change. (b) Miter of a microstrip 90° corner to reduce discontinuity from corner capacitance

When a pair of microstrip lines form a junction of high and low impedance lines, the exposed section of the low impedance (wider) line is electrically longer than its physical length and requires a capacitive end correction. Equation (C.18) may be used to compute an approximate correction length for removal. It is assumed that the low impedance (wider) line has a width of W_2 and the narrower high impedance line is of width W_1. Δ_s is the step correction:

$$\Delta_s \approx \Delta L \left(1 - \frac{W_1}{W_2}\right) \tag{C.18}$$

When a microstrip transmission line forms a sharp right angle bend, excess capacitance appears at the corner of the line. A reflection will be produced by the corner capacitance. To compensate for this effect, the corner is often mitered. This can be seen in Figure C.5. Figure C.5(a) shows a microstrip transmission line with a right angle bend and a step in width. Figure C.5(b) illustrates the use of a miter to reduce the corner capacitance and reduce the reflection it causes. The miter fraction m is defined as

$$m = 1 - \frac{b}{\sqrt{W_1^2 + W_2^2}} \tag{C.19}$$

A value of 0.6 is a good choice for m. When $(W_1 = W_2)$ (C.19) reduces a definition conventionally found in the literature [11].

C.3 Dielectric covered microstrip line

In Subsection 6.4.6, it is shown that radiation from microstrip transmission lines increases with substrate thickness. One avenue of mitigation is to use thin substrate microstrip transmission lines with a higher dielectric to reduce microstrip transmission

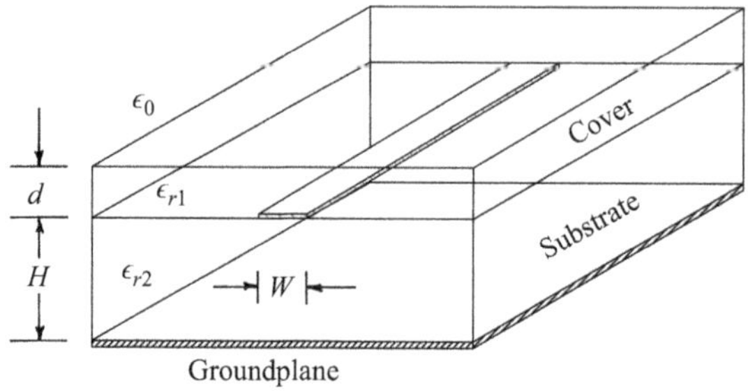

Figure C.6 Microstrip transmission line with dielectric cover geometry

line radiation, placing a lower dielectric permittivity layer on the line and embedding the transmission line conductor between the two layers.

The analysis presented here was formulated by Bahl and Stuchly [12]. The design geometry is presented in Figure C.6.

The substrate layer has a thickness of H and a relative dielectric permittivity of ϵ_{r2}. The cover layer has a thickness of d with a relative dielectric permittivity ϵ_{r1}. Free space is assumed above the top dielectric layer. The width of the microstrip transmission line conductor is W.

The equations used to compute a quasi-static solution for the effective relative dielectric permittivity, ϵ_e, and characteristic impedance, Z_0, are evaluated by computing the capacitance per unit length with and without the dielectrics present:

$$\frac{1}{C} = \frac{1}{\pi\epsilon_0}\int_0^{\infty} \frac{\left(1.6\left\{\frac{\sin\left(\frac{\beta W}{2H}\right)}{\left(\frac{\beta W}{2H}\right)}\right\} + \frac{2.4}{\left(\frac{\beta W}{2H}\right)^2}\cdot\left\{\cos\left(\frac{\beta W}{2H}\right) - \frac{2\sin\left(\frac{\beta W}{2H}\right)}{\left(\frac{\beta W}{2H}\right)} + \frac{\sin^2\left(\frac{\beta W}{4H}\right)}{\left(\frac{\beta W}{4H}\right)^2}\right\}\right)^2}{\left[\epsilon_{r1}\frac{\epsilon_{r1}\tanh\left(\frac{\beta d}{H}\right)+1}{\epsilon_{r1}+\tanh\left(\frac{\beta d}{H}\right)} + \epsilon_{r2}\coth(\beta)\right]\cdot\beta}\,d\beta \tag{C.20}$$

where C_0 is the capacitance no dielectric present, C is the capacitance with dielectric present, $\epsilon_e = C/C_0$, $Z = 1/cC_0$, $Z_0 = Z/\sqrt{\epsilon_e}$, $V_p = c/\sqrt{\epsilon_e}$, V_p is the phase velocity, and c is the speed of light in a vacuum, ϵ_0 is the permittivity of free space, ϵ_{r1} is the relative permittivity of dielectric cover, and ϵ_{r2} is the relative permittivity of dielectric substrate.

The integral of (C.20) can be numerically computed using Gaussian quadrature using the techniques in Appendix B.1. The results of example computations for a covered microstrip transmission line with a high dielectric substrate and low dielectric cover are found in Table C.2.

Table C.2 Covered 50 Ω microstrip transmission line with increasing cover thickness and constant substrate thickness

50 Ω covered microstrip transmission line **Cover – ϵ_{r1} Substrate – ϵ_{r2}**					
ϵ_{r1}	ϵ_{r2}	H	d	W	ϵ_e
3.00	10.2	762 μm	126 μm	715 μm	7.03
3.00	10.2	762 μm	254 μm	699 μm	7.17
3.00	10.2	762 μm	508 μm	682 μm	7.33
3.00	10.2	762 μm	762 μm	675 μm	7.40
3.00	10.2	762 μm	1524 μm	666 μm	7.48
3.00	10.2	762 μm	3048 μm	663 μm	7.52

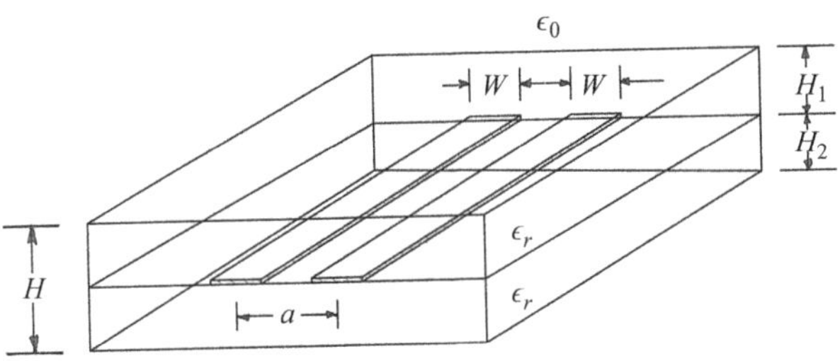

Figure C.7 Twin strip transmission line geometry

C.4 Twin strip transmission line

The twin strip transmission line is a printed version of the twin lead transmission line. Generally, it consists of two parallel strips on a dielectric slab. This transmission line is also known as the *coplanar strip transmission line* or *parallel-strip line*. This transmission line is sometimes used to drive balanced printed antennas.

The general version of the geometry of a twin strip transmission line is shown in Figure C.7. This type of transmission line generally does not have an upper dielectric layer (i.e. $H_1 = 0$). Both upper and lower dielectric layers have the same relative dielectric permittivity. The center to center spacing of the strips is denoted as a. The width of each strip is equal to W.

The capacitance of this structure has been analyzed by Yamashita and Yamazaki [13]. The computation of the capacitance of this structure with C and without the dielectric present C_0 allows one to evaluate the characteristic impedance Z_0, guide wavelength λ, and relative effective dielectric permittivity ϵ_e. The assumptions for this analysis are that the strip separation is small compared with a free space wavelength ($a \ll \lambda_0$). The thickness of the conductor t is much smaller than the thickness H, and H is much smaller than a free space wavelength ($t \ll H \ll \lambda_0$).

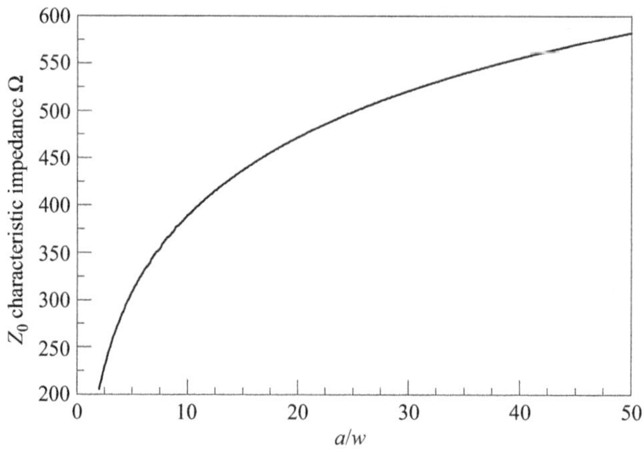

Figure C.8 Computed characteristic impedance (Z_0) of a twin strip transmission line with $\epsilon_r = 2.55$, $H_1 = 0$ and $H_2 = 1.0$ mm

The capacitance of this structure is given by

$$\frac{1}{C} = \frac{1}{\pi \epsilon_r \epsilon_0} \int_0^{\infty} \frac{\tilde{F}(\beta)}{\beta} \left| \frac{f(\beta)}{Q} \right|^2 d\beta \tag{C.21}$$

$$\tilde{F}(\beta) = \frac{[\epsilon_r \cosh(\beta H_1) + \sinh(\beta H_1)][\epsilon_r \cosh(\beta H_2) + \sinh(\beta H_2)]}{[\epsilon_r^2 + 1] \sinh(\beta H) + 2\epsilon_r \cosh(\beta H)} \tag{C.22}$$

$$\left| \frac{f(\beta)}{Q} \right|^2 = \frac{4}{\beta W} \sin\left(\frac{\beta W}{2}\right) \sin\left(\frac{\beta a}{2}\right) \tag{C.23}$$

The characteristic impedance Z_0 may be computed from

$$Z_0 = \frac{120\pi \epsilon_0}{(CC_0)^{\frac{1}{2}}} \tag{C.24}$$

The guide wavelength is

$$\lambda = \left(\frac{C_0}{C}\right)^{\frac{1}{2}} \lambda_0 \tag{C.25}$$

The relative effective dielectric permittivity is

$$\epsilon_e = \left(\frac{C}{C_0}\right) \tag{C.26}$$

The integration of (C.21) may *not* be computed accurately using Gaussian quadrature. The author computed small integration intervals with Richardson's extrapolation [14] and added them up to evaluate the integration.

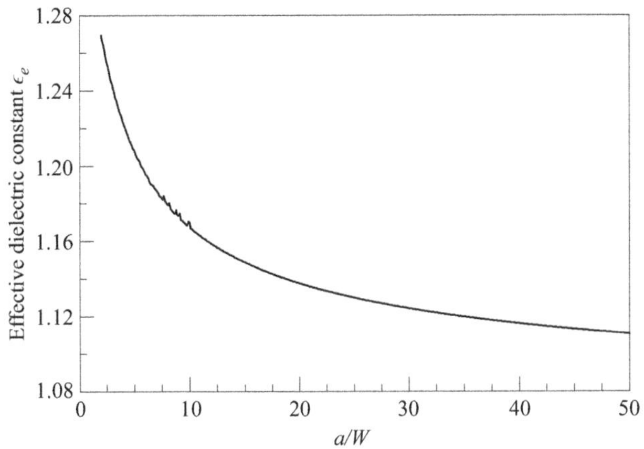

Figure C.9 Computed relative effective dielectric permittivity (ϵ_e) of a twin strip transmission line with $\epsilon_r = 2.55$, $H_1 = 0$ and $H_2 = 1.0$ mm

In Figure C.8, the computed results for the characteristic impedance of a twin strip transmission line with respect to a/W are shown. We can see that the Z_0 of the twin strip transmission line becomes larger as the separation between strips compared with the strip width becomes large. This is consistent with the twin lead transmission line.

In Figure C.9, are the computed results for the effective dielectric permittivity of a twin strip transmission line with respect to a/W. We can see the ϵ_e of the twin strip transmission line becomes smaller as the separation between strips compared with the strip width becomes large.

Alternative analysis methods for Twin Strip transmission line analysis may be found in the literature [15,16].

C.5 Parallel plate transmission line

The geometry of a parallel-plate transmission line is shown in Figure C.10. Two metal plates of width W, are separated by a dielectric sheet of height H, with a relative dielectric permittivity of ϵ_r. This transmission line can support a TEM transmission line mode. The electric field is $\hat{y}$ directed with the magnetic field $\hat{x}$ directed. The electric field is across the separated metal plates with a distribution similar to those of a capacitor. We may use the formula for a parallel-plate capacitor to express the capacitance per unit length (C) of this transmission line as

$$C = \epsilon_0 \epsilon_r \frac{W}{H} \text{ F/m} \tag{C.27}$$

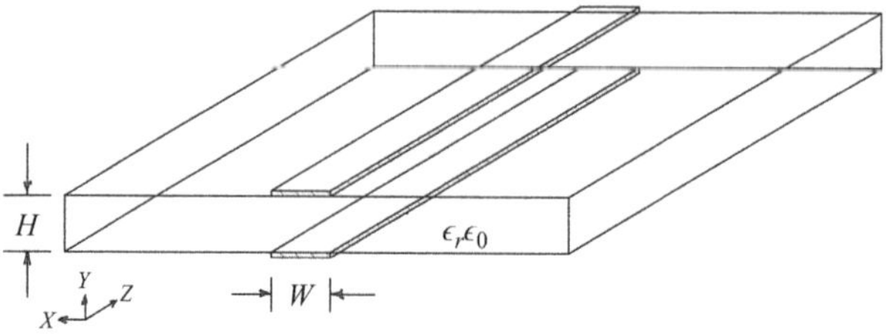

Figure C.10 Geometry of a parallel-plate transmission line

The phase velocity v_p of a TEM wave is

$$v_p = \frac{c}{\sqrt{\epsilon_r}} = \frac{1}{\sqrt{\mu_0\epsilon_0\epsilon_r}} = \frac{3.0 \cdot 10^8}{\sqrt{\epsilon_r}} \text{ m/s} \tag{C.28}$$

For the case of a lossless transmission line the phase velocity is given by

$$v_p = \frac{1}{\sqrt{LC}} \text{ m/s} \tag{C.29}$$

We may use (C.29) to compute the inductance per unit length of this parallel-plate transmission line:

$$L = \frac{1}{v_p^2 C} = \mu_0 \frac{H}{W} \text{ H/m} \tag{C.30}$$

We may now use the derived capacitance per unit length (C.27) and inductance per unit length (C.30), to write an expression for the characteristic impedance of the parallel-plate transmission line:

$$Z_0 = \sqrt{\frac{L}{C}} = \sqrt{\frac{\mu_0}{\epsilon_0\epsilon_r}}\frac{H}{W} = \frac{377}{\sqrt{\epsilon_r}}\frac{H}{W} \ \Omega \tag{C.31}$$

We assume that $W \gg H$ for the capacitance per unit length expression (C.27) and that the fringing fields at the edges are a negligible contribution to the total capacitance. This assumption is essentially an open-circuit or magnetic wall boundary condition at $\pm\frac{W}{2}$ (coordinate system at the center of the width W of the transmission line and centered between the plates):

Analysis:

$$Z_0 = \frac{377}{\sqrt{\epsilon_r}}\frac{H}{W} \ \Omega \tag{C.32}$$

Synthesis:

$$\frac{W}{H} = \frac{377}{\sqrt{\epsilon_r} Z_0} \tag{C.33}$$

It is possible to write a relationship for ϵ_e. This done by again using the phase velocity equation:

$$\epsilon_e = \left(\frac{c}{v_p}\right)^2 = \left(\frac{\left[\frac{1}{\sqrt{LC_0}}\right]}{\left[\frac{1}{\sqrt{LC}}\right]}\right)^2 = \frac{C}{C_0} \tag{C.34}$$

where C_0 is the capacitance per unit length of the transmission line with the same geometric configuration with the dielectric replaced by free space/air ($\epsilon_r = 1$). It follows that

$$\epsilon_e = \frac{\epsilon_0 \epsilon_r \frac{W}{H}}{\epsilon_0 \frac{W}{H}} = \epsilon_r \tag{C.35}$$

This computation assumes no fringing fields, which is a good approximation when $W/H \gg 1$.

References

[1] Wu, T.T., "Theory of the Microstrip," *Journal of Applied Physics*, March 1957, Vol. 28, No. 3, pp. 299–302.

[2] Schneider, M.V., "Microstrip Lines for Microwave Integrated Circuits," *Bell System Technical Journal*, May–June 1969, Vol. 48, No. 5, pp. 1421–1444.

[3] Lev, J.J., "Synthesize and Analyze Microstrip Lines," *Microwaves & RF*, January 1985, pp. 111–116.

[4] March, S., "Microstrip Packaging: Watch The Last Step," *Microwaves*, December 1981, pp. 83–94.

[5] March, S., "Microstrip Errata: 'Microstrip Packaging: Watch The Last Step'," *Microwaves*, February 1982, p. 9.

[6] Pucel, R.A., Massé, D.J., and Hartwig, C.P., "Losses in Microstrip," *IEEE Transactions on Microwave Theory and Techniques*, June 1968, Vol. 16, No. 6, pp. 342–350.

[7] Pucel, R.A., Massé, D.J., and Hartwig, C.P., "Errata: 'Losses in Microstrip'," *IEEE Transactions on Microwave Theory and Techniques*, Vol. 16, No. 12, December 1968, p. 1064.

[8] Kajfez, D., and Tew, M.D., "Pocket Calculator Program for Analysis of Lossy Microstrip," *Microwave Journal*, December 1980, pp. 39–48.

[9] Wheeler, H.A., "Transmission Line Properties of a Strip on a Dielectric Sheet on a Plane," *IEEE Transactions on Microwave Theory and Techniques*, Vol. 25, No. 8, August 1977, pp. 631–647.

[10] Hammerstad, E.O., and Bekkadal, F., *A Microstrip Handbook*, ELAB Report, STF 44 A74169, N7034, University of Trondheim, Norway, 1975.

[11] Edwards, T.C., *Foundations for Microstrip Engineering*, John Wiley, New York, 1981.

[12] Bahl, I.J., and Stuchly. S.S., "Analysis of a Microstrip Covered with a Lossy Dielectric," *IEEE Transactions on Microwave Theory and Techniques*, February 1980, Vol. MTT-28, No. 2, pp. 104–109.

[13] Yamashita, E., and Yamazaki, S., "Parallel-Strip Line Embedded in or Printed on a Dielectric Sheet," *IEEE Transactions on Microwave Theory and Techniques*, Vol. 16, No. 11, November 1968, pp. 972–973.

[14] Booton, R.C., Jr., *Computational Methods for Electromagnetics and Microwaves*, John Wiley & Sons, 1999.

[15] Di Paolo, F., *Networks and Devices Using Planar Transmissions Lines*, CRC Press, 2000, Chapter 11.

[16] Yildiz, C., "New and Very Simple Synthesis Formulas for Coplanar Strip Line," *Microwave and Optical Technology Letters*, January 2005, Vol. 44, No. 2, pp. 199–202.

Appendix D
Antenna topics

D.1 Friis transmission formula

An equation that describes the Path Loss between two antennas was introduced by Friis [1]. The formula was introduced in terms of the effective area of the transmitting and receiving antenna:

$$P_r = \frac{A_t A_r}{\lambda^2 R^2} P_t \tag{D.1}$$

where P_r is the power available at the output port of the receiving antenna, P_t is the power fed into the input port of the transmitting antenna, A_r is the effective area of the receiving antenna, A_t is the effective area of the transmitting antenna, R is the distance between the antennas, and λ is the wavelength.

The power delivered to the terminals of an antenna is defined as

$$P_r = SA_e \tag{D.2}$$

where S is the power density of a plane wave incident on the antenna in W/m^2 and A_e is the effective area of the antenna m^2.

The gain of an antenna in terms of effective area is

$$G = \frac{4\pi}{\lambda^2} A_e \tag{D.3}$$

The Friis equation may now be written in terms of antenna gain:

$$P_r = G_t G_r \left(\frac{\lambda}{4\pi R} \right)^2 P_t \tag{D.4}$$

where G_t is the transmit antenna gain and G_r is the receive antenna gain.

The relationship between gain over an isotropic source (G) and that gain expressed in decibels with respect to an isotropic source is

$$G_{\mathrm{dBi}} = 10 \log_{10} G \tag{D.5}$$

It is often useful to write the Friis equation using logarithms to determine the Path Loss:

$$\text{Path Loss (dB)} = 20 \log_{10}(FR) - 27.558 - G_r\,(\mathrm{dB}) - G_t\,(\mathrm{dB}) \tag{D.6}$$

where F is the frequency in MHz and R is the distance between antennas in meters.

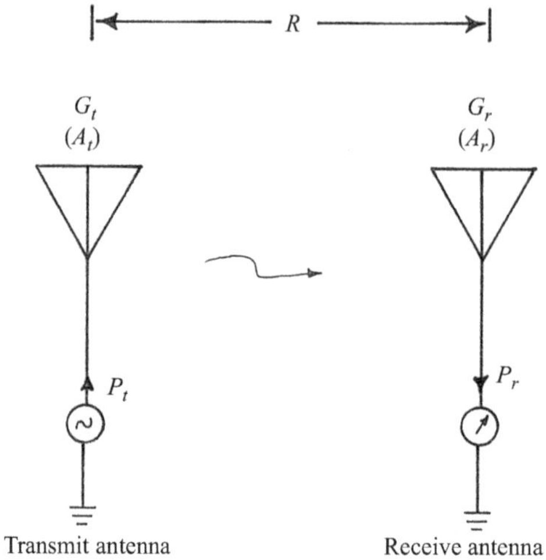

Figure D.1 Friis transmission formula geometry

The Friis transmission formula assumes that the value of R is sufficiently large so a plane wave is incident at the receive antenna. To satisfy this requirement, the separation distance should be at least:

$$R > \frac{2d^2}{\lambda} \tag{D.7}$$

where d is the largest linear dimension of either antenna. The formula also assumes free space.

D.2 Wireless link range versus power input

It is often of interest to estimate the effective range of a communications link. If we assume that the noise environment remains constant at any given distance, this allows us to write an equation for the maximum distance of a communications link R_0, for a given power input P_{t0}, which just meets the systems signal to ratio requirement:

$$P_r = G_t G_r \left(\frac{\lambda}{4\pi R_0} \right)^2 P_{t0} \tag{D.8}$$

If we wish to increase the range of the communications link to a distance R_1, which may be defined as a multiple of R_0:

$$R_1 = \Delta R_0 \tag{D.9}$$

Table D.1 Power increase (in dB) versus increased link distance factor Δ

dB	Δ
3.01	$\sqrt{2}$
6.02	2
9.54	3
12.04	4

We may write a second equation using the Friis transmission formula:

$$P_r = G_t G_r \left(\frac{\lambda}{4\pi \Delta R_0} \right)^2 P_{t_1} \tag{D.10}$$

The two equations are equated so the receive power is identical in both cases, which after canceling common terms becomes

$$\frac{P_{t_1}}{P_{t_0}} = \Delta^2 \tag{D.11}$$

$$\text{Power}_{\text{dB}} = 10 \log_{10} \left(\frac{P_{t_1}}{P_{t_0}} \right) = 20 \log_{10}(\Delta) \tag{D.12}$$

To double the distance, we use $\Delta = 2$. Equation (D.12) shows that the power into the antenna used for transmission must be increased by 6.02 dB to double the range. Table D.1 shows the relationship between the antenna gain (or system gain) and the increase in the link distance factor. If the gain of an antenna increases by 6.02 dB, then the distance a link will be maintained is twice the original distance. When increased by 9.54 dB, the link distance will be increased by a factor of 3. This indicates that the signal-to-noise ratio remains constant.

D.3 Decibels

The decibel provides a logarithmic comparison of two power levels. A power level is assigned as a reference value P_r, and this value is used for comparison to a measured or computed power level P. The decibel (dB) is defined as

$$\text{dB} \equiv 10 \log_{10}(P/P_r) \tag{D.13}$$

The decibel is a convenient method of expressing power because of its logarithmic definition, which transforms the mathematical operations of multiplication and division to addition and subtraction. Powers and roots become multiplication and division. Addition and subtraction must be done with the numerical values prior to

Table D.2 Decibel to power ratio estimates

dB	P/P_r
3	2
6	4
9	8
10	10
13	20
20	100
26	400
30	1 000
39	8 000
40	10 000

conversion to decibels or they must be converted from decibels to numerical values. This may be accomplished with

$$P/P_r = 10^{(\mathrm{dB})/10} \tag{D.14}$$

When a power level reference value is tied to a known absolute reference value, such as a milliwatt, the power in decibels is generally referred to as *decibels with respect to a milliwatt* and written dBm. When 1 W is used as P_r, this is called *decibels with respect to a watt* and written dBW. The gain value of an antenna is referenced with respect to an isotropic antenna's power level and known as *decibels with respect to an isotropic radiator* and written dBi. Acoustic sound pressure levels are also expressed using decibels.

The definition of a decibel in (D.13) may be expressed in terms of units which equivalently express power. The power dissipated by a resistor in a given electric circuit is V^2/R. We can choose a reference voltage V_r and resistance R_r. When substituted into (D.13), we obtain:

$$\begin{aligned} \mathrm{dB} &= 10\log_{10}\left[\frac{V^2/R}{V_r^2/R_r}\right] \qquad \text{(D.15)} \\ &= 20\log_{10}(V/V_r) - 10\log_{10}(R/R_r) \end{aligned}$$

For the special case of $R = R_r$, this expression reduces to

$$\mathrm{dB} = 20\log_{10}(V/V_r) \tag{D.16}$$

Decibels allow one to express large dynamic ranges in an accessible manner. When power is expressed in decibels, one can readily determine the approximate change in power. A change of 3 dB is close to a factor of 2 in power (1.995), and a 10 dB change is a factor of 10 in power. A 13 dB difference is a factor of 10 in power multiplied by a factor of 2 or approximately 20 (19.95).

Table D.2 has estimates of the power ratios for a number of decibel values. These estimates are very useful in engineering work for a quick view of the magnitude of quantities. The exact values are given by (D.14). For instance 39 dB is estimated to be a power ratio of 8000 ($= 10 \cdot 10 \cdot 10 \cdot 2 \cdot 2 \cdot 2 = 10^{39.03/10}$). The actual value is $10^{39/10} = 7\,943.38$ which is only a 0.71% difference.

D.3.1 Historical origin of the decibel (dB)

The origin of the decibel is rooted in the telephone industry. It was known that the sound level a person could hear was related to the length of the transmission line between a person speaking and the person listening at the telephone receiver. This loss was quantified with a unit called a "mile of standard cable" or MSC. This unit was in use at least as early as 1904. The rating of a telephone circuit was performed by using a standard circuit with a listener at one end, and switching in known amounts of cable, until the performance of the standard and equipment under test were deemed "equivalent." The equivalent loss overcome by a new set of equipment might be 16 miles or 16 MSC.

The mile of standard cable had a resistance of 88 Ω, with a capacitance of 54 nF per mile. The limitations of this measurement method became obvious, and a mathematical definition in terms of power was proposed. In 1924, the new unit was dubbed *The Transmission Unit* or TU. The TU was chosen so two amounts of power would differ by $10^{0.1}$ and any two amounts of power differ by N units when they are in the ratio of $10^{N(0.1)}$. The number of transmission units would become

$$N = 10 \log_{10} P_1/P_2 \tag{D.17}$$

The unit was designed so it would be *Approximately Equal in Effect on Volume to a "Mile of Standard Cable."* The definition of the transmission unit was tied to the perceptual tests of volume using miles of standard cable.

> A Consideration of the above qualifications and of the various units suggested, led to the adoption of $10^{0.1}$ as the most suitable ratio on which to base the unit of transmission efficiency. The transmission unit is logarithmic, distortionless, is based on a power ratio and its relation to that ratio is a simple one. Its effect on the transmission of telephonic power corresponding to speech sounds is about 6% less than that of one mile of standard cable [5].

In 1929, the transmission unit, TU, was renamed the decibel after Alexander Gram Bell, by the Bell System:

> The Bell System has adopted the name "decibel" for the "transmission unit," based on a power ratio of $10^{0.1}$. This is in accordance with the terminology for the decimal unit, with the prefix "deci" being the usual one for indicating a one-tenth relation. For convenience, the symbol "db" will be employed to indicate the name decibel [6].

The original suggestion of "db" has been altered to dB.

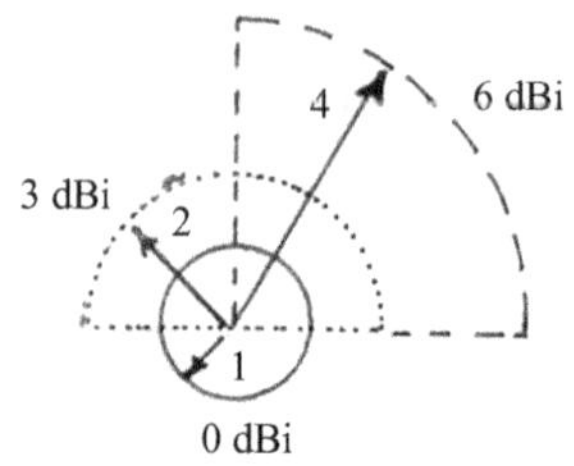

Figure D.2 Isotropic antenna, hemispherical antenna, and quarter sphere antenna

D.4 Antenna gain and directivity

An antenna with a radiation pattern which is constant (independent of angle) is an *isotropic* antenna. An isotropic antenna that radiates 100% of the power it accepts at its driving point (P_0) is generally used as a reference antenna to which physically realizable antennas are compared. The radiation from an isotropic antenna may be described mathematically as

$$S = \frac{P_0}{4\pi R^2} \tag{D.18}$$

where S is in power per unit area or *power density*.

The input power accepted by the driving point of an isotropic antenna is spread equally over a sphere with a surface area of $4\pi R^2$. If the radiation pattern of an isotropic source is altered so it radiates only over an upper hemisphere, but is still constant over that hemisphere, but zero otherwise, the same amount of power (P_0) is distributed over half of the area of the isotropic case. This means that the power over this radiation hemisphere is twice as large as the isotropic case:

$$S = 2\frac{P_0}{4\pi R^2} \; (0 < \phi < 2\pi, 0 < \theta < \pi) \tag{D.19}$$

If the upper hemisphere is again divided in half so the radiation is allowed to radiate equally over one-fourth of the area of the full (isotropic) sphere and is zero elsewhere, the amount of power will be four times the power of an isotropic antenna:

$$S = 4\frac{P_0}{4\pi R^2} \; (0 < \phi < \pi, 0 < \theta < \pi) \tag{D.20}$$

In each case, we are able to use a factor to describe the increase in the radiation level experienced in terms of radiation over a sphere. We can rewrite this with a value of D:

$$S = D\frac{P_0}{4\pi R^2} \; (0 < \phi < \pi/2, 0 < \theta < \pi) \tag{D.21}$$

We have restricted the radiation of the antenna to one-eighth of a sphere ($0 < \phi < \pi/2, 0 < \theta < \pi$) in the equation above, then $D = 8$. We will refer to this factor as the directivity of an antenna.

The antennas discussed so far to illustrate directivity are not physically realizable. When measuring an antenna we will have a varying function $D(\phi, \theta)$ which describes the radiation pattern:

$$S = D(\theta, \phi)\frac{P_0}{4\pi R^2} \tag{D.22}$$

The maximum value of $D(\theta, \phi)$ is assumed to be the *directivity* of an antenna.

Physical antennas have losses which decrease the total amount of power radiated with respect to the input power P_0. The pattern of an antenna is assumed not to be affected by the losses inherent in its construction. The *gain* of an antenna (G) is related to its directivity by the antenna's efficiency (e):

$$G = eD \tag{D.23}$$

$$S(\theta, \phi) = eD(\theta, \phi)\frac{P_0}{4\pi R^2} = G(\theta, \phi)\frac{P_0}{4\pi R^2} \tag{D.24}$$

As with directivity, the gain of an antenna is assumed to be the maximum value of $G(\theta, \phi)$ over the radiation sphere. Directivity is a property of the pattern shape alone. At one time, directivity was often called *pattern gain*, and when losses were included it became *antenna gain*. Methods to estimate the directivity of antennas have been developed and are very useful for approximating expected gain before measurement. The gain of an antenna is in terms of power and may be written in decibels with respect to an isotropic antenna:

$$G_{\text{dBi}} = 10\log_{10}(G) \tag{D.25}$$

Gain is generally assumed to be with respect to an isotropic antenna. In some instances, antennas are compared with a dipole as a reference (dBd). This reduces the gain in dBi by 2.14 dB.

One may multiply both sides of equation (D.24) by R^2 to obtain the *radiation intensity* U:

$$U(\theta, \phi) = G(\theta, \phi)\frac{P_0}{4\pi} \tag{D.26}$$

The total power radiated P_r by an antenna is the surface integral of the radiation intensity over the entire radiation sphere:

$$P_r = \int_0^{2\pi}\int_0^{\pi} G(\theta, \phi)\frac{P_0}{4\pi}\sin\theta\; d\theta\; d\phi \tag{D.27}$$

The ratio of the total amount of power radiated by an antenna divided by the amount of power accepted by the antenna's driving point is the efficiency e of an antenna. Dividing (D.27) by P_0, we obtain:

$$e = \frac{P_r}{P_0} = \int_0^{2\pi}\int_0^{\pi} \frac{G(\theta, \phi)}{4\pi}\sin\theta\; d\theta\; d\phi \tag{D.28}$$

Losses in the materials out of which an antenna is constructed reduce the total radiated power.

Directivity may be more formally defined in terms of radiation intensity:

$$\text{Directivity} = \frac{\text{maximum radiation intensity}}{\text{average radiation intensity}} = \frac{U_{\text{maximum}}}{U_0} \tag{D.29}$$

The average radiation intensity of an antenna may be obtained by integrating the radiation intensity over the entire radiation sphere and then dividing by the area of a sphere in steradians which is 4π:

$$U_0 = \frac{1}{4\pi}\int_0^{2\pi}\int_0^{\pi} U(\theta,\phi)\sin\theta \, d\theta \, d\phi \tag{D.30}$$

The average radiation intensity is assumed to have accounted for all the power in an electromagnetic wave (i.e. total power in two orthogonal polarizations).

D.5 Attenuation and voltage standing wave ratio

In some wireless applications, a considerable amount of cable may exist between the driving point of an antenna and the signal source which feeds it. If the attenuation due to a length of the cable is known, its effect on voltage standing wave ratio (VSWR) may be approximated as if it were a lumped attenuator [2]. Generally, a length of cable produces ripple in the measured VSWR, and an attenuator does not. The attenuation of a cable is frequency dependent, whereas over its operating bandwidth an attenuator is approximately constant. When the attenuation introduced is known, (D.31) may be used to predict the new input VSWR, if the original driving point VSWR is known [3]:

$$\text{VSWR}_{\text{input}} = \frac{\sqrt{\left(\frac{\text{VSWR}_{\text{DP}}-1}{\text{VSWR}_{\text{DP}}+1}\right)^2 10^{-(A/5)}} + 1}{1 - \sqrt{\left(\frac{\text{VSWR}_{\text{DP}}-1}{\text{VSWR}_{\text{DP}}+1}\right)^2 10^{-(A/5)}}} \tag{D.31}$$

where A is the attenuation of an attenuator in dB (positive), VSWR_{DP} is the VSWR at the driving point of an antenna (or load) and $\text{VSWR}_{\text{input}}$ is the VSWR following the attenuator.

D.5.1 Example 1

A resistive microwave load of 150 Ω is connected to a 50 Ω system which produces a 3.0:1 VSWR. What will the VSWR measured be if a 7 dB attenuator is placed between the load and the 50 Ω system? Using D.31, we obtain:

$$\text{VSWR}_{\text{input}} = \frac{\sqrt{\left(\frac{3-1}{3+1}\right)^2 10^{-(7/5)}} + 1}{1 - \sqrt{\left(\frac{3-1}{3+1}\right)^2 10^{-(7/5)}}} = 1.22$$

In terms of Return Loss, (D.31) becomes

$$\text{RL}_{\text{input}} = [\text{RL}_{\text{DP}} + 2A] \tag{D.32}$$

D.5.2 Example 2

A Return Loss of 2.922 dB (6.0:1 VSWR) is produced by a resistive load. A 7 dB attenuator is used to pad the load. What is the Return Loss measured at the input of the attenuator?

$$\mathrm{RL}_{\mathrm{input}} = 2.922\,\mathrm{dB} + 2 \cdot 7.0\,\mathrm{dB} = 16.92\,\mathrm{dB}$$

D.6 Return Loss and Reflection Loss

The term *Return Loss* has an apocryphal origin. In July 1924 *Bell System Technical Journal*, W.H. Martin states on p. 407, when discussing the Transmission Unit, which would later be called the decibel:

> ... A power ratio of 2 represents a gain of 3.01 units, and a power ratio of 0.5 corresponds to a loss of 3.01 units.

This may offer a clue as to how a loss could become defined as a positive quantity, but loss as written above would imply a minus sign. It would be a loss of 3.01 units or −3.01 units. Alternatively, a gain of 3.01 units would be +3.01 units.

The term Return Loss has been defined by the IEEE as

$$\mathrm{RL} = 10\log_{10}\left(\frac{P_{incident}}{P_{reflected}}\right) \tag{D.33}$$

where $P_{incident}$ is the incident power propagating toward the driving point of an antenna under test (AUT) and $P_{reflected}$ is the power reflected back toward the power source.

When $P_{reflected} < P_{incident}$ Return Loss is a *positive quantity*. The Return Loss is the ratio of power propagating toward a mismatch and that reflected back expressed in decibels. For a passive device, we can express the ratio in terms of reflection coefficient Γ:

$$\left(\frac{P_{in}}{P_{ref}}\right) = \left|\frac{1}{\Gamma^2}\right| \tag{D.34}$$

When substituted into (D.33), we obtain:

$$\mathrm{RL} = 10\log_{10}\left|\frac{1}{\Gamma^2}\right| = -20\log_{10}|\Gamma| \tag{D.35}$$

The term Return Loss is constantly expressed as a negative quantity in technical papers, rather than as a positive one as its definition requires. The term was not in general use before 1960 [4]. Its definition is counter-intuitive, has become pedantic, and is at-odds with its general use. To avoid the lack of utility the definition has introduced, the negative value of this quantity is presented in many different ways in technical papers and texts as a work-around. Examples are $|S_{11}|_{\mathrm{dB}}$, $10\log_{10}|\Gamma^2|$, and Log Mag.

It is more natural to use the incident power as the decibel reference:

$$\mathrm{RL} = 10\log_{10}\left(\frac{P_{reflected}}{P_{incident}}\right) \tag{D.36}$$

Many authors instinctively use Return Loss this way. If the reference power for the decibel definition of Return Loss had been chosen as the incident power, and the measured power level returned defined as the measured quantity, which makes intuitive sense, the sign would be negative for Return Loss values, as it often seen in general use, and Return Loss would be defined thus:

$$\mathrm{RL} = 20\log_{10}|\Gamma| \tag{D.37}$$

The use of many alternative labels for the same quantity in technical papers indicates that many authors have an understanding of the definition of Return Loss, but find the negative value more intuitive and useful. This intuitive confusion has led many to use the negative value of Return Loss in articles and textbooks which has produced an inconsistency in nomenclature [4].

The term *Reflection Loss* has been used by engineers and is defined in a manner consistent with reflection coefficient expressed in decibels, but provides a more meaningful description of the measurement:

$$\text{Reflection Loss} = 20\log_{10}|\Gamma| \tag{D.38}$$

I would like the IEEE to revisit its definition of Return Loss and change it to become consistent with (D.38), but this is very unlikely, despite its obvious awkwardness and pedantic origins. As the terms Reflection Loss and Return Loss have identical abbreviations: RL, please substitute Reflection Loss for any labels that were intended to be Return Loss and are not consistent with its current definition. If you see Return Loss on a plot that by definition should be Reflection Loss, in my view there is no error as the definition should be changed to become intuitive.

In the 3rd edition of this book, I will attempt to adopt the use of Reflection Loss for reflection coefficient expressed in decibels. This will provide a unifying term for $|S_{11}|_{\mathrm{dB}}$, $10\log_{10}|\Gamma^2|$, etc., that we currently need as alternatives to a negative value of current definition of Return Loss.

D.7 Attenuators

There are instances, such as some RFID receivers, where the VSWR of an antenna can have a significant effect on the sensitivity of a receiver. Despite an excellent match, an antenna with more efficiency loss and a better VSWR can outperform a well-matched antenna. One can maximize the overall sensitivity of the system by introducing an optimum amount of attenuation near the driving point of the antenna. Attenuators are often called pads or matching pads [2,7].

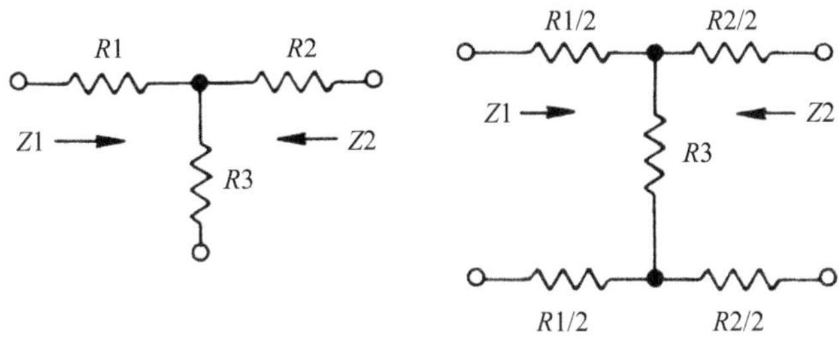

Figure D.3 A Tee attenuator for unbalanced (grounded) use is shown on the left. On the right is an H attenuator for balanced applications

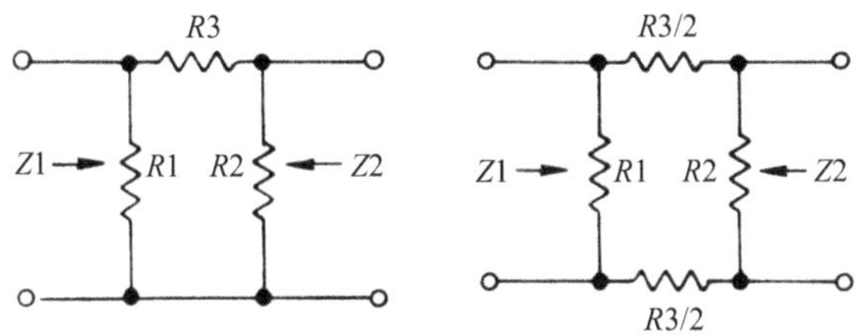

Figure D.4 On the left is an unbalanced Pi attenuator. On the right is a balanced O attenuator

On the left side of Figure D.3 is a Tee attenuator. This is often used with microstrip and coplanar waveguide. When using twin strips or other balanced transmission line, a balanced H attenuator should be used. The governing equations are

$$N = 10^{-\mathrm{RL}/10} \tag{D.39}$$

$$R3 = \frac{2\sqrt{NZ1 \cdot Z2}}{N-1} \tag{D.40}$$

$$R2 = Z2\left(\frac{N+1}{N-1}\right) - R3 \tag{D.41}$$

$$R1 = Z1\left(\frac{N+1}{N-1}\right) - R3 \tag{D.42}$$

In some instances, a Tee attenuator may not be appropriate in an unbalanced application. A Pi attenuator is an alternative unbalanced attenuator, and it is the left-hand graphic in Figure D.4. On the right-hand side is a balanced O attenuator which is an alternative to the H attenuator.

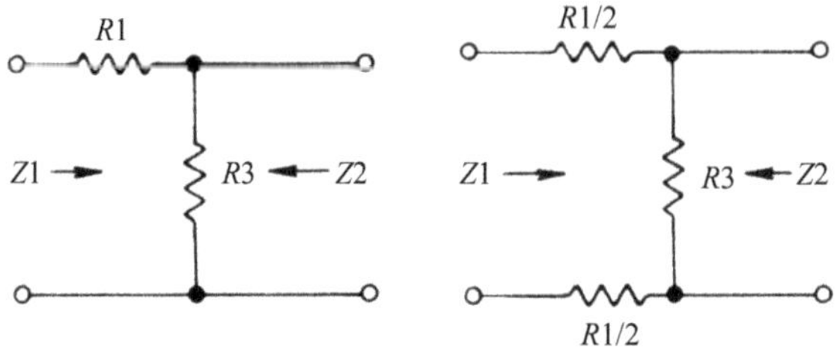

Figure D.5 Minimum loss attenuator for an unbalanced application (left) and minimum loss attenuator for a balanced system (right)

The design equations are

$$R3 = \frac{N-1}{2}\sqrt{\frac{Z1 \cdot Z2}{N}} \tag{D.43}$$

$$\frac{1}{R1} = \frac{1}{Z1}\left(\frac{N+1}{N-1}\right) - \frac{1}{R3} \tag{D.44}$$

$$\frac{1}{R2} = \frac{1}{Z2}\left(\frac{N+1}{N-1}\right) - \frac{1}{R3} \tag{D.45}$$

There are situations where one needs to provide a transition from a transmission line of a given characteristic impedance to another which is not the same. A minimum loss attenuator is often introduced to provide a smooth transition. Figure D.5 has the topology for minimum loss attenuators for unbalanced and balanced cases.

The equations are

$$R1 = Z1\sqrt{1 - \frac{Z2}{Z1}} \tag{D.46}$$

$$R3 = \frac{Z2}{\sqrt{1 - \frac{Z2}{Z1}}} \tag{D.47}$$

$$N = \left(\sqrt{\frac{Z1}{Z2}} + \sqrt{\frac{Z1}{Z2} - 1}\right)^2 \tag{D.48}$$

$$\text{Attenuator Loss (dB)} = 10 \log_{10} N \text{ with } (Z1 > Z2)$$

D.7.1 Example 3

Television and cable industries generally use 75 Ω coaxial cable, but most test equipment is 50 Ω. A minimum loss pad is generally placed between the 50 Ω test equipment and the 75 Ω device under test. Using (D.46), we obtain $R1 = 43.30$ Ω, (D.47) gives

$R3 = 86.60\ \Omega$. The value of N produced by (D.48) is 3.73205 for an insertion loss of 5.72 dB.

References

[1] Friis, H.T., "A Note on a Simple Transmission Formula," *Proceedings of The I.R.E. (IEEE)*, Vol. 34, No. 5, May 1946, pp. 254–256.

[2] *Reference Data for Radio Engineers*, Howard W. Sams, 1982, Chapter 11.

[3] Rizzi, P.A., *Microwave Engineering Passive Circuits*, Prentice Hall, 1988, p. 267.

[4] Bird, T.S., "Definition and Misuse of Return Loss," *IEEE Antennas and Propagation Magazine*, April 2009, Vol. 51, No. 2, pp. 166–167.

[5] Martin, W.H., "The Transmission Unit and Telephone Transmission Reference Systems," *Bell System Technical Journal*, July 1924, Vol. 3, No. 3, pp. 400–408.

[6] Martin, W.H., "Decibel—The Name for the Transmission Unit," *Bell System Technical Journal*, Vol. 8, No. 1, January 1929, pp. 1–2.

[7] Lewart, C., *Science and Engineering Programs for the IBM PC*, Micro Text Publications, New York, NY, 1983, Chapter 15.

Appendix E
Impedance matching techniques

E.1 The λ/8 transmission line transformer

In 1967, Donald Steinbrecher at MIT published a single page paper on the use of $\lambda/8$ transmission line sections [1]. He stated in this work: "Any load impedance can be transformed to a real impedance by a $\lambda/8$ transformer whose characteristic impedance is equal to the magnitude of the load impedance."

If one has an impedance of $Z_L = R_L + jX_L$, this can of course be written in polar form as $|Z_L|\angle\theta$. When a $\lambda/8$ transmission line section is attached to this load, and its characteristic impedance Z_0 is equal to $|Z_L|$, the input impedance becomes pure real.

The requirement that $Z_0 = |Z_L|$ can often produce Z_0 values which are impractically large for microstrip realization in many cases, but should be considered as an option, especially for lower impedances. A quarter-wave impedance transformer may be used with the eighth-wave transformer to match a complex impedance to a desired resistance value.

E.1.1 Example: Combined λ/8 and λ/4 transformer matching

As an example of a complex to real impedance transformation using a λ/8 transmission line section, we will use the load impedance presented in Figure E.1. The circuit consists of a 55.76 Ω resistor and a 3.01 nH inductor at 2.715 GHz. The load impedance is $Z_L = 55.76 + j51.36\ \Omega$ which in polar form is $|75.765|\Omega\angle 42.678°$.

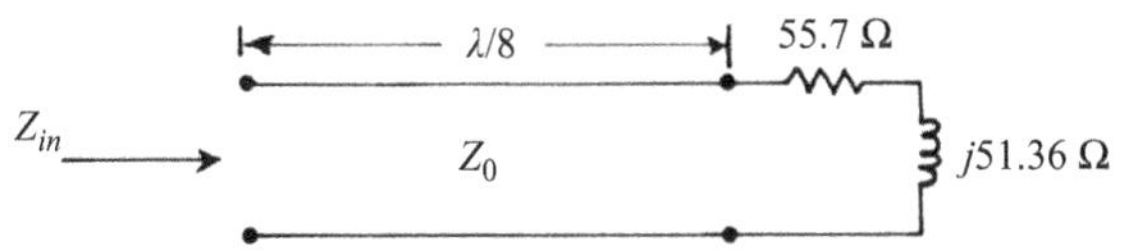

Figure E.1 The example load impedance of 55.76 + j51.36 Ω is connected to an eight-wave transformer, which will produce an input impedance which is pure real, when the characteristic impedance of the λ/8 matching transformer is equal to the magnitude of the load impedance

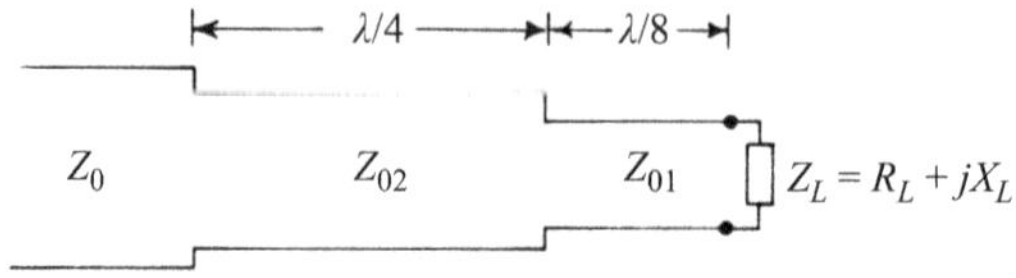

Figure E.2 A quarter-wave impedance transformer may be used with an eighth-wave impedance transformer to match a load to a desired system impedance [1]

The lossless transmission line equation is

$$Z_{in} = Z_0 \frac{Z_L + jZ_0 \tan \beta L}{Z_0 + jZ_L \tan \beta L} \tag{E.1}$$

In the case of a $\lambda/8$ section of transmission line, this reduces to

$$Z_{in} = Z_0 \frac{Z_L + jZ_0}{Z_0 + jZ_L} \tag{E.2}$$

We use a $\lambda/8$ matching transformer with $|Z_0| = Z_{01} = 75.765\ \Omega$. The input impedance becomes

$$Z_{in} = 75.765 \frac{55.7 + j51.36 + j75.765}{75.765 + j(55.7 + j51.36)}$$

$$Z_{in} = 173.0 - j89.6 \cdot 10^{-6}\ \Omega$$

$$Z_{in} \approx 173.0\ \Omega$$

We now need to transform this pure real impedance value to 50 Ω in order to match the system reference impedance. Because the input impedance is now pure real, this can easily be done using a quarter-wave transformer as shown in Figure E.2.

$$Z_{02} = \sqrt{50 \cdot R_{in}} = 93.0\ \Omega$$

The transmission line impedances in this example are all very realizable in most microstrip transmission line media using common substrates.

E.1.2 Dual λ/8 transmission line transformer

Two $\lambda/8$ transformers, and a single $\lambda/4$ transformer, may be used to conjugate a source and load impedance with a total section length of $\lambda/2$. This is shown in Figure E.3.

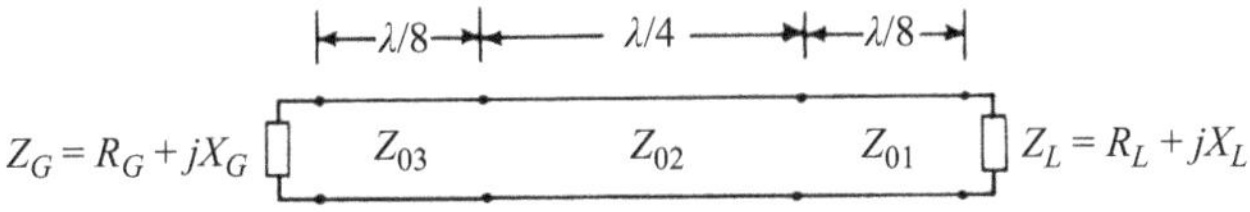

Figure E.3 Two λ/8 and one λ/4 transformer may be used to conjugate match source and load impedances simultaneously [1]

The following relations are used to accomplish this matching:

$$Z_{01} = |Z_L| \tag{E.3}$$

$$Z_{02} = \sqrt{\frac{R_L R_G |Z_L||Z_G|}{(|Z_L| - X_L)(|Z_G| - X_G)}} \tag{E.4}$$

$$Z_{03} = |Z_G| \tag{E.5}$$

E.2 Q matching with λ/8 transmission line transformer

There may be a design where one would like to minimize the amount of physical area used to match a load. It is possible to use a $\lambda/8$ transformer with lumped element Q matching, in place of the quarter-wave transformer in Figure E.2 to decrease the footprint of the matching network.

Q matching is based on the transformation of a series to a parallel circuit. In our example, we start with a pure real load created by a λ/8 transformer, add parallel reactance to produce the desired real part of 50 Ω in the series equivalent circuit, and then add a second reactance in series to cancel the series reactance, leaving a desired pure real resistance of 50 Ω.

The use of Q matching is detailed by White [2].

E.2.1 Example: Combined λ/8 transformer and Q matching

We will use the load of Example E.1.1 with the $\lambda/8$ transformer. At the design frequency, the input impedance at the input of the $\lambda/8$ transformer becomes a real impedance value of 173 Ω. At the design frequency, we can view this network as equivalent to a single resistive load of 173 Ω shown in Figure E.4.

We now use Q matching to transform the impedance level from 173 Ω (rather than using a quarter-wave transformer) to the system reference impedance (50 Ω).

The equations governing the two situations (series and parallel) are presented in Figure E.5. We can use the parallel relationship, which transforms the resistance between parallel and series circuits, to compute the circuit Q:

$$R_P = R_S(1 + Q^2) \tag{E.6}$$

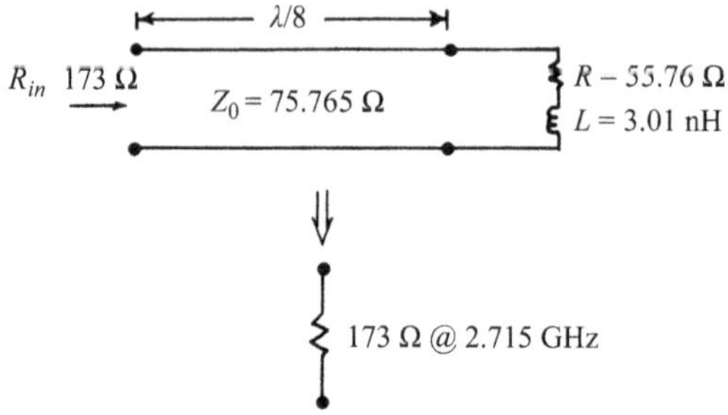

Figure E.4 The λ/8 transformer with impedance transformation to pure real can be viewed as a single resistive element

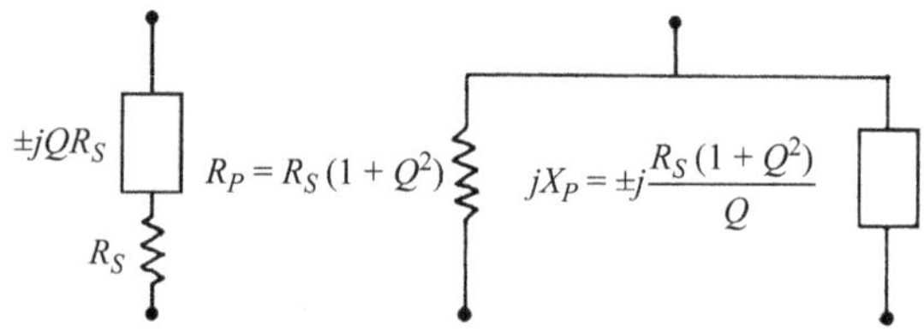

Figure E.5 Series to parallel equivalent circuits used in the Q matching method

We know $R_P = 173\ \Omega$, and we desire a series resistance of $R_S = 50\ \Omega$. Using (E.6), we obtain the circuit Q:

$$Q = \sqrt{\frac{173}{50} - 1} = 1.568$$

We now determine what shunt reactance is required in the parallel circuit to produce a series circuit with a real part of 50 Ω. The reactance is

$$X_p = \frac{R_s(1 + Q^2)}{Q} \tag{E.7}$$

$$X_p = \frac{50(1 + (1.568)^2)}{1.568} = 110.288\ \Omega$$

If we choose a shunt inductor as our parallel circuit reactance, we compute the inductance from the inductive reactance $|X_L|$ as 6.45 nH

$$L = \frac{|X_L|}{2\pi f} = \frac{110.288}{2\pi(2.715 \cdot 10^9)} = 6.45\ \text{nH}$$

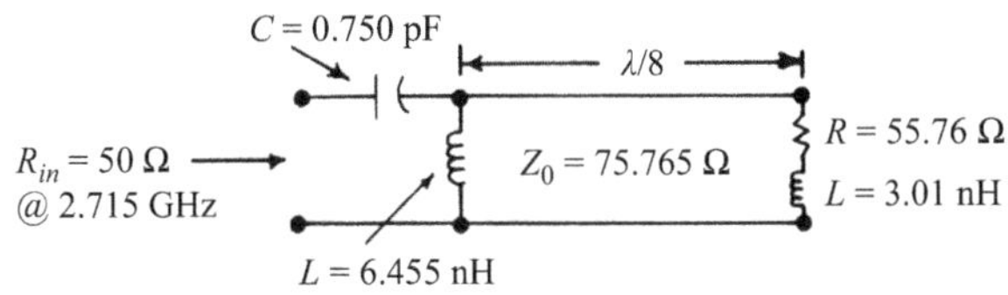

Figure E.6 Impedance matching design using λ/8 transformer with Q matching in place of a λ/4 transformer

The reactance of the series circuit is determined from the circuit Q

$$X_s = QR_s = 1.568 \cdot 50 = 78.4\ \Omega$$

We will need a series capacitive reactance $|X_c|$ to cancel the series inductive reactance, which gives us a capacitance value of 0.750 pF. This is a very small capacitance to realize in an actual circuit, but may be used for illustration

$$C = \frac{1}{2\pi f |X_c|} = \frac{1}{2\pi (2.715 \cdot 10^9)78.4} = 0.75\ \text{pF}$$

The final matching circuit is shown in Figure E.6. One can choose to use a shunt capacitor with a series inductor as an alternative Q matching design. In some instances, this may produce a more desirable topology or component values.

A circuit simulator was used to analyze these two designs for comparison. The results are in Figures E.7 and E.8. The original impedance is designated with triangles on the Smith chart and the negative return loss plot. The matched impedance with an eighth- and quarter-wave transformer is marked with squares, and the eighth-wave transformer with an LC Q matching circuit is designated with diamonds. The matching results are nearly identical. One should keep in mind that in general the Q of transmission line sections is larger (lower loss) than discreet components. The Q of an inductor may produce larger losses in a circuit than desired. This is of course a tradeoff between size and matching efficiency. In some cases, one may be able to use Q matching directly for an even simpler matching circuit design.

E.3 Single section series T-line impedance transformer

The single section transformer, Figure E.9, cannot match all complex loads, but its simplicity makes it attractive as a candidate method for a simple matching network. This network is detailed in an online monograph by Orfanidis [3]. In order to match a complex load $Z_L = R_L + jX_L$, with a single section impedance transformer of characteristic impedance Z_0, and normalized length $L_1 = l_1/\lambda_1$ (l_1 is the physical length and

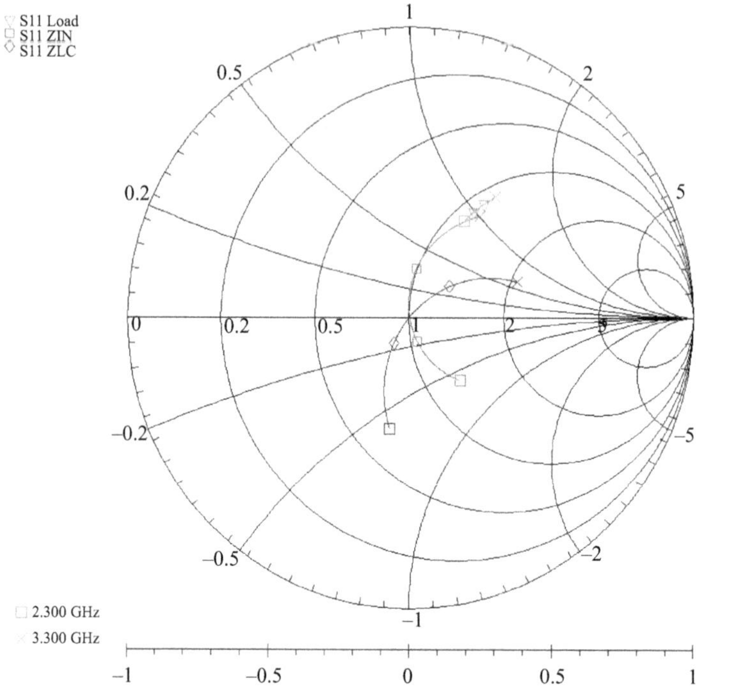

Figure E.7 *Smith chart plot of the original impedance (triangle), λ/8 with λ/4 matching (square), and λ/8 with Q matching circuit (diamond)*

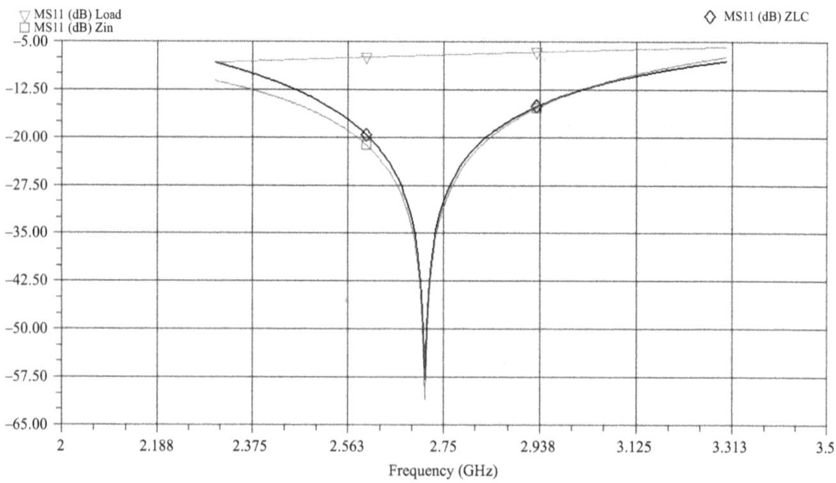

Figure E.8 *Negative return loss plot of the original impedance (triangle), λ/8 with λ/4 matching (square), and λ/8 with Q matching circuit (diamond)*

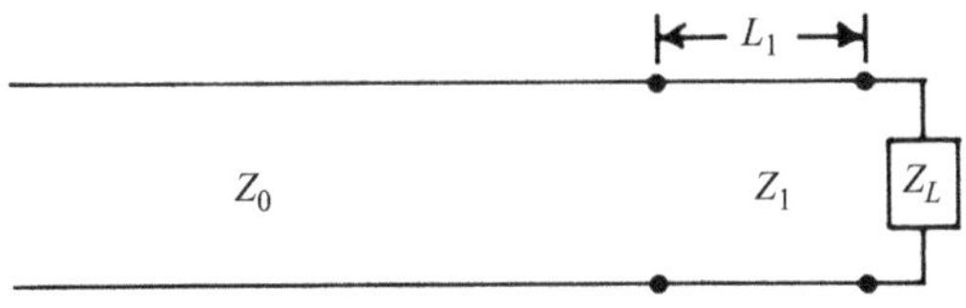

Figure E.9 Single section series impedance transformer

λ_1 is the transmission line wavelength) to a transmission line system with a reference value of Z_0 (usually 50 Ω) either of these conditions must be satisfied:

$$Z_0 < R_L \quad \text{or} \quad Z_0 > R_L + \frac{X_L^2}{R_L} \tag{E.8}$$

There is always a solution when Z_L is a pure real impedance. This suggests that it might be useful to consider using this type of transmission line matching solution with a $\lambda/8$ transformer.

When the conditions are met for a single line match, the transformer impedance and length are given by

$$Z_1 = \sqrt{Z_0 R_L - \frac{Z_0 X_L^2}{Z_0 - R_L}} \tag{E.9}$$

$$L_1 = \frac{1}{2\pi}\tan^{-1}\left[\frac{Z_1(Z_0 - R_L)}{Z_0 X_L}\right] \tag{E.10}$$

E.3.1 Example: Single section impedance match

$$Z_1 = \sqrt{50 \cdot 55.76 - \frac{50(51.36)^2}{50 - 55.76}} = 160.269\ \Omega$$

$$L_1 = \frac{1}{2\pi}\tan^{-1}\left[\frac{160.269(50 - 55.76)}{50 \cdot 51.36}\right] = -54.924 \cdot 10^{-3}$$

The solution is negative and normalized to the wavelength. We obtain a realizable solution by adding one-half wavelength of transmission line to obtain $L = 0.4451\lambda$. Orfanidis has MATLAB® code available online of the solution of these equations for a single section series transmission line transformer, which uses the mod() function to produce a physically realizable length: L1 = mod(L1, 0.5).

In Figures E.10 and E.11, we have a comparison of the different methods used to match in the example antenna impedance of Example E.1.1 so far. The single line section has a number of disadvantages compared with the other methods. It is longer than the other methods in this case, and so it will take up more physical area to realize, and the single line section has reduced impedance bandwidth compared with the other methods.

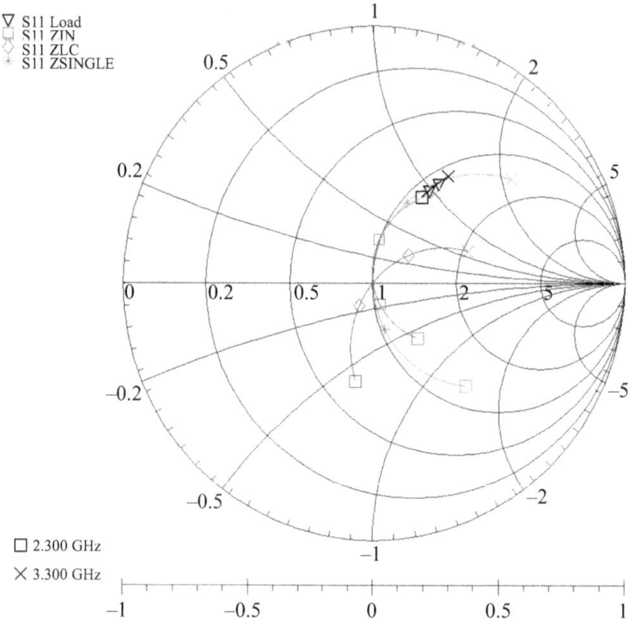

Figure E.10 Smith chart plot of the original impedance (triangle), λ/8 with λ/4 matching (square), λ/8 with Q matching circuit (diamond), and with a single line section (star)

There may be circumstances where this solution will be shorter than other options, and so one should always check to see if it is a realistic possibility.

E.4 Bramham–Regier two-section impedance transformer

In 1959, P. Bramham invented a two-section impedance transformer which is useful for matching the characteristic impedances between two transmission lines. Its total length is generally λ/6 which makes it a compact solution to this problem over a narrow frequency range [4]. F.A. Regier expanded the solution so that a complex terminating impedance can be matched to a transmission line [5]. The length of the two sections are of the same length as a single quarter-wave transformer, but match a complex load to a real driving point impedance.

The complex load has a section of transmission line which is of the desired system impedance Z_0, which is generally 50 Ω, with a length L_0. The next section has a characteristic impedance which is restricted by the design equations and designated Z_{01} with a length L_1. The value of Z_{01} is normalized to the system impedance and designated as n

$$n = Z_{01}/Z_0$$

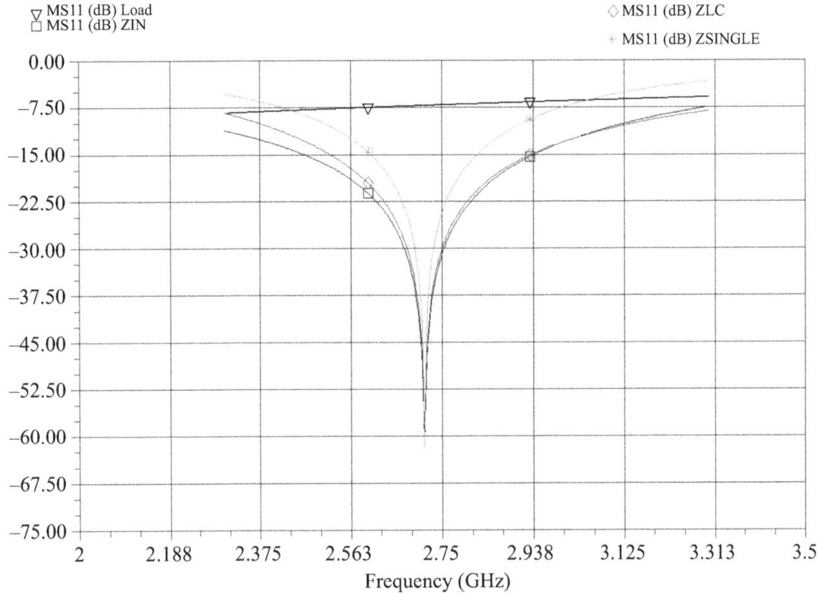

Figure E.11 Negative return loss plot of the original impedance (triangle), λ/8 with λ/4 matching (square), λ/8 with Q matching circuit (diamond), and with a single line section (star)

The value of n must always be greater or less than one for most practical designs. The load impedance Z_L is normalized to the system impedance and designated $\overline{Z}_L$:

$$\overline{Z}_L = Z_L/Z_0 = \overline{R}_L + j\overline{X}_L$$

After a value of the transformer impedance n has been chosen, one must calculate the value B. If B is imaginary, then a physical solution is not possible, and one must adjust n accordingly:

$$B = \pm\sqrt{\frac{k}{m-k}}$$

$$k = \frac{(\overline{R}_L - 1)^2 + \overline{X}_L^2}{\overline{R}_L}$$

$$m = \left(n - \frac{1}{n}\right)^2$$

The positive value for B is chosen; it produces the shortest physical solution. The value A may now be evaluated with

$$A = \frac{\left(n - \frac{\overline{R}_L}{n}\right)B + \overline{X}_L}{\overline{R}_L + \overline{X}_L nB - 1}$$

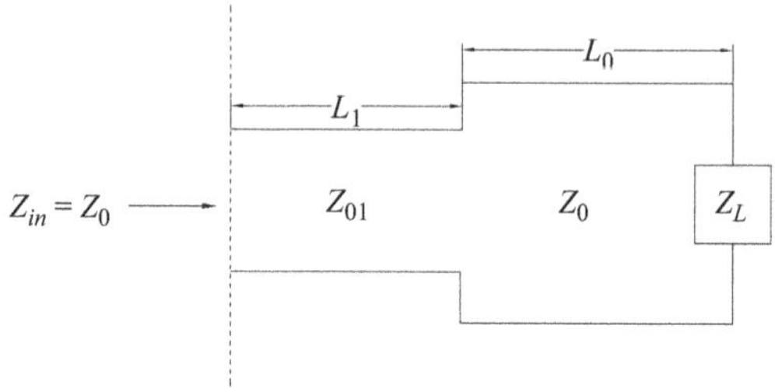

Figure E.12 Two section Bramham–Regier (BR) transmission line transformer design

The values A and B are related to the line lengths:

$$A = \tan(\beta_0 L_0)$$

$$B = \tan(\beta_1 L_1)$$

where β_0 and β_1 are the phase constants for each section of transmission line with guide wavelengths of λ_0 and λ_1, respectively.

The line lengths are determined with

$$L_0 = \tan^{-1}(A)\frac{\lambda_0}{2\pi}$$

$$L_1 = \tan^{-1}(B)\frac{\lambda_1}{2\pi}$$

E.4.1 Example: BR transmission line transformer design

First, the value of n is determined so that the value of $m > k$ such that the value of B is not complex. We will again match the same load impedance as is used in the earlier examples:

$$Z_L = 55.7 + j51.36\ \Omega$$

When it is normalized to $Z_0 = 50\ \Omega$, we obtain

$$\overline{Z}_L = 1.114 + j1.0272$$

It follows that $k = 0.958829$ and $B = 0.861745$, which allows us to determine $A = 1.20502$. The line lengths are determined to be $L_0 = 0.139755\,\lambda_0$, and $L_1 = 0.113203\,\lambda_1$. Assuming that the transmission line dielectric permittivity is unity (free space), and the wavelength is 110.5 mm (2.715 GHz), the line lengths are

$$L_0 = 15.44 \text{ mm}$$

and

$$L_1 = 12.51 \text{ mm}$$

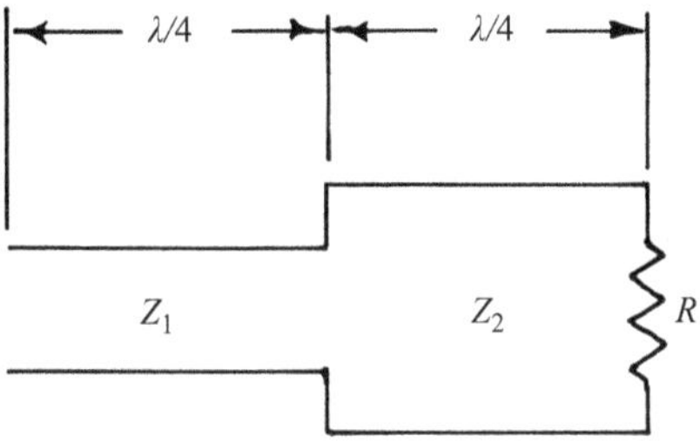

Figure E.13 Two-section Chebyshev quarter-wave transmission line transformer design

E.5 Two-section Chebyshev impedance transformer

If one has the available physical space to implement a more sophisticated impedance transformer than often exists in many wireless applications, one could combine a $\lambda/8$ transformer with a two $\lambda/4$ section Chebyshev impedance transformer as shown in Figure E.13.

The normalized equations for the design of a two-section Chebyshev impedance transformer are [6]:

$$Z_1 = (RS)^{\frac{1}{4}} \tag{E.11}$$

where R is the normalized load resistance, S is the maximum allowable standing wave ratio $S{:}1$, and Z_1 is the normalized characteristic impedance of the input (first section) of the transformer. After obtaining Z_1, we can then calculate Z_2 using:

$$Z_2 = \frac{R}{Z_1} \tag{E.12}$$

where Z_2 is the normalized impedance of the section connected to the normalized load resistance R. The normalizing resistance is the system reference impedance, which is 50 Ω ($Z_0 = 50$ Ω) in this example.

E.5.1 Example: λ/8 with two-section Chebyshev transformer

Using (E.11) and (E.12), we compute the normalized values to match the impedance. We will once again use the RL circuit from Example E.1.1 as an example load to match. From previous computations, we know that the input resistance of the eighth-wave transformer is 173.0 Ω:

$$R = 173.0/50.0 = 3.46$$

we will allow a maximum VSWR of 1.7:1 ($S = 1.7$)

$$Z_1 = (3.46 \cdot 1.7)^{\frac{1}{4}} = 1.5573$$

$$Z_2 = 3.46/1.5573 = 2.221$$

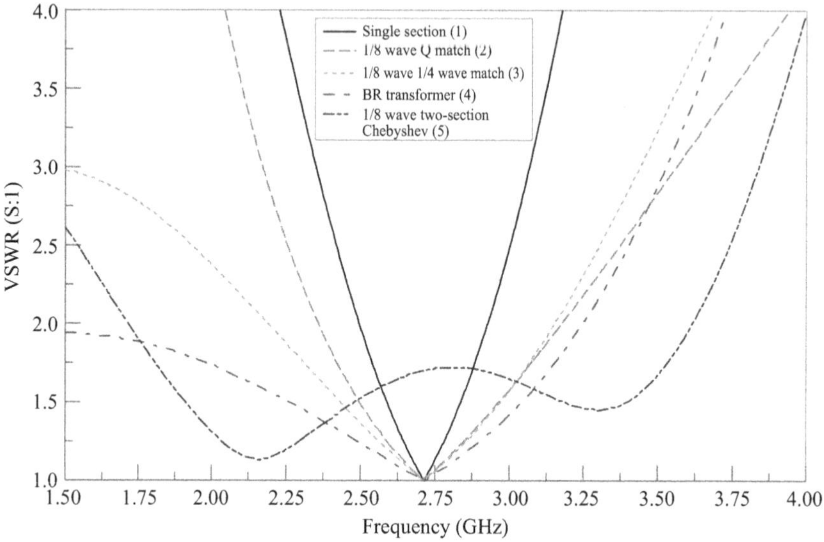

Figure E.14 VSWR plot of the different matching solutions examined previously. (1) Single transmission line impedance matching, (2) λ/8 transformer with LC Q matching, (3) λ/8 transformer with λ/4 transformer matching, (4) BR transformer, (5) λ/8 transformer with two-section Chebyshev λ/4 transformer matching

Multiplying the values of Z_1 and Z_2 by 50 Ω, we obtain our final design parameters:

$$Z_1 = 78.0\ \Omega$$

$$Z_2 = 111.1\ \Omega$$

The transformer was implemented and analyzed with a circuit simulator as done previously. All the previous matching solutions, and the two-section Chebyshev Transformer, with eighth-wave series transformer, is presented in Figure E.14. We see a considerable increase in bandwidth when using a two-section Chebyshev transformer. Of course what one gains in bandwidth is traded off with line length, and larger mismatch over the band.

In many designs, the length of the matching section is of considerable importance. Table E.1 shows a comparison of the lengths needed for the implementation of these alternative matching schemes.

One can see that there is a considerable trade-off in length versus bandwidth performance. This examination of matching options is by no means exhaustive, and there are more options for matching in the literature [7].

Table E.1 Matching methods versus required line lengths

Matching method	Line length
Single section of transmission line	49.18 mm
$\lambda/8$ with Q-matching	16.00 mm
$\lambda/8$ with quarter-wave transformer	41.44 mm
BR transformer	27.95 mm
$\lambda/8$ with a two-section Chebyshev transformer	69.06 mm

E.6 Bode-Fano limits/matching overview

A classic textbook about the subject of impedance matching is *Microwave Filters, Impedance-Matching Networks, and Coupling Structures* [8]. In a footnote on p. 5 of this text, we read

> Simple matching networks can give very great improvements in impedance match, and as the number of matching elements is increased the improvement per additional element rapidly becomes smaller and smaller. For this reason fairly simple matching networks can give performance which comes close to the theoretically optimum performance for an infinite number of impedance-matching elements.

This is good news for realizing minimum footprint matching networks with maximum matching bandwidth. There are fundamental trade-offs one must keep in mind when designing a matching network. These are known as the Bode-Fano criteria. Pozar has a good discussion of impedance matching limitations which we will offer as a set of gloss points [9]:

1. A broader-band match can only be achieved at the expense of a worse match over the larger frequency range.
2. The matching bandwidth cannot be perfectly matched except at a single point, which means only a finite number of frequencies can be matched in with a perfect match.
3. High Q circuits are harder to match than low Q circuits.

As the bandwidth of an impedance match is increased, the level of match decreases (i.e. more mismatch is introduced). Some people refer to this as a gain bandwidth product. This is an approximate, but fairly accurate way to look at this restriction. In order to keep a constant gain bandwidth product, one must give up match for bandwidth, or bandwidth for match.

A dominant single mode (linearly polarized) rectangular microstrip antenna has a driving point impedance which looks very much like an RLC circuit. It was shown in Chapter 4 that by using an impedance matching network with a 2:1 VSWR matching requirement, one is limited to four times the bandwidth of an unmatched antenna.

In Section E.5, the design of a two-section Chebyshev impedance transformer was illustrated. This improved the (2:1 VSWR) bandwidth considerably over other methods described. It has been proverbially believed that a Chebyshev impedance transformer is the optimum compromise between the mismatch ripple in the matched bandwidth and the maximum bandwidth achievable. In 1981, Carlin and Amstutz challenged this assumption [10]. The researchers provide examples of loads with which they first use a Chebyshev impedance transformer as a matching network and then proceed to use a method they call the *Real Frequency Technique* to match the load with better results. Clearly, a Chebyshev match is not the optimum possible. The technique is very mathematically intensive and a "Simplified Real Frequency Technique" has been proposed. It too is mathematically intensive. This simplified method has been detailed by Gerkis [11].

Increasingly, there is a demand to create dual-band antennas which may require dual-band matching networks. Unfortunately, there appears to be very little literature extant on the subject. A paper on the design of dual-band Chebyshev impedance transformers by Castaldi *et al.* has a useful discussion of dual-band matching [12]: The lack of any developed analytical methods for matching two separate frequency bands leaves only computer optimization as a viable solution method.

There are many pitfalls one can encounter when designing a matching network. In some cases, one can design a matching network which has components with an effective series resistance (ESR) which produces large energy dissipation. There are instances when an antenna may appear to be matched by a circuit, but if the antenna is removed from this matching circuit, the measured return loss does not change drastically, indicating that the network is acting more as an attenuator than a matching circuit. A matching circuit is a resonant circuit and has its own Q, which can mask the fact that the circuit is acting more like a resonant attenuator, rather than a matching circuit.

The design and implementation of a matching network is only as good as the quality of the physical inductors and capacitors used. The accuracy of any circuit simulation is dependent upon how well these physical devices mirror the theoretical elements used in the circuit simulator. One needs to characterize the lumped elements with measurement or non-ideal models to obtain good results. Often deviations seen between analysis and measurement are from the use of ideal resistors, inductors, and capacitors in a circuit simulator, when the physical elements are non-ideal. Kielkowski provides a detailed discussion of the importance of device characterization to successful circuit design [13].

The attentive reader may have noted that the load impedance used for illustration of the previous methods ($Z_{in} = 55.76 + j51.365\ \Omega$) has a real part which is very close to 50 Ω and that a single series capacitor could be used to match the driving point impedance. In practice, this was done to match the driving point impedance of a tapered omnidirectional microstrip antenna [14]. A surface mount 1 pF capacitor was used as a "via" through the circuit board in series with the antenna's driving point impedance to cancel the series inductive reactance present and produce an acceptable match.

References

[1] Steinbrecher, D.H., "An Interesting Impedance Matching Network," *IEEE Transactions on Microwave Theory and Techniques*, June 1967, Vol. 15, No. 6, p. 382.

[2] White, J.F., *High Frequency Techniques*, John Wiley and Sons, 2004, pp. 67–74.

[3] Orfanidis, S.J., *Electromagnetic Waves and Antennas*, www.ece.rutgers.edu/orfanidi/ewa/.

[4] Bramham, P., "A Convenient Transformer for Matching Co-axial Lines," *CERN 59-37*, November 1959.

[5] Regier, F.A., "Impedance Matching with a Series Transmission Line Section," *Proceedings of The IEEE*, Vol. 59, No. 7, July 1971, pp. 1133–1134.

[6] Wheeler, G.J., *Microwave Engineer's Handbook*. Vol. 1, Artech House, 1971, 17 pp.

[7] Hall, L.T., Hansen, H.J., Davis, B.R., and Abbott, D., "Performance Analysis of a Series Transformer for Complex Impedance Matching," *Microwave and Optical Technology Letters*, June 2005, Vol. 45, No. 6, pp. 491–494.

[8] Matthaei, G., Young, L., and Jones, E.M.T., *Microwave Filters, Impedance-Matching Networks, and Coupling Structures*, Artech House, 1980 (Reprint).

[9] Pozar, D.M., *Microwave Engineering*, McGraw Hill, 1990, pp. 325–327.

[10] Carlin, H.J., and Amstutz, P., "On Optimum Broad-Band Matching," *IEEE Transactions on Circuits and Systems*, May 1981, Vol. CAS-28, No. 5, pp. 401–405.

[11] Gerkis, A.N., "Broadband Impedance Matching Using the 'Real Frequency' Network Synthesis Technique," *Applied Microwaves and Wireless*, July/August 1988, pp. 26–36.

[12] Castaldi, G., Fiumara, V., and Pinto, I.M., "A Dual-Band Chebyshev Impedance Transformer," *Microwave and Optical Technology Letters*, October 2003, Vol. 30, No. 2, pp. 141–145.

[13] Kielkowski, R., *Spice Practical Device Modeling*, McGraw Hill, 1995, Chapter 2.

[14] Bancroft, R., and Bateman, B., "An Omnidirectional Microstrip Antenna with Low Sidelobes," *Microwave and Optical Technology Letters*, July 2004, Vol. 42, No. 1, pp. 68–69.

Appendix F
Baluns for printed antennas

F.1 Transmission line theory

In Appendix E, a number of impedance matching methods were introduced. The notion of impedance matching has underlying assumptions which can obscure important physical aspects of creating microwave transitions. For instance, suppose one were to take a coaxial cable with a characteristic impedance of 50 Ω and connect it to a twin lead transmission line, which also has a characteristic impedance of 50 Ω. We have an impedance which is identical at each side of the boundary. Should we have any concern that a mismatch could occur?

From an electromagnetic fields standpoint, the electric and magnetic fields of a coaxial transmission line are significantly different from the fields of a twin lead transmission line. In order to guarantee that no mismatch exists at the 50 Ω coaxial cable and 50 Ω twin lead interface, one needs an electromagnetic mode transformer. The coaxial transmission line is often referred to as an *unbalanced* transmission line because its outer shield is at ground potential, and the inner shield potential is above the ground potential. In the case of a twin lead transmission line, the two conductors have a symmetric potential above and below ground potential. The twin lead transmission line is seen to be balanced ($\pm$), and the coaxial cable is referred to as unbalanced $(0, +)$. When the two are connected, the outer shield of the coaxial cable mode is at ground potential, and the twin lead mode is not.

We have described the mismatch that can occur between two TEM transmission lines which have identical characteristic impedance, but mismatched fields. When examined in terms of *distributed circuit analysis*, it is clear that a mismatch can occur.

Transmission line theory rests on a number of assumptions which are well described by Chipman [1].

Chipman writes

> The distributed circuit analysis of uniform transmission lines, begun by William Thomson (Lord Kelvin) in 1855 and completed by Oliver Heaviside about 1885, is described by the following postulates.
>
> **Postulate 1** The uniform system or line consists of two straight parallel conductors.

> The adjective "uniform" means that the materials, dimensions and cross-sectional geometry of the line and its surrounding medium remain constant through the length of the line. Typically a signal source is connected at one end of the system and a terminal load is connected at the other end, ...
>
> This postulate does not require that the two conductors be of the same material or have the same cross-sectional shape. The analysis is therefore valid for a conductor of any material and cross section enclosing another conductor of any material and cross section, for a wire parallel to any conducting plane or strip or strip, and for many other useful constructions in addition to the simple example of two parallel wires of circular cross section and of the same diameter and material. ...

The postulate requires that the cross-section geometry through a transmission line remains constant. A coaxial cable has an outer conductor of a given diameter and an inner conductor of a given diameter, and a given dielectric material remain the same along the coaxial cable. One can have two different materials, such as an outer conductor of iron and an inner conductor of brass, as long as the transmission line is made of these materials without exception along its length (i.e. the cross-section geometry and materials remain the same). If a coaxial line were made of a copper inner and outer conductor for 1 m, then the outer shield became iron with a brass inner conductor for 1 m, with the same geometry and dielectric permittivity; this is viewed as two separate uniform transmission lines connected together. This distinction can be of great importance in practical cases. Chipman states:

> In general twists or bends in a transmission line violate the "uniformity" postulate and create effects not explainable by the distributed circuit theory. these effects will be negligible if the rate of the twist or bend does not exceed about one degree in the length of line comparable with the separation of the line conductors.
>
> The uniformity postulate is also violated by any discontinuities in a line, such as the termination points of an otherwise uniform system, or the point of connection between two uniform lines that differ physically in some respect. In the vicinity of discontinuities like these, phenomena will occur which are not in accord with distributed circuit theory. The anomalous behaviour is usually confined to distances not greater that a few times the separation of the line conductors, on either side of the discontinuity.

Postulate 2 The currents in the line conductors flow only in the direction of the length of line.

Postulate 2 essentially restricts us to TEM mode transmission lines.

Postulate 3 At the intersection of any transverse plane with the conductors of a transmission line, the instantaneous total currents in the two conductors are equal in magnitude and flow in opposite directions.

Postulate 4 At the intersection of any transverse plane with the line conductors, there is a unique value of potential difference between the conductors at any instant, which is equal to the line integral of the electric field along all paths in the transverse plane, between any point on the periphery of one of the conductors and any point on the periphery of the other.

Postulates 3 and 4 also restrict the mode to a TEM type along the transmission line.

Postulate 5 The electrical behavior of the line is completely described by four distributed electric circuit coefficients, whose value per unit length of line are constant everywhere on the line. These electric circuit coefficients are resistance and inductance uniformly distributed as series circuit elements along the length of the line, together with capacitance and leakage conductance uniformly distributed as shunt circuit elements along the length of the line.

In the case of a coaxial cable connected to a twin lead transmission line of identical characteristic impedance, we can see that Postulate 1 is violated because the cross-sectional geometry changes. The value of inductance and capacitance per unit length is different for coaxial cable and twin lead transmission line, which violates Postulate 5. Distributed circuit analysis indicates that we would need some type of circuit to transition between the two different transmission lines—even though they both have 50 Ω characteristic impedance.

The mode transformer which converts the electromagnetic field and transforms impedance from an unbalanced transmission line to a balanced one is proverbially known as a *balun* [2]. There are numerous cases in antenna engineering where an unbalanced feed is required. This is true for printed dipoles and printed loop antennas.

F.2 L-C lattice balun

At frequencies which are below 1 GHz, a lumped element lattice type balun is often of great utility. It can be used to feed printed loops and printed dipoles from microstrip transmission line.

The balun is essentially a bridge with two capacitors and two inductors, which produce a ±90° phase shift [3]. The design equations for this balun are given by

$$\omega = 2\pi f \tag{F.1}$$

$$Z_c = \sqrt{R_i \cdot R_L} \tag{F.2}$$

$$L = \frac{Z_c}{\omega} \tag{F.3}$$

$$C = \frac{1}{Z_c \cdot \omega} \tag{F.4}$$

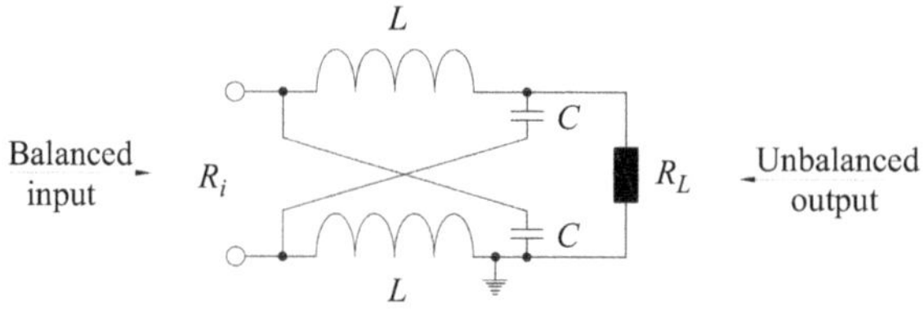

Figure F.1 Lumped element lattice balun

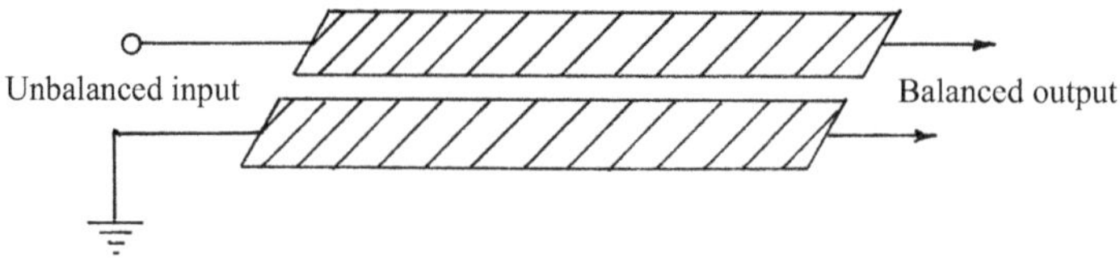

Figure F.2 Coupled microstrip transmission line balun

These design equations allow one to create a 50 Ω coaxial cable to 50 Ω twin lead transition in a straightforward manner. Assuming the design frequency is 1 GHz we obtain: $L = 7.96$ nH and $C = 3.18$ pF.

For the case of a classic 75 Ω coaxial cable to 300 Ω twin lead at 489.25 MHz (Channel 17) we obtain: $L = 48.80$ nH and $C = 2.17$ pF.

The lattice balun has the advantage that any input or output resistance may be arbitrarily matched. However, A lumped element balun can have lower utility at high microwave frequencies. The design of lattice baluns of higher order may be found in the literature [4].

F.3 Coupled microstrip transmission line balun

A simple balun, fabricated using a microstrip transmission line, is a pair of quarter wavelength long edge coupled microstrip lines. As is seen in Figure F.2, a microstrip transmission line feeds a single unbalanced input, with the other input side port shorted to ground. The coupling from the input microstrip line to the coupled side encourages a differential (odd) mode on the strips. There are no closed form CAD equations for this design, so the implementation is experimental.

An example of this balun would be for the excitation of an electrically small RFID loop. In Figure F.3, a graphic of the currents on an RFID near field coupling loop is shown, which is fed with a coupled microstrip transmission line balun. The inset shows the uniform current on the small loop, which is driven using the balun.

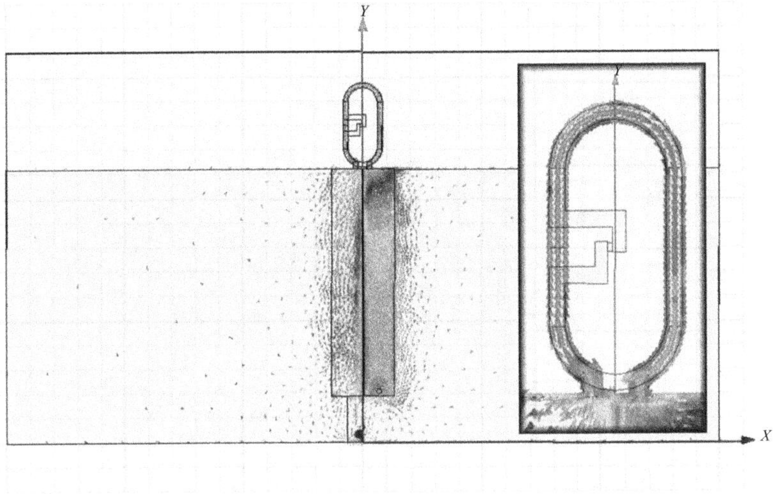

Figure F.3 RFID near field coupling loop using a coupled microstrip transmission line balun. The inset shows current on loop with RFID tag present 2 mm above the loop

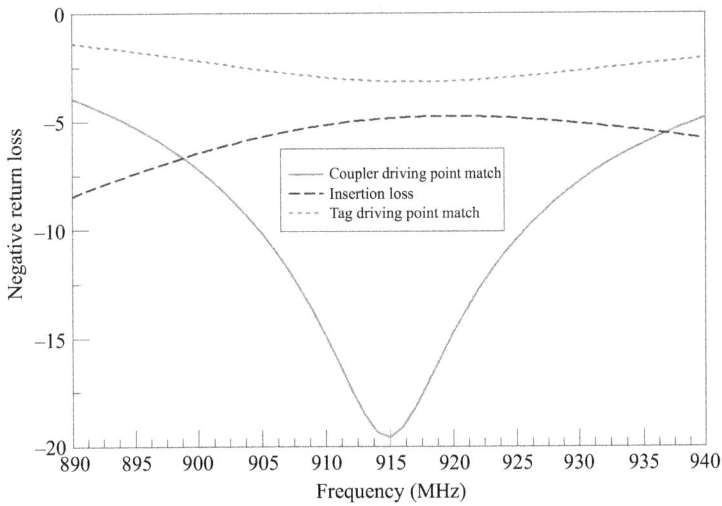

Figure F.4 Input match to coupler, insertion loss from coupler to RFID tag, and input match of the tag

Figure F.4 shows a graph of the driving point match at the input of the coupler (microstrip line at the bottom of Figure F.3). The insertion loss of the coupler to tag, viewed as a two port device, is plotted along with the driving point match of the tag. The tag is 2 mm above the coupling loop. The design clearly demonstrates the usefulness of a coupled microstrip transmission line balun.

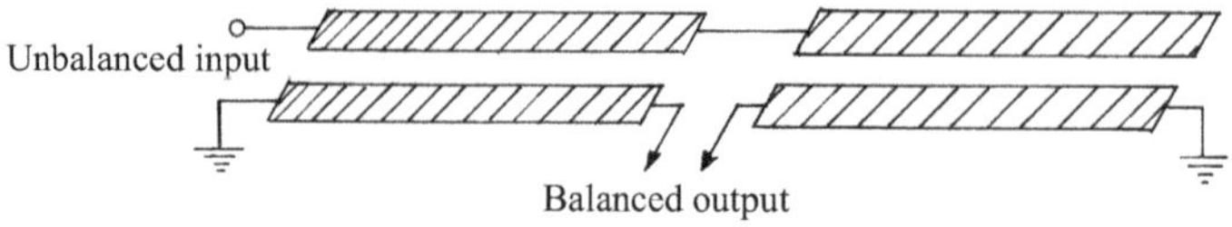

Figure F.5 Illustration of a printed Marchand balun

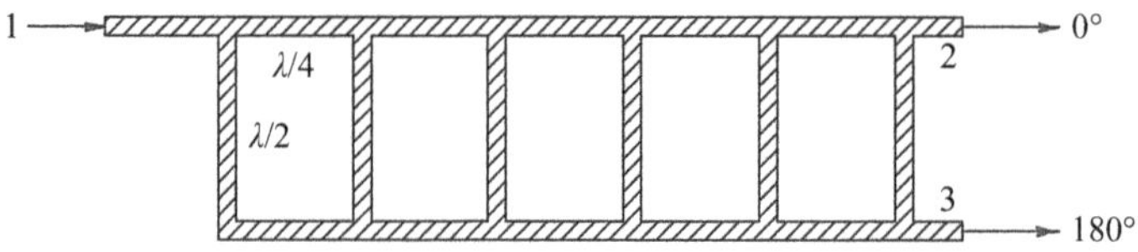

Figure F.6 Microstrip branchline (ladder) balun

F.4 Microstrip transmission line Marchand balun

Better performance than that obtained from a coupled microstrip transmission line balun can be realized using a microstrip transmission line version of a Marchand balun [5]. This design is less sensitive to a low coupling ratio than the coupled microstrip transmission line balun of Section F.3.

The printed Marchand balun is shown in Figure F.5. Each coupled section is one-quarter wavelength long at the design frequency. Although the performance of a printed Marchand balun is superior to a coupled microstrip transmission line balun, its topology is also larger.

Microstrip Marchand baluns can be modified to present a 50 Ω load to the common mode (unbalanced) and remain transparent to the differential mode (balanced) at their output port [6]. In the case of antennas, this design could be used to enforce the desired current on a printed antenna and therefore the desired radiation pattern.

F.5 Microstrip branchline (ladder) balun

Microstrip branchline baluns can produce considerable bandwidth. A microstrip branchline balun was used in the design of the printed dipole antenna of Section 7.8. The printed dipole, fed with a two section microstrip branchline balun, produces a 25% 2:1 VSWR impedance bandwidth.

This balun is fabricated from N sections of quarter-wavelength parallel microstrip lines, with half-wavelength shunt lines, as shown in Figure F.6 [7]. The bandwidth of the balun increases with increasing numbers of sections. After six sections are added, no significant bandwidth improvement occurs. This type of balun design is also designed in an empirical manner.

References

[1] Chipman, R.A., *Transmission Lines*, Shaum's Outline Series in Engineering, McGraw Hill, 1968.

[2] Hu, S., "The Balun Family," *Microwave Journal*, September 1987, pp. 227–229.

[3] www.rfic.co.uk.

[4] Kuylenstierna, D., and Linner, P., "Design of Broad-Band Lumped-Element Baluns with Inherent Impedance Transformation," *IEEE Transactions on MTT*, December 2004, Vol. 52, No. 12, pp. 2739–2745.

[5] Marchand, N., "Transmission–line Conversion Transformers," *Electronics*, December 1944, Vol. 17, pp. 142–146.

[6] Kooho, J., Campbell, R.L., Hanaway, P., *et al.*, "Marchand Balun Embedded Probe," *IEEE Transactions on Microwave Theory and Techniques*, Vol. 56, No. 5, May 2008, pp. 1207–1214.

[7] Al Basraoui, M., and Shastry P., "Wideband Planar Log-Periodic Balun," *International Journal of RF & Microwave Computer Aided Design*, Vol. 11, No. 6, November 2001, pp. 343–353.

Index

www.ingramcontent.com/pod-product-compliance
Lightning Source LLC
LaVergne TN
LVHW011915200926
848484LV00002B/626